Practical Concepts for
Capstone Design Engineering

Frederick Bloetscher, Ph.D., P.E., LEED-AP
Daniel Meeroff, Ph.D., E.I.

J.ROSS
PUBLISHING

Copyright © 2015 by J. Ross Publishing, Inc.

ISBN 978-1-60427-114-0

Printed and bound in the U.S.A. Printed on acid-free paper
10 9 8 7 6 5 4 3 2 1

Library of Congress Cataloging-in-Publication Data

Bloetscher, Frederick.
  Practical concepts for Capstone design engineering / by Frederick
Bloetscher and Daniel Meeroff.
        pages cm
  Includes bibliographical references and index.
  ISBN 978-1-60427-114-0 (hardcover : alk. paper) 1. Engineering design.
2. Communication in engineering design. I. Meeroff, Daniel, 1973- II.
Title.
  TA174.B56 2015
  620′.0042—dc23
                              2014048609

    Direct all inquiries to J. Ross Publishing, Inc., 300 S. Pine Island Road, Suite #305, Plantation, Florida 33324.

Phone: (954) 727-9333
Fax: (561) 892-0700
Web: www.jrosspub.com

Table of Contents

Preface ix
About the Authors xiii
Web Added Value™ xv

Chapter 1. Introduction to Capstone Design 1
1.1 The Capstone Design Process 2
1.2 Course Objectives 3
1.3 Project Selection 4
1.4 Course Management Structure 4
1.5 Group Selection 5
1.6 Course Delivery Structure 6
1.7 Getting Started 8
1.8 Deliverables 8
1.9 Assessment 10
1.10 Last Words 14
1.11 References 14

Chapter 2. Career Opportunities and Leadership 17
2.1 Self-Assessment 17
2.2 Types of Job Opportunities 19
2.3 Branding 20
2.4 Create the Group Design Firm 24
2.5 Teaming Skills 24
2.6 References 28
2.7 Assignments 29

Chapter 3. The Profession and Ethical Conduct 31
3.1 Engineering Ethics 32
     3.1.1 Where Do Ethics Come From? 33
     3.1.2 The Philosophers Weigh In 34
     3.1.3 Creeds, Codes, and Canons 35

3.2 Ethical Issues in Engineering 36
    Case Study 1. Licensure in Multiple States 37
    Case Study 2. Practicing without a License 37
    Case Study 3. Design Defect 38
    Case Study 4. Failing to Seal the Documents 38
    Case Study 5. Sealing Documents That Are Not Final 38
    Case Study 6. Misleading Testimony about a Design 39
3.3 Licensure 39
3.4 References 43
3.5 Assignments 43

Chapter 4. Getting the Design Contract 49
4.1 Building a Capital Project 53
4.2 What Owners (Should) Look for from Consulting Engineers 55
4.3 What Consultants Do Not Need 56
4.4 The Typical Public Sector Proposal Process 56
     4.4.1 Scope of the Project 57
     4.4.2 Requirements of Proposers 58
     4.4.3 Evaluation of Proposals 61
4.5 The Typical Private Sector Proposal Process 63
4.6 Stages in the Design Process 63
     4.6.1 Conceptual Design 64
     4.6.2 Predesign 64
     4.6.3 Preliminary Design 65
     4.6.4 Final Design 65
4.7 Construction Documents 66
4.8 Scheduling and Project Delivery 66
4.9 References 69
4.10 Assignments 69

Chapter 5. Communication Skills for Engineers 71
5.1 Overview of the Engineering Writing Style 71
5.2 General Document Development and Outlining 72
5.3 Proper Grammar 75
5.4 Reference Citations 75
    5.4.1 Journal Citations 76
    5.4.2 Conference Proceedings and Symposium Citations 76
    5.4.3 Book Citations 76
    5.4.4 Report Citations 76
    5.4.5 Unpublished Material Citations 77
    5.4.6 Web Page Citations 77
    5.4.7 Thesis and Dissertation Citations 77
5.5 Persuasive Writing 77
5.6 Engineering Graphics 78
    5.6.1 Numerical Tables 79
    5.6.2 Figures 80

5.7   Proofreading Strategies 85
5.8   Fonts 86
5.9   Margins 86
5.10 Typical Engineering Documents 86
      5.10.1 Meeting Minutes 87
      5.10.2 Emails and Informal Notes 88
      5.10.3 Memoranda 91
      5.10.4 Standard Business Letters 91
      5.10.5 Progress Reports 93
      5.10.6 The Basis of Design Report 96
      5.10.7 Technical Memoranda 98
      5.10.8 Interim and Final Technical Reports 100
5.11 Public Speaking 103
      5.11.1 Visual Aids 109
      5.11.2 Questions and Answers 112
      5.11.3 Evaluating Presentations 113
5.12 References 116
5.13 Assignments 116
5.14 Appendix 119
      5.14.1 Grammar 119
              5.14.1.1 Pronoun Antecedents 119
              5.14.1.2 Subject-Verb Agreement 119
              5.14.1.3 Plural Nouns 121
      5.14.2 Punctuation 121
              5.14.2.1 Comma Use 122
              5.14.2.2 Colon Use 122
              5.14.2.3 Semicolon Use 123
              5.14.2.4 Hyphen Use 123
      5.14.3 Capitalization 124
      5.14.4 Common Spelling Errors 124
      5.14.5 Misused Words 125
      5.14.6 Abbreviations 125
      5.14.7 Numbers 126
      5.14.8 Figurative Language Use 127
      5.14.9 Voice 127
      5.14.10 Gender Issues 127
      5.14.11 Writing Pitfalls to Avoid 128
              5.14.11.1 Redundancies 128
              5.14.11.2 Informal Language 129
              5.14.11.3 Inventing New Words or Phrases 129
              5.14.11.4 Double Negatives 129

Chapter 6. Alternative Analysis 131
6.1 Application to Design Projects 133
6.2 Selection Criteria 134
6.3 Scoring System 136

6.4 Alternative Selection Matrix 136
6.5 Sensitivity Analysis 137
6.6 Automobile Purchase Example 138
6.7 Reference 140
6.8 Assignments 140

Chapter 7. High-Performance Construction 143
7.1 Why Build Green? 144
7.2 Agencies That Evaluate Green Building Performance 146
    7.2.1 International Organization for Standardization 146
    7.2.2 U.S. Environmental Protection Agency 147
    7.2.3 U.S. Green Building Council® 149
7.3 LEED® Certification 150
    7.3.1 Requirements 150
        7.3.1.1 Sustainable Sites 151
        7.3.1.2 Water Efficiency 151
        7.3.1.3 Energy and Atmosphere 151
        7.3.1.4 Materials and Resources 152
        7.3.1.5 Indoor Environmental Quality 152
        7.3.1.6 Innovation and Design Process and Regional Priority Credits 153
7.4 Triple Bottom Line 153
7.5 References 154
7.6 Assignments 155

Chapter 8. Environmental Site Assessment 157
8.1 Scope 157
8.2 The Environmental Professional 160
8.3 Site Reconnaissance 161
    8.3.1 Exterior Reconnaissance 161
    8.3.2 Interior Reconnaissance 166
8.4 Records Review 169
8.5 Interviews 180
8.6 Evaluation and Report 181
8.7 Nonscope Considerations 182
8.8 Phase II and III 187
8.9 References 187
8.10 Assignments 188

Chapter 9. The Site Plan Development Process 189
9.1 Community Plans and Codes 190
9.2 Site Development 191
9.3 Easements, Rights-of-Way, and Setbacks 193
9.4 Utilities, Parking Requirements, and Roadwork 194
9.5 Building Code Requirements and Functionality 204
9.6 Assignments 209

Chapter 10. The Floor Plan Development Process 211
10.1 Building Program 211
10.2 Floor Planning 213

Chapter 11. Engineering Economics 231
11.1   Interest Rates 234
11.2   Single-Payment Present Worth 235
11.3   Future Value or Single-Payment Compound Amount 236
11.4   Annual Worth 239
11.5   Future Worth Given an Annuity 241
11.6   Gradients 243
11.7   Shifted Annuities 246
11.8   More about Interest Rates 249
11.9   Dealing with More Complex Cash Flow Diagrams 252
11.10 Comparing Options 257
        11.10.1 Break-Even Analysis 258
        11.10.2 Annual Worth Analysis 259
11.11 Inflation Adjustment 260
11.12 Depreciation 264
11.13 A Word of Caution 268
11.14 References 268
11.15 Assignments 268
Appendix 11A: Interest Tables 272

Chapter 12. Preliminary Site Design and Nonstructural Concepts 279
12.1 Roof Systems 280
12.2 On-site Stormwater Drainage 288
12.3 Potable Water Systems 303
12.4 Sanitary Sewer Systems 314
12.5 Heating, Ventilation, and Air Conditioning 320
12.6 Parking Considerations 326
12.7 Transportation 328
12.8 Landscaping 334
12.9 References 336

Chapter 13. Structural Design Concepts 337
13.1 Load and Resistance Factor Design 340
13.2 Types of Loads 341
        13.2.1 Dead Loads 341
        13.2.2 Live Loads 342
        13.2.3 Wind Loads 343
        13.2.4 Roof Loads 346
                13.2.4.1 Wind Loads 347
                13.2.4.2 Rain Loads 347
                13.2.4.3 Snow Loads 348

13.2.5 Earthquake Loads 350

13.2.6 Other Loads 351

13.3 Structural Design Concepts 351

13.3.1 Concept of Tributary Area 351

13.3.2 One-Way and Two-Way Slabs 353

Example 1. One-Way Slab Design of Flexure Reinforcement 356

13.3.3 Beams and Girders 358

Example 2. Design of Continuous Steel Beam for Flexure by LRFD 359

13.3.4 Columns 360

Example 3. Design of Steel Columns 361

13.3.5 Walls 362

Example 4. Concrete Shear Wall Design 365

13.3.6 Lateral Load Analysis 368

13.3.7 Serviceability 369

13.3.8 Structural Detailing 371

13.4 Foundation Design Concepts 373

13.4.1 Shallow Foundations 374

13.4.2 Strip Footer Foundations 377

Example 5. Strip Footer 379

13.4.3 Shallow Footer Foundations 380

Example 6. Size of Footer 383

13.4.4 Pile Foundations 385

13.5 References 388

Chapter 14. Cost Estimating 389

14.1 Purpose and Process 389

14.2 Stages of Cost Estimating 390

14.3 Bidding Process 392

14.4 Asset Management 393

14.5 Life Cycle Analysis 395

14.6 References 396

14.7 Assignments 396

Chapter 15. Conclusion 397

Index 399

Preface

Training the next generation of engineers and engineering educators is a national priority, but with declining enrollment in undergraduate engineering programs and declining interest in engineering education, the resulting critical shortage of qualified workers has intensified the need to strengthen the professional competency of the graduates entering the engineering workforce despite a robust demand in the workplace. In the long term, the growth and development of human civilization will rely on the ability to resolve challenges using science, technology, engineering, math, and creativity. The changing materials and environments, along with the infrastructure challenges, are just the surface of the expansive need for competent engineering graduates.

Proposals to enhance engineering design education have included the development of design expectations across the curriculum, team-based learning activities, and assessments to gauge student attainment of outcomes, but the key obstacle for undergraduate students is transitioning from traditional lecture-based coursework to more realistic, practice-oriented training. The key is to stimulate creativity and critical thinking to solve real-world problems by putting engineering skills into practice in the classroom. It is precisely these two components that are missing from the traditional engineering curriculum, which emphasizes the regurgitation of equations and repetition of standard problem sets, neither of which reflects the real world. The capstone design experience during the undergraduate student's senior year provides an excellent opportunity to transition more smoothly from the classroom setting to the workplace environment. Professional engineering is a process, one that first requires a full understanding of the problem and associated challenges (due diligence) and a means to define the problem in a context that can lead to the second aspect—the actual design of the solution.

There is widespread agreement on the value of offering a capstone design course that involves real-world projects, industry partnerships, and student teamwork, but often the stumbling block is time (on the part of both faculty and students) and the need to find or create real-world projects. Help from nonacademics is critical to the success of such a capstone course sequence. To enrich the learning experience of fundamental coursework in the undergraduate program, this textbook was created to provide a framework of the critical components and provide a guide for students and faculty as they navigate through the design process. This book results from over 10 years of teaching the capstone course, integration of experiences from engineering consultants and clients about the expectations of engineering graduates, and comments resulting from the adoption of the basic tenets of the class at several other universities.

Although the capstone design course is found in almost all accredited chemical, civil, environmental, industrial, and mechanical engineering programs, there are no existing textbooks that present the breadth of topics covered here. This text is designed to be adopted for a one- or two-semester sequence in engineering design. For example, at Florida Atlantic University and the University of Miami, the civil engineering capstone sequence is offered as a two-semester course. The capstone courses typically represent the first time that engineering students are exposed to professional practice in their academic program. A two-semester course permits adequate time for the transition from repetitive problem solving to thinking through situations where many of the variables are uncertain or unknown.

However, relying solely on industry to lead the project is unrealistic due to the time commitments for reading and evaluating reports, meeting with students, mentoring, scaffolding progress, and assessing presentations. It is not uncommon to dedicate 20 hours per week or more to a class like this, far more than adjunct professors or industry professionals can devote. Commitment of full-time faculty with the experience and/or willingness to include external concepts is critical to a successful student learning experience. The key is finding the faculty members who are interested in the process, have the experience, and are willing to dedicate the time. Senior design never should be a class that is a burden for someone to teach. The value to students will diminish as a result. Having registered professional engineers involved as a part of the class is extremely valuable to the learning experience and for future employment opportunities.

This textbook will be most meaningful if the student has completed fundamental civil engineering coursework. For prerequisites, students should obtain department approval, and it is recommended that students complete the following coursework prior to enrolling in the capstone course: introductory transportation engineering, soil mechanics, applied hydraulics or fluid mechanics, materials science, structural analysis, surveying, computer-aided design, and introductory environmental engineering. For the second portion of the course, students should have completed the following coursework: steel and/or concrete structures, foundation design, transportation engineering design, environmental engineering design, and hydrology or drainage design. Because the students create their own design solution, they should take the course in consecutive semesters, with the same project and teammates to encourage the students to engage in the project results more fully. Changing groups or adding a student from a prior course often disrupts group dynamics and the design process of the group. This should be avoided. Students should complete the two semesters consecutively, in the same groups.

We should note that in the years since this (required) capstone design sequence was initially offered at the previously mentioned institutions, the course has been consistently rated as the one of the best student experiences in exit interviews, the course that students learn the most in, and the most time-consuming class they took during their academic careers. There is a correlation there. The students get interested in certain aspects of the engineering design process and then aggressively pursue them. The key is to spark interest in students, provide general guidance from faculty and professionals, and allow students to learn and pursue ideas and complex solutions on their own. The results can be inspirational, and student knowledge increases exponentially from the junior year. From the authors' experiences, several projects that were undertaken as part of this course were actually constructed, including a LEED® Gold-certified library (Figure 1), the world's first LEED Gold-certified nanofiltration water treatment plant (Figure 2), a LEED Platinum engineering building

Figure 1 City of Dania Beach, FL LEED Gold library

Figure 2 City of Dania Beach, FL LEED Gold nanofiltration water treatment plant

at Florida Atlantic University, a LEED Gold environmental center, and several public school buildings, in addition to proposed city parks, mixed-use facilities, hotels, and transit stations.

For students whose academic goal is to become a civil engineer, the capstone design course should be the most exciting and valuable learning experience before entering into professional practice. Because there is not a comprehensive textbook that can address the holistic design issues, this textbook was designed to be different from traditional textbooks that cover only the fundamentals of engineering. Instead, it centers on key skills needed to complete a capstone project. The intent is to permit students to transition from purely academic work to solving actual problems in a more realistic setting. The student projects should be real and, whenever possible, developed by working professionals who interact with students and participate as mentors or jury members during the course. This approach of integrating students, faculty, design professionals, clients, consultants, and regulators is the best way we have found to bridge the gap between the classroom and the profession, and the results are always astounding.

About the Authors

Frederick Bloetscher, Ph.D., P.E., LEED-AP, DWRE, is currently an associate professor at Florida Atlantic University in Boca Raton, where his focus is on water resources, water supply sustainability, stormwater (including sea level rise), and wastewater disposal issues. He received his bachelor's degree in civil engineering from the University of Cincinnati and earned his master of public administration degree from the University of North Carolina at Chapel Hill. His Ph.D. is in civil engineering from the University of Miami. His areas of interest include water and wastewater resource management, membrane processes, utility management and finance, groundwater, and waste disposal options. He is also the president of Public Utility Management and Planning Services, Inc. (PUMPS). PUMPS is a consulting firm dedicated to the evaluation of utility systems, needs assessments, condition assessments, strategic planning, capital improvement planning, grant and loan acquisition, interlocal agreement recommendations, bond document preparation, consultant coordination, permitting, and implementation of capital improvement construction. Dr. Bloetscher was previously an adjunct faculty member at the University of Miami in Coral Gables, the former utility director and deputy director for several large water and sewer systems, and a former city manager in North Carolina. He is the former Chair for the Water Resource Division Trustees, Groundwater Resource Committee and Education Committee for the American Water Works Association. He is a LEED-AP and holds professional engineering licenses in nine states.

Dr. Bloetscher has been nominated for the Teacher of the Year award a number of times by his students and has received two university-wide leadership awards, plus two national leadership awards. Dr. Bloetscher co-teaches the first semester of the capstone design course at Florida Atlantic University and leads the second, where the planning and conceptual design of green building construction is turned into preliminary plans, specifications, and basis of design reports.

In 2012, Dr. Bloetscher and Dr. Meeroff received the National Council of Examiners for Engineering and Surveying (NCEES) Award for Connecting Professional Practice and Education for their work on the Dania Beach nanofiltration facility, which is the first LEED Gold water treatment facility in the world. Dr. Bloetscher was the LEED administrator for the project.

Daniel E. Meeroff, Ph.D., E.I., is associate chair and professor in the Department of Civil, Environmental & Geomatics Engineering at Florida Atlantic University (FAU). His area of specialization is environmental engineering, more specifically, water and wastewater engineering, water chemistry, solid/hazardous waste management, sustainable building strategies, and pollution prevention. Dr. Meeroff is the founder and director of the Laboratories for Engineered Environmental Solutions (Lab.EES) at FAU (http:\\labees.civil.fau.edu). He earned his bachelor's degree in environmental science from Florida Tech and his master's and Ph.D. degrees in civil/environmental engineering from the University of Miami. In 2011, Dr. Meeroff was selected by the students for the Excellence and Innovation in Undergraduate Teaching Award at FAU and has been nominated for the Teacher of the Year award numerous times by his students. In 2014, the Engineer's Council recognized Dr. Meeroff as the John J. Guarrera Engineering Educator of the Year, and the student body at FAU selected him as the Distinguished Teacher of the Year, the highest teaching honor at the university. In the field of green building design, Dr. Meeroff co-teaches the first semester of the innovative capstone design course at FAU.

*Free value-added materials available from
the Download Resource Center at www.jrosspub.com*

At J. Ross Publishing, we are committed to providing today's professor and student alike with practical tools that enhance the learning experience. That is why we offer free ancillary materials available for download on this book and all participating Web Added Value™ publications. These online resources may include interactive versions of material that appears in the book or supplemental templates, worksheets, models, plans, case studies, proposals, spreadsheets, and assessment tools, among other things. Whenever you see the WAV™ symbol in any of our publications, it means bonus materials accompany the book and are available from the Web Added Value Download Resource Center at www.jrosspub.com.

Downloads available for *Practical Concepts for Capstone Design Engineering* include instructor material such as final exams, course syllabi, grading rubrics, and PowerPoint® slide shows that correspond with specific topics within the book. All materials can be modified for classroom instruction.

Introduction to Capstone Design

As noted in the preface, training the next generation of engineers and engineering educators is a national priority (National Academy of Sciences 2005). Declining interest in engineering education and the consistent erosion of the number of required credits for a degree have created a critical shortage of qualified professionals, yet the complexity of societal issues has intensified the need to increase and expand the professional competency of the engineering workforce for the next generation (Seymour 2001). Students need to learn more, but have less time to do so, and currently much of that time is devoted to rote problem solving as opposed to actual engineering. The key obstacle for undergraduate students is transitioning from traditional lecture-based coursework to more realistic, practice-oriented training. As a result, there is widespread agreement on the value of offering a capstone design course that involves real-world projects (Padmanabhan and Katti 2002), industry partnerships (Kumar 2000), and student teamwork in preparation for entering the engineering workforce (Todd et al. 1995).

The Accreditation Board for Engineering and Technology (ABET) calls for accredited undergraduate engineering degree programs to have a capstone design experience in which fourth-year students work in teams for one or two semesters on a practical design project. The purpose of a capstone course is to provide students with a culminating engineering design experience that allows them to apply the fundamental coursework and skills learned during their engineering curriculum to solve an engineering problem in a way that incorporates appropriate engineering standards with multiple realistic constraints. Capstone projects should be industry-related, which helps to bridge the gap from the university environment to the professional ranks.

Engineering infrastructure projects typically are designed to last for long periods of time, and their life cycle impacts on the environment are not fully explored during typical undergraduate coursework. By integrating concepts such as teamwork, mentoring, life cycle analy-

1

sis, and environmental stewardship into the preliminary design phase, engineering graduates will be better prepared to actively play a role in improving the condition of both the built environment and the natural environment, putting them in position to make lasting contributions to building a sustainable future. One way to accomplish this is by introducing the concept of high-performance building design and green engineering into the capstone course.

The goal of a capstone design course, or sequence, must be to encourage students to use their creativity, innovation, curiosity, and educational foundations to solve complex, real-world problems. To enrich the learning experience of fundamental coursework in the undergraduate program, it is necessary to expose students to practical applications of the basic subdisciplines of the engineering curriculum, including elements of the key courses they have taken prior to their senior year. In addition, classes that students typically will take as seniors, such as engineering economics, construction management, and other technical electives, will be important.

1.1 The Capstone Design Process

For students whose academic goal is to become a professional engineer, the capstone design course should be the most exciting and valuable learning experience before entering into practice. According to ABET, many subjects make up the engineering disciplines, and students need an understanding of all of those subjects to be successful in their careers and to obtain their professional licenses. The focus of this book is different from traditional textbooks in that this one is designed to integrate many aspects of the professional practice experience, instead of just covering the fundamentals of engineering. The capstone course and the resulting project center around skills that involve the successful design development of a real-world project such as a commercial/institutional high-performance building, with ancillary issues such as environmental impacts, transportation, resiliency against natural disasters, flood protection, compliance with local ordinances, application and interpretation of building codes, and concepts that limit options such as economics, local politics, and even historical preservation. These multiple realistic design constraints are similar to those that professionals have to address every day in their careers. The project should be real or based on a real project. It may actually be in the design process when assigned, and, whenever possible, the design professionals working on the project should be a part of the course in some meaningful way.

In the first phase of the capstone project, students perform a project needs assessment, conduct site reconnaissance, and develop a conceptual design that often includes floor plans and site plans. In the second phase, the project is taken from the conceptual stage to a set of preliminary design drawings. A professional engineer should be engaged to approve acceptable conceptual plans, which serve as the basis for developing a set of preliminary design drawings (using AutoCAD® and Revit® or other three-dimensional building information modeling software), along with all of the design calculations, modeling results, and support documents that comprise a basis of design report. Throughout the course, student teams will present their progress to an invited jury of outside professionals (e.g., department advisory council members, practicing engineers, regulators, owners, and other interested stakeholders) to provide feedback and comments whenever possible. This interaction with working professionals is critical for the growth of student engineers into practice. The intent is to transition students from purely academic work to solving actual problems in a more realistic setting.

1.2 Course Objectives

Students are expected to learn how to approach complex challenges, discover solutions, and deal with multiple realistic design constraints while incorporating engineering design standards. It is important to realize that there is no solution manual for capstone projects. While often there are many different answers that will work in the real world, designs are considered elegant, clever, and innovative instead of partially correct. The idea is to expose students to the thought process of how an engineer arrives at an appropriate solution. Creativity, ingenuity, and innovation are important and always must be encouraged. Professionals will find that students have great ideas and often challenge the status quo, which is a good thing for the profession and society.

Participants in the course are expected to communicate effectively in a professional manner, both in written and presentation format. The following is an example of the stated objectives in a capstone course syllabus:

◆ Develop design project teams and deliverables acceptable to a sponsor or client
◆ Develop effective communication, teaming, and leadership skills
◆ Develop an understanding of professional practice issues, such as involvement in professional societies, licensing, ethics, and continuing education
◆ Develop a practical understanding of the application of engineering economics
◆ Integrate prior engineering coursework to develop feasible solutions while incorporating appropriate engineering standards with multiple realistic constraints (see Figure 1.1)

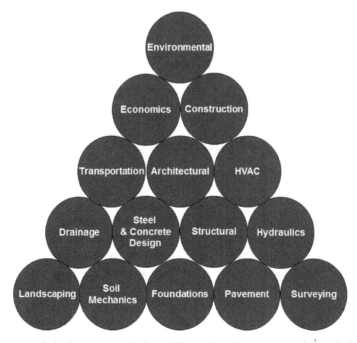

Figure 1.1 Integrated design approach that utilizes all prior coursework for design of a high-performance building project

1.3 Project Selection

A suitable design project will incorporate as many of the aspects of the engineering curriculum as possible. The local zoning board, community redevelopment board, members of the department industry council, and even alumni are great sources for capstone projects. Locally significant construction projects may be easier to coordinate compared to theoretical exercises or projects that are located great distances away. Local governments often start planning years in advance for projects, which makes them perfect for students to work on. Students should physically visit potential development sites to learn how to gather data in the field. This is *not* a satellite image activity done by computer. Students need to experience the sites with their own senses. Realization that the work will be evaluated by professionals in the field or the actual owner of the project adds to the richness of the learning experience.

It is important to make sure that the capstone project is not too focused on one single aspect of engineering, such as site development or highway design, which may alienate students interested in other engineering disciplines. For example, design of a multistory building encompasses nearly all of the elements of engineering training, with the possible exception of certain aspects of environmental engineering. That is why it is important to place an emphasis on high-performance building design by requiring students to meet sustainability criteria such as the LEED® checklist from the U.S. Green Building Council® or other systems such as BREEAM, CEEQUAL, PERSI, and Green Globes, among many others. Examples of suitable projects that have been successfully pursued by students of the authors include college dormitories, hospital buildings, libraries, water and wastewater treatment plants, digester gas energy capture projects, school buildings, mixed-use buildings, hotels, train stations, local multimodal transportation projects, office buildings, apartments, airport terminals, recreation centers, civic buildings, and parks.

When developing a capstone project for this course, each student team should be provided with (1) a scope of work that outlines the client's needs, (2) key technical information such as a geotechnical report near the job site with results of a soil boring log and soil bearing capacity tests, and (3) a contact person at the job site (from either the ownership group, the construction management team, or the architect's office) to assist in coordinating site visits, interviews, and information requests.

1.4 Course Management Structure

A capstone design course works best when it is team taught by faculty members (representing the fundamental academic engineering perspective), consulting engineers (representing the practical engineering perspective), and government officials (representing the owner's perspective or the regulatory perspective), along with contributions from a variety of outside lecturers to provide both academic and real-world connections. It is vital that at least one (and preferably as many as possible) of the members of the instructional team is a licensed professional engineer. By focusing on project-based learning through the development of a high-performance building for example, students also will learn how to practice responsible stewardship. Thus, students will be better prepared to deal with an evolving job market in an ever-changing world with an increasing human population, energy and water limitations, adaptations to climate change, and economic and social inequities.

The role of the instructional team is to act as the principal engineers in charge of the students' fictional consulting firms. They also serve a guidance function by providing the project background, interpreting scope issues, acting as an initial facilitator or liaison between the student groups and the client, providing critical reviews and feedback, assigning change orders, and enforcing class policies and procedures. The instructors also have the final say in conflict resolution.

1.5 Group Selection

Prior to becoming eligible for the capstone design course, students typically must obtain department approval. It is recommended that the following prerequisites be completed *prior to* enrolling in the course:

1. Introductory transportation engineering
2. Soil mechanics
3. Applied hydraulics or fluid mechanics
4. Materials science
5. Structural analysis
6. Surveying
7. Computer-aided drafting
8. Introductory environmental engineering

In other words, students should be eligible to register for the Fundamentals of Engineering (FE) exam within 6 months of registering for the class.

For the second portion of the course, it is strongly recommended that the following coursework be completed before starting:

1. Steel and/or concrete structures
2. Foundation design
3. Transportation engineering design
4. Environmental engineering design
5. Water resources, hydrology, or drainage design

By the first class meeting, each student should submit a resume and brief personal statement that identifies his or her engineering interests, past and current employment, career goals, perceived strengths and weaknesses, and geographic information. This last piece of information is critical because the class may be made up of widely diverse commuter students or the course may be delivered through a long-distance learning platform; hence, appropriate arrangements can be made to facilitate effective student teams.

Several lessons have been observed through the process of team selection. Groups should not be assigned randomly, nor should students be allowed to select their own groups. Using the information provided in the resume and personal statement, teams of three to five students each are assigned based on several key pieces of information, including ability to communicate effectively, leadership abilities, and grade point average (GPA). It is important to have students with different interests in each group. For instance, if all four members of a team are interested in transportation, then who will do the structures or the surveying? It also is important to consider group dynamics. For instance, a team of four members with weak leadership personalities will tend to struggle to get work done. Finally, GPA should be considered as well. If a high-GPA student is paired with low-GPA students, only one student may want to do all the work, and the other three will not get as much value out of the experience.

Once the groups are established, a project manager is selected by the team members. The duties of the project manager are to oversee and coordinate team activities, create and maintain a collegial atmosphere, ensure professional-quality work is submitted on time, build team consensus, resolve conflicts within the team, delegate and assign tasks to individual team members and micro-teams, approve time sheets, and maintain the group's focus on the project goals. It is not uncommon for the individual who serves as project manager to change over the course of the project, depending on team needs or personal issues. The other team members' roles are to attend group meetings, show initiative, take responsibility, contribute thoughtfully, give and receive useful feedback, deliver on commitments, avoid friction between team members, assist teammates by checking each other's work, and do *more* than their fair share of the effort.

1.6 Course Delivery Structure

The course employs two overlapping teaching methodologies: lecture and professional practice sessions. The lectures focus on introducing certain aspects of the engineering profession that will be relevant to the assignments. The first several weeks of the course include critical lectures to convey the appropriate background concepts and lay the framework and expectations for the rest of the class. These topics include the following:

- ◆ **Technical communication skills**—A premium is placed on professional-quality writing, outlining, effective graphical presentations, persuasive writing, creativity, and logical thinking. The focus here is to be able to present engineering concepts and ideas to different audiences using computer-based tools and other visual aids, in both written and oral formats. Separate modules are presented for preparing engineering progress reports, responses to requests for proposal, technical memos, and construction documents.
- ◆ **Project management skills**—Strategies for working effectively in a team setting and developing leadership skills, project management tools, and scheduling capabilities are addressed.
- ◆ **Alternative analysis skills**—Methods to systematically analyze engineering alternatives to come up with the preferred option using objective criteria are presented.
- ◆ **Site assessment skills**—Guidelines for conducting a Phase I Environmental Site Assessment based on the most recent version of ASTM 1527 are covered.
- ◆ **Site planning skills**—Concepts involving zoning, easements, setbacks, rights-of-way, parking requirements, landscaping issues, accessibility, and drainage, as well as water, sanitary sewer, stormwater, reclaimed water, cable TV, natural gas, telephone, electrical, and wireless utilities are presented. This topic also should include a brief discussion of conceptual development of programming for the overall site layout and floor plans.
- ◆ **Engineering economics skills**—The time value of money, preparation of preliminary cost estimates, quantity takeoffs, and bidding are discussed.
- ◆ **Ethical conduct**—The concepts of engineering ethics and the professional responsibilities of professional engineers are presented.

In the beginning of the second phase of the course, a series of technical lectures are presented which focus on specific design considerations and approaches, such as structural building concepts; foundation concepts; plumbing concepts; roadway design; drainage strategies; lift station design; heating, ventilation, and air conditioning (HVAC) design; and roof concepts. These brief presentations often are delivered by technical advisors in their specific area of expertise. This serves to introduce the teams to an appropriate design approach for their projects, and it also introduces the technical advisors or outside faculty members to the student teams for later consultation during the semester as mentors.

Beyond the lectures are a series of professional practice sessions that involve presentations of progress toward developing the capstone design; results of investigations; research and data collection; and development of technical memos that describe the basis of design in detail. These professional practice sessions allow the instruction team to foster development of specific engineering skills through guided exploration and to periodically monitor progress. The presentation format can facilitate this growth and help the students make progress throughout each semester toward attaining predefined goals. Each presentation is essentially a milestone in a cleverly scaffolded approach that forces students to make manageable progress toward the final deliverable. The approach is gradual progress toward a basis of design document by dividing up the work into manageable components so that the project grows from the initial design charrettes to a professional-quality report complete with calculations, drawing sets, specifications, and details. Staged presentation times can reduce the tendency of students to go into "crisis mode" to complete the work. Trying to prepare complex calculations and presentations for a client at the last minute does not work in the real world. The goal is to alter student work habits. A jury of faculty, representatives from industry, the client, and peers evaluates the presentations and written reports and provides feedback for revision (see Figure 1.2). Although the students are not yet professionals, they should become accustomed to filling out time sheets for compensation and learn how to manage their time wisely. Instead of money, students are "paid" in grades.

Figure 1.2 Students sharing presentation with faculty

1.7 Getting Started

Once the project scope has been assigned, each team must come up with a set of well-defined goals and a plan to accomplish them. This plan typically addresses accountability, leadership structure, communication among members, strategies to encourage innovation and sharing of ideas, means to manage and resolve conflicts, and well-defined roles and responsibilities of each individual team member. This begins the process of developing an organizational chart in preparation for responding to a request for qualifications. The next step is to come up with a name and logo for the fictitious consulting firm, along with a shared vision statement to guide decision-making priorities.

Soon after the teams have developed their fictitious design firms, they must begin their initial work by exploring the concept of sustainability and high-performance building design. Teams are encouraged to seek out design inspirations and innovative technologies that they might want to include in their designs at this point. The goal is to establish some key components that will set a team's design apart from competitors. This will be the first opportunity to present in front of the instructional staff and peers. Buoyed by critical evaluations from their first presentation, each student team performs a needs assessment for the project, develops a well-defined scope of work, performs site reconnaissance and engineering due diligence, and then develops floor plans and site plans to create a basis of design report. Then students should have their conceptual plans evaluated by a professional engineer, and each team should prepare a set of preliminary design drawings (in AutoCAD) with all the calculations, modeling results, and support documents for structural, geotechnical, water, sewer, stormwater, and transportation aspects of the project in the second phase of the class.

1.8 Deliverables

By the end of the first phase, each team is required to submit its basis of design notebook, along with a complete set of preliminary plans. Preparation of all interim deliverables should have this final work goal in mind. During this first phase, the following deliverables are presented as pieces of the *Final Predesign Notebook*:

1. **General introduction**—Details the team's interpretation of the project scope, design goals, and objectives.
2. **Response to request for qualifications**—Answers the question "Why should you hire our team?" and requires the students to brainstorm ideas and concepts that they plan to incorporate in their proposed design. It also requires the team members to outline their strengths, detail their project management plan, come up with a realistic schedule for accomplishing the work, and refine their professional resumes.
3. **Phase I Environmental Site Assessment**—Answers the question "What are the existing recognized environmental conditions on the site?" and requires the students to investigate the existing site to identify any current or historical recognized environmental conditions, past activities, impacts of development, long-term sustainability, and due diligence. During this exploration, each team conducts site reconnaissance, interviews, and record reviews with federal, state, and local regulatory agencies.

4. **Preliminary site plan**—Answers the question "What is proposed to be built on the site and inside the building?" and focuses each team on developing a preliminary site plan and floor plan for the project. Once again, the scope of work is restated in relation to the design goals, site constraints, and opportunities for innovation. Using this framework, the existing site conditions are presented and a set of viable alternatives are analyzed. The proposed site plan is then presented, along with solutions for stormwater, drainage, parking, accessibility, and utilities, as well as preliminary cost estimates and "green" features. Final floor plans also are presented for approval.

During the second phase, each team should submit a series of progress reports, culminating in the *Preliminary Design Briefing Report*, which includes technical design memoranda for each of the following topics:

1. **Structural plan**—Includes justification for the structural layout, detailed calculations for critical loads, locations of structural elements, and details for all connections, as well as appropriate drawings and specifications.
2. **Foundation plan**—Includes justification for the foundation layout, analysis of the soil borings, a geotechnical report, and a grading plan, as well as appropriate drawings, specifications, and detailed calculations.
3. **Transportation plan**—Includes a breakdown of parking requirements, modeling traffic impacts, appropriate horizontal/vertical curves, pavement design, and cross sections. This plan also addresses access and egress issues, along with Americans with Disabilities Act compliance, and includes appropriate drawings and specifications.
4. **Drainage plan**—Includes justification for the drainage layout including plans to retain all stormwater on-site, for example. Calculations for runoff routing, retention, storage, and treatment, as well as appropriate drawings and specifications, also are included.
5. **Plumbing and HVAC plan**—Includes an analysis of the fixture units, meter sizing, and sizing of pipelines within the building using EPANET or similar simulation software, along with appropriate drawings and specifications. It also includes a concept for design of the HVAC system, with capacity of air handlers, cooling/heating units, control systems, and ductwork.
6. **Roof plan**—Includes evaluation of alternatives to remove stormwater from the roof of the building in a timely fashion, particularly if vegetated roofs (or "green roofs") are utilized. Students use the appropriate building code and their knowledge of applied hydraulics to design slopes, scuppers, overdrains, downspouts, and roof sealing materials to prevent roof leaks. Appropriate drawings and specifications also are included.
7. **Utilities plan**—Includes a capacity analysis for sizing potable water, sanitary sewer, reclaimed water, stormwater pipelines, and irrigation. Design drawings include plan and profile views.
8. **Landscaping plan**—Many local governments have landscaping ordinances. In many cases, these policies may restrict regional water usage and encourage xeriscaping concepts, drought-tolerant species, local flora, etc. Therefore, students are required to create a landscaping plan that is in compliance with all local and regional regulations.

9. **Energy model**—For long-term operation and sustainability of the building, students must run an energy model to determine what options they may be able to build into the structure to reduce energy consumption.
10. **High-performance building checklist**—Includes an analysis of the potential "green" features captured by the design team, as well as the required documentation.
11. **Preliminary cost analysis**—Outlines the capital costs for delivering the building project, including equipment, materials, labor, installation, design development, and contingencies.

1.9 Assessment

Homework assignments that cover the various lecture topics are collected approximately biweekly. Homework subjects include technical communication skills, resume writing, interpretation of codes, and analysis of engineering alternatives. A final examination is given on lecture materials, largely focused on engineering economics, technical communication skills, alternative analysis, and site planning. In this exam, students are encouraged to use the *FE Reference Handbook* and FE-approved calculator. Another component of the final grade is a subjective score assigned to level of professionalism, as exhibited by the quality of in-class discussion and interviews with instructional staff outside of class. Before grades are assigned at the end of each phase, it is recommended that involvement in at least one professional society function, event, conference, or competition be required. The purpose is to encourage students to develop networking opportunities with practicing professionals outside of the educational setting (and get jobs). Also, the reputation of the institution within the local community is enhanced when students participate in civic engagement activities with their future peers.

Interpersonal relationships with teammates are an important component of the group dynamics; therefore, students submit time sheets for approval signatures (from the project manager and one of the principals) on a biweekly basis. At the end of each phase, students are asked to reflect on the performance and contributions of their team members as well as submit a self-evaluation, as part of the *Final Predesign Notebook*.

By design, the largest portion of the grade should be allocated to the professional practice assignments. These are comprised of (1) timed 20-minute oral presentations with 10 to 15 minutes for answering questions and feedback and (2) written reports, each of which is assigned equal weight. Group members generally receive the same score on presentations and on submittals, except in very unusual circumstances (e.g., a team member does not participate). The oral presentations allow the student groups to experience what it is like for consultants to report to their clients. Minimum requirements for the content of each deliverable are explicitly spelled out in the syllabus. Student teams make final presentations to a jury consisting of members of the department advisory committee, faculty, invited professionals, alumni, and interested stakeholders. This final presentation is treated like a milestone event in the career of the engineering student. For many, it will be the final act before graduation. The popularity of the final presentation with local industry representatives is a testament to their understanding the importance of participating in the education of their future employees.

The jury for the final presentation scores the technical content, knowledge of the subject matter, effectiveness of visuals, organization of the presentation, coordination and participation of team members, professionalism, and responses to questions on a scale of 1 to 5. A sample grading sheet is shown in Table 1.1, which divides the evaluation into the following sections: content, organization, delivery, discussion, and overall impression. The ability of students to field questions during and after the presentation also is scored. A rubric is quite helpful in letting the students know the level of expectation for each of the presentations. A sample rubric is shown in Table 1.2.

Because students are "paid" in grades, the assignments should consist of evaluations based on the professional quality of the team-based interim professional practice assignments (pre-

Table 1.1 Example of a grading sheet used to evaluate student presentations

Please indicate a score based on a scale of 1 to 5 where: 5 = Excellent 4 = Good 3 = Fair 2 = Weak 1 = Poor				
	Team 1 Name	Team 2 Name	Team 3 Name	Team 4 Name
CONTENT				
Subject matter (technical content)				
Knowledge of subject (understanding of issues demonstrated)				
ORGANIZATION				
Introduction (objectives clearly presented)				
Continuity (facts presented in a logical sequence, transitions well)				
Conclusion (reasonable summary and recommendation presented)				
DELIVERY				
Schedule (effective use of time, well prepared, rehearsed)				
Body language (eye contact, no distracting or annoying mannerisms)				
Visuals (clear, free of clutter, effective, related to discussion items)				
DISCUSSION				
Question and answer (answers reflect understanding of topic)				
OVERALL IMPRESSION				

Table 1.2 Example of a rubric for student presentations

	Excellent	Good	Fair	Poor	Unacceptable
☑ Content	All team members display professional level of knowledge of subject matter, with no important content left out and no incorrect material presented.	All team members display professional level of knowledge of subject matter, with minor amount of subject material left out or minor amount of incorrect material presented.	Majority of team members display professional level of knowledge of subject matter, with minor amount of subject material left out or minor amount of incorrect material presented.	Some team members display professional level of knowledge of subject matter, with minor amount of subject material left out or minor amount of incorrect material presented.	No team members display professional level of knowledge of subject matter, with minor amount of subject material left out or minor amount of incorrect material presented.
◆ Subject matter	All important topics are covered during the presentation, with no essential elements missing or misrepresented.				
◆ Knowledge of subject	Each member of the team demonstrates an understanding of the essential topics presented.				
☑ Organization	Presentation has a strong introduction, an effective body of material that supports the conclusion, and a strong ending.	Presentation has deficiencies in only one of the following: introduction, body, or conclusion.	Presentation has deficiencies in two of the following: introduction, body, or conclusion.	Presentation has deficiencies in all of the following: introduction, body, and conclusion.	Presentation is missing introduction, body, or conclusion.
◆ Introduction	Presentation starts strong with scope and objectives clearly presented.				
◆ Continuity	Facts are presented in a logical sequence and presentation transitions effectively between speakers.				
◆ Conclusion	Finishes strong with reasonable summary and/or recommendations presented, as justified from the body of the presentation.				
☑ Delivery	Presentation is effective in terms of rhythm, visuals, and presenters' body language.	Presentation has deficiencies in only one of the following: rhythm, visuals, or presenters' body language.	Presentation has deficiencies in two of the following: rhythm, visuals, or presenters' body language.	Presentation has deficiencies in all of the following: rhythm, visuals, and presenters' body language.	Presentation is clearly not rehearsed, visuals are unprofessional, and/or presenters' body language is unprofessional.

	Excellent	Good	Fair	Poor	Unacceptable
◆ Rhythm	Presentation demonstrates effective use of time, presenters seem well prepared, and presentation appears rehearsed.				
◆ Visuals	Visuals are effective, free of clutter, related to the discussion, and meaningful.				
◆ Body language	Presenters maintain eye contact with the audience and are free of any distracting or annoying mannerisms.				
☑ Discussion	All questions are fielded professionally, confidently, and correctly while avoiding defensive or argumentative responses.	Majority of questions are fielded professionally, confidently, and correctly while avoiding defensive or argumentative responses.	Some questions are fielded professionally, confidently, and correctly while avoiding defensive or argumentative responses.	Only one question is fielded professionally, confidently, and correctly while avoiding defensive or argumentative responses.	None of the questions are fielded professionally, confidently, and correctly while avoiding defensive or argumentative responses
◆ Question and answer session	Answers supplied reflect an understanding of the topic.				
☑ Overall impression	Presentation addresses all important subject matter, demonstrates conceptual understanding of the content, and responds to the purpose of the report; slides are cohesive, clear, concise, and organized well; presentation has many strengths; delivery is professional; question and answer session shows excellent engineering judgment.	Presentation addresses most of the important subject matter, demonstrates conceptual understanding of the content, and responds to the purpose of the report; majority of the slides are cohesive, clear, concise, and organized well; presentation has strengths; delivery is professional; question and answer session shows good engineering judgment.	Presentation addresses some of the important subject matter, demonstrates conceptual understanding of the content, and responds to the purpose of the report; some of the slides are cohesive, clear, concise, and organized well; presentation has few strengths; delivery is professional; question and answer session shows some engineering judgment.	Presentation addresses little of the important subject matter, demonstrates conceptual understanding of the content, and responds to the purpose of the report; some of the slides are not cohesive, clear, concise, or organized well; presentation requires major revision; delivery is not professional; question and answer session shows lack of engineering judgment.	Presentation is completely unprofessional.

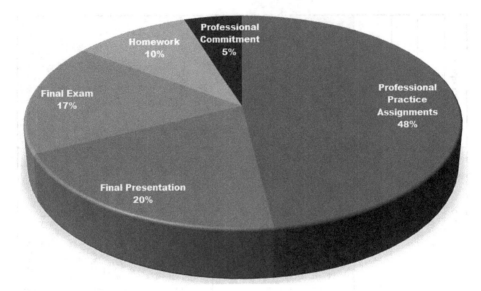

Figure 1.3 Example of grading breakdown structure

sentations and reports) and the final presentation and report. Individual performance also is evaluated on exams, homework, and other assignments such as class discussions, critiques, feedback, and attendance at professional meetings (professional commitment). An example of a grading breakdown structure is shown in Figure 1.3.

1.10 Last Words

The course content and delivery of a capstone design class are very different from traditional lecture courses in which students are trained to work out problems in the back of the book. Because the capstone class focuses on practice-oriented design, the course functions to tie all of the fundamental concepts together in a realistic project that will better prepare students for the upcoming challenges in their careers. The opportunity to work on a real project that will actually be constructed is valuable exposure to engineering practice. The demands of the course require the students to develop time management skills as they learn to work successfully in a team environment on their own. The open-ended nature of the course allows students to explore creative alternatives, and the participation of industry partners acts to channel that creativity so as to be grounded in practical feasibility and cost competitiveness.

1.11 References

Kumar, S. (2000). "Industry participation in a capstone design course." *Proceedings of the ICEE 2000 Conference,* Taipei, Taiwan.

National Academy of Sciences (2005). *Rising above the Gathering Storm: Energizing and Employing America for a Brighter Economic Future,* Committee on Science, Engineering, and Public Policy (COSEPUP) Division on Policy and Global Affairs, statement delivered by Chairman N. Augustine to the Committee on Science, U.S. House of Representatives, October 20, 2005, National Academy Press, Washington, DC.

Padmanabhan, G., and Katti, D. (2002). "Using community-based projects in civil engineering capstone courses." *Journal of Professional Issues in Engineering Education and Practice*, 128(1), 12–18.

Seymour, E. (2001). "Tracking the processes of change in US undergraduate education in science, mathematics, engineering, and technology." *Science Education*, 86(1), 79–105.

Todd, R., Magelby, S.P., Sorensen, C.D., Swan, B.R., and Anthony, D.K. (1995). "A survey of capstone engineering courses in North America." *Journal of Engineering Education*, 84(4), 165–174.

Career Opportunities and Leadership

As an engineer, there will be many potential career opportunities in the job market and many opportunities to grow and develop. Engineers are in high demand because there are many types of engineering jobs available from which to choose, such as regulators (state, federal, and local governments), equipment manufacturing representatives, contractors, public works and utility managers (both public and private), engineering consultants, project managers, business owners, and academicians. The current infrastructure crisis and aging nature of the workforce will ensure a favorable job market for entry-level engineers for many years to come. Students should ask themselves if they want to be involved with project management, design, manufacturing and sales, construction, instruction, regulation, or higher education. There are exciting, interesting, and challenging job opportunities in each field.

It is encouraging to note that graduates from engineering programs rarely have much difficulty finding a job related to civil, environmental, construction, or geomatics engineering, especially if their communication skills are excellent. The importance of good communication skills cannot be overstated. Careers depend on them, so they will be discussed more later.

2.1 Self-Assessment

Each student needs to decide where his or her interests lie. No one should choose for an individual. Will the career path lead to the private sector or the public sector? If a student wants to have extensive hands-on training, then working for a large private firm or perhaps a government utility, transportation department, public works entity, or regulatory agency may have more appeal. Most large entities offer some degree of mentoring and oversight to allow an entry-level employee to grow into a professional engineer under the guidance of a more experienced engineering supervisor.

In contrast, a student who craves independence may be attracted to a small to medium-size private firm or agency. Keep in mind that engineers need a license to practice engineering (or to even call themselves engineers in some jurisdictions), so before starting a business, it is important to gain experience. Without the professional engineering license, starting a business will be extremely difficult. The downside to joining a smaller company is that there is less opportunity to receive mentoring, but the excitement of working on many different types of projects tends to keep boredom from repetitive work at a minimum.

Because engineers generally have many opportunities to advance and increase their salaries, especially after obtaining their professional engineering license, students should not focus too much on starting salary when first joining the workforce. Keep in mind when getting that first job that the starting salary should not be the only criterion. It is vastly more important to enjoy the job and the tasks being performed so that the job does not feel like work. A job should be more than that—it should be interesting and challenging, with varied types of roles and responsibilities. There also are other considerations in any evaluation of a prospective job: working hours, travel requirements, retirement programs, lifestyle, training opportunities, and opportunities for personal growth and career advancement.

It may be helpful to do some soul searching and ask some important questions. Remember to be honest here:

◆ What do I want to do first?
◆ What do I want during the next 3 to 5 years?
◆ Where do I want to be in 10 years?
◆ Where do I want to be in 30 years?

Develop a plan and implement it. For example, in the third year of engineering school, students should be pursuing internship opportunities, should be holding leadership positions in a professional society, should be performing well academically in the junior-level coursework (this is the core engineering curriculum), and should be identifying and refining personalized engineering interests and career goals. By the fourth year, students should be updating their resumes, actively conducting a job search and attending career fairs, interviewing, networking with practicing engineers through professional societies, attending professional meetings regularly, contemplating graduate school, and preparing for (and passing) the Fundamentals of Engineering exam.

Young engineering graduates are well positioned to be successful, but there are important concepts to keep in mind during the transition from a supportive academic world to the harsh reality of the professional world, where difficult business decisions are made often. Bill Gates, former CEO of Microsoft Corporation, has established a series of rules about life for potential graduates in America, which apply to engineering graduates as well (Sykes 1995):

Rule 1: Life is not fair—get used to it!

Rule 2: The world doesn't care about your self-esteem. The world will expect you to accomplish something BEFORE you feel good about yourself.

Rule 3: You will NOT be a vice-president until you earn it.

Rule 4: If you think your teacher is tough, wait until you get a boss.

Rule 5:   No job is beneath your dignity. Your grandparents had a different word for burger flipping and house cleaning—they called it opportunity.

Rule 6:   If you mess up, it's not your parents' fault, so don't whine about your mistakes; learn from them.

Rule 7:   Before you were born, your parents weren't as boring as they are now. They got that way paying your bills, cleaning your room, and listening to you tell them how idealistic you are. So before you save the rain forest from the blood-sucking parasites of your parents' generation, try delousing the closet in your own room.

Rule 8:   Your school may have done away with winners and losers, but life HAS NOT.

Rule 9:   Life is not divided into semesters. You don't get summers off, and very few employers are interested in helping you FIND YOURSELF. Do that on your own time.

Rule 10: Television is NOT real life. In real life, people actually have to leave the coffee shop and go to real jobs.

Rule 11: Be nice to nerds. Chances are you'll end up working for one someday.

2.2 Types of Job Opportunities

In the public sector, opportunities exist at various levels of government, such as:

◆ Federal government (e.g., Environmental Protection Agency, Department of Energy, Corps of Engineers, Federal Highway Administration, etc.)

◆ State government (e.g., departments of environmental protection, water management districts, departments of transportation, etc.)

◆ Local governments including cities, counties, and special districts (e.g., water and sewer departments, public works divisions, utility authorities, permitting offices, growth management, etc.)

◆ Others (e.g., universities, NGOs)

Public sector jobs usually have greater job security and more stable salaries, but this does not mean that the compensation is necessarily lower than in the private sector. Public sector positions traditionally have had strong retirement programs, long hours (which limit flexibility), and more repetitive or mundane types of work (offering less challenging projects for young engineers), although the opposite can also be true—lots of challenging opportunities. Often, public sector employers are looking for people who want to "step up" or get ahead quickly. Young engineers can help advance their careers quickly under such conditions. However, some public sector jobs have a graduation grade point average (GPA) requirement of 3.0. If a graduate's GPA is lower than this, then job opportunities may be restricted.

In the private sector, opportunities exist in many different types of companies, such as:

◆ Developers
◆ Consulting firms
◆ Contractors

- ◆ Manufacturing and sales
- ◆ Vendors
- ◆ Private utility companies (e.g., telephone, power, cable, water, and sewer)
- ◆ Privately held not-for-profits and NGOs if not public
- ◆ Research and development (e.g., pharmaceutical companies, petroleum/mining companies)

Private sector jobs generally have different advancement potential and greater earning potential in those companies that are successful in garnering work. The downside will be the number of hours, which can vary widely depending on the amount of work, and less job security, which is more dependent on the economy rather than performance and merit.

Key items of interest when sorting out potential employment opportunities include:

- ◆ **Organization**—Is the organizational structure conducive to future success and vertical advancement?
- ◆ **Profession**—Does the job description align with personal career goals and interests? If not, the chance of success diminishes.
- ◆ **Geographical location**—How willing is a job candidate to move from his or her current location? Is the candidate willing to work in a particular region or climate zone? What about lifestyle? This choice may put an artificial limit on potential job opportunities. For instance, it may be necessary to take a lower paying job in order to stay in a preferred geographic location or to get better hands-on training.
- ◆ **Personal guidance**—An engineer needs to have someone in his or her early career, preferably a professional engineer, who can act as a mentor on the path to licensure. It also is helpful to have an employer that is open to training opportunities to expand a young engineer's knowledge of the field.
- ◆ **Growth**—Getting good guidance will help an engineer develop professionally—especially to get his or her license. Things like involvement in professional societies can expand opportunities. Some firms will even pay for graduate studies. This is a very good opportunity to enhance future career marketability.

Now that the ground rules are known, it is time to start creating and marketing a personal brand to provide the best odds for success in the profession.

2.3 Branding

A student who wants to be successful needs to acquire experience. This means networking and having a resume that stands out, having excellent communication skills, and making a good first impression. This is known as primacy (which will be discussed in Chapter 5). The resume is the first critical piece of branding. Often this document is the first opportunity to make a good impression with a potential employer, as most job applications require a resume and cover letter. The resume should be concise, easy to read, free of errors, and completely honest. For the most part, it contains a summary of the applicant's experience, educational background, any other pertinent qualifications, and contact information. The key to writing an effective resume is being aware that the reader probably will scan it for only a few seconds before making a yes or no decision (Kaplan 2012). Using eye-tracking technology, 30 professional recruiters were sampled over a 10-week period, and it was found that recruiters spend most of the 6 seconds they look at a resume on the applicant's name, educational history, current position, and most recent past position (Evans 2012). The design and layout of the resume, then, should focus on these key elements in the top third of the first page so as not to be overlooked.

The major elements of a resume for engineers are described in more detail as follows:

◆ **Contact information**—The applicant's name should appear at the top of the page, centered in large type, followed by permanent address, telephone number, and email. Students in college generally have a temporary address as well, so it is important to make sure that any mail sent to the address on the resume will be received. Otherwise, a job offer could go unnoticed. Make sure that the email address does not have an offensive alias. It might be a good time to create a professional email account with your full name. Then make sure to check the account daily so as to respond to any request for an interview in a timely fashion. The telephone number likely will be a personal cellular phone, so make sure that the voice mail message is professional because recruiters will not be amused by a childish message. Another tip is not to take up too much unnecessary space with all of the contact information; just make sure it is at the top and it is correct. Add a link to your online profile, but not Facebook®.

◆ **Objective**—This optional section describes the career goal of the applicant. It states what kind of job the goal is. The more specific to the field of engineering, the more likely it is the human resources manager will be able to redirect the resume to the appropriate hiring department. This statement, if included, should be targeted specifically to the job description in the advertisement. For example, if the job is for an entry-level traffic safety specialist, then the objective should include the words transportation engineering. Do not make the objective generic or fluffy, like "seeking a challenging, high-paying job with a Fortune 500 company with ample opportunities for advancement."

◆ **Education**—This section describes postsecondary education in reverse chronological order, that is, the most recent first. Do not include high school here. List all universities or colleges and the degrees earned, or if pending include the expected graduation date. Include other institutions attended beyond high school, even if no degree was obtained. Recent graduates or soon-to-be recent graduates typically will include minors, certificates, relevant coursework, and specific computer programming or software skills. With respect to GPA, listing it on the resume is optional, and it should be included only if good. A 2.3 GPA is a deterrent. Interviewers can simply request the transcripts if they must know the GPA.

◆ **Licensure**—It is assumed that engineering students and recent graduates will take and pass the Fundamentals of Engineering exam, so make sure to include a section that provides the license number, the state, and the date granted. Then update this section after passing the Principles and Practice of Engineering exam.

◆ **Skills**—This section should include things such as computer software knowledge, instrument experience, field sampling, surveying, foreign languages, and other relevant skills.

◆ **Experience**—This section describes the candidate's work experience, listed in reverse chronological order. For each position listed include the dates, the job title, the employer and address, and a brief description of the duties using action words. Do not be afraid to list nonengineering-related work because gaps in the work timeline are flagged more often than unrelated work. For example, if a previous or current job is in retail, then focus on describing any engineering-related activities, such as leadership roles, bookkeeping, supervision, training, report writing, management, technical equipment operation, productivity optimization, client interactions, volunteer work, other responsibilities, etc. Make sure to describe past job duties in the past tense using action words like "supervised three interns" or "increased productivity by 17% in the first year." When describing current work experience, use the present tense active voice, like "designing an upgrade for a 15-million-gallon-per-day reverse osmosis water treatment plant." Devote some time to crafting this section carefully because it is one of the sections that is critically reviewed by potential employers. Be specific and be concise, while remaining focused on engineering-related skills that will translate to the job in question.

◆ **Other information**—This optional section is reserved for items like military service, honors, awards, memberships, interests, hobbies, language fluency, etc. It is unlikely that this section will be heavily weighted by prospective recruiters, but it may help to show how well rounded the candidate is or break the ice with the interviewer.

◆ Do *not* include birth date, marital status, family information, religious denomination, or similar information. There are laws regarding what employers can ask about these issues. Including them in a resume puts employers in a compromised position and indicates the applicant is unaware of basic personnel practices, which can result in disqualification from consideration.

Remember that developing a resume is not a one-time job. It should be updated continuously with every new accomplishment, new skill, and new career milestone. No two resumes will ever be alike, but a sample of an acceptable resume format is shown in Figure 2.1.

In the era of social media, traditional resumes do not differentiate people, except in exceptional cases, so students need to take proactive measures. Dan Schawbel (2009) wrote a book called *Me 2.0*, which outlines how people and companies should create themselves as a "brand" to let others know who they are, what they stand for, and what their beliefs are. The concept is called branding, and the first impression often is the one that sticks with others. As a result, it is important that a personal "brand" be effective.

Employers will find out how much a student cares about the engineering profession by performing Internet searches by name and reviewing comments made on social networks. Anything negative that exists online can and probably will be found by a prospective employer. Consultants, entrepreneurs, and job creators are increasingly using social networking sites like LinkedIn® to find suitable candidates for employment. This is the modern way to find a job and market online. CareerBuilder.com estimates that 53% of hiring managers use social networks as a background check (CareerBuilder.com 2010), so it is absolutely critical to manage personal online content!

Start developing an online profile by registering for popular social media networks; for example, create a Google® account (http://google.com/profiles) or a LinkedIn account. An individual's full name should be used to create a personal account. Nicknames can give the wrong impression or be confusing to potential employers. Fill out all of the sections that relate to the following pieces of information: professional summary, experience, education, recommendations, personal (professional) websites, interests, keywords, and groups. Always include a recent photo (professional quality in formal business attire). Make the profile public, and start a personal network by linking to other social media sites, like Twitter®,

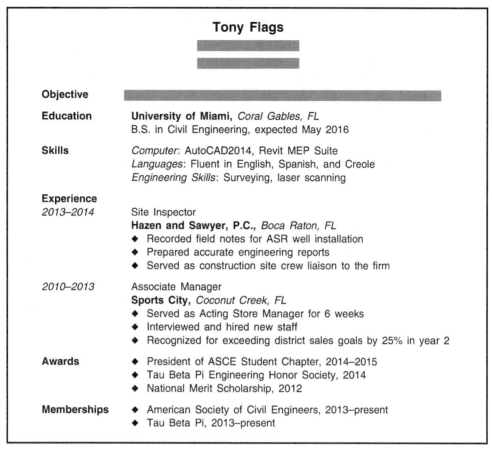

Figure 2.1 Sample of an acceptable resume format

Facebook, etc., to locate personal contacts such as classmates, faculty, coworkers, employers, etc. Create a custom URL using your full name. Update and improve the profile, and take advantage of site tools like "improve your profile" (be sure to follow all directions). Make sure to write a strong personal value statement. Give and receive recommendations. These kinds of firsthand endorsements are highly valued by recruiters.

Conduct an Internet search of yourself to see what online content is associated with your name and review the top 10 results. There are too many horror stories about what is revealed online. For example, one student was surprised to discover that the first item that appeared in an Internet search of his name was a newspaper article about a drug-related crime. Another found out her DUI mug shot and subsequent mug shots for violating her DUI probation were two of the top four photos that show up under images when searching for her full name. Imagine how prospective employers viewed her candidacy after finding those! You might be surprised by what comes up, but note that this is what potential recruiters will be using to form their first impressions. The Internet is forever, but steps can be taken to remove undesirable content.

Branding identifies students by who they are, why they should be considered for a job, or why they might be qualified to do certain kinds of work. Students will use branding in class, their careers, and with professional organizations. It is the entrée into the network of professionals.

2.4 Create the Group Design Firm

One challenge students find difficult to cope with is that the future is now. Students need to practice what they should do in the future while in the relatively safe environment of the capstone class. The focus will be on the design team and the expectations for each of its members. Armed with a loosely defined project scope, the first step is for students to introduce themselves to their new teammates and exchange contact information. The next order of business is to select or elect a project manager and then come up with a creative company name that adds meaning to the group. Make it memorable, and try to avoid dull or uninspired names like initials (e.g., J.S.B., Inc. for John, Sally, and Bob), any trademarked names (e.g., Pepsi®, Ace®, etc.), sports franchises, or college mascots.

An example of a clever name is *Catalyst Engineering Consultants*. A catalyst is an instrument of change, and the team wanted to be known to outsiders for their fresh new ideas. A catalyst allows the reactants to come together to speed up the process, indicating that the team works together with the client to get the job done right and on time. Plus, a catalyst is never used up in a reaction, so it is sustainable, signifying that the team considers sustainability concepts as a priority.

Along with a thoughtful concept for the team name, a vision statement will solidify what things are important to the team and how they will impact decision making. Ideally, these core values should connect back to the goals of the project. The statement clearly and concisely conveys the direction of the organization and its priorities. A vision statement is not a slogan. It powerfully communicates the team's intentions and motivates the team toward successful project completion. Some examples include:

> Systematically solving society's situations with safe, sustainable, and superior structures.

> We embrace innovative ideas beyond your imagination with safety, efficiency, and the sustainability of our natural environment as our priorities for a better future.

Although these may sound a lot like slogans, they also convey useful information, like what the major priorities of the team are.

2.5 Teaming Skills

Acceptance of others and having empathy are the first steps toward coming together, which is the beginning of teamwork. It is important to remember to listen respectfully and carefully to what others have to say. Be aware and be perceptive: there may be something that will be useful to help keep the team working together effectively or solve a team problem. In a team environment, "self-promotion" is counterproductive. The goal is to become more than the sum of the individual parts. Students must suppress any desire to control a situation to gain advantage or personal prestige because eventually this will result in breaking down the team effort and limiting the spontaneity and creativity that can lead the team toward innovative solutions.

Teams that discuss the issues, share opinions, resolve disagreements, and gain an understanding of all the differing points of view offered are teams that function well and perform effectively. Therefore, conditions need to be established whereby every team member has an opportunity to speak and be heard. Some team members may be uncomfortable expressing their opinions, questioning others, or defending points of disagreement in group settings. They likely will distance themselves for fear of confrontation or humiliation. If this happens, one or more important points of view will be silent or lost. It therefore makes sense to create some important ground rules for teamwork:

1. **Speaking**—Only one person speaks at a time. Speak so everyone can hear. Make sure that everyone can hear clearly. Never allow the loudest voice in the group to seize control of the proceedings.
2. **Listening**—Give the speaker full attention. Stay receptive to what others have to say and remain open-minded to new ideas. Listen without making assumptions or judgments. Acceptance of others and having empathy are important prerequisites. Do not engage in side conversations while someone is speaking. Turn off mobile devices. Be aware and be perceptive. Take a moment to understand the argument and comprehend the meaning of the words before reacting.
3. **Using time wisely**—For all team meetings, rehearsals, presentations, and appointments, make a commitment to be on time, start on time, and end on time, as mutually agreed upon. Showing up on time is not enough; be prepared to get the work done in the time allotted.
4. **Focusing**—Stay on target, addressing what needs to be accomplished "in the now" at each team meeting. Use an agenda that states the goal of each meeting at the outset to avoid wasting valuable time. If the discussion goes off course, bring the team back to the goal of the agenda item.
5. **Being open to the outcome**—Keep an open mind about what the outcome of the meeting might be. It could be very different and more refreshing than anticipated.
6. **Honoring personal commitments and trusting in team decisions**—If the team has made a decision, then move forward. There is no sense in continuing to argue a point that has already been decided. If something is not working out, then be flexible enough to admit mistakes and move on from there.
7. **Always making the best effort possible**—It is unethical to do anything else. A team that aims for mediocrity inevitably will miss that goal. Setting the standard too low is a recipe for failure. Set the bar high, and aim to be the best. Show competitive spirit and strive to reach the top. Always play to win!

The best teams are made up of individuals who go out of their way to make each other look good. An effective team has many open lines of communication and meets frequently to discuss progress, plan for future deliverables, evaluate concepts, make decisions, work on deliverables, check each other's work, rehearse presentations, and edit written submittals. Effective team meetings can be very helpful in keeping the project on schedule. However, if team meetings are ineffective, they can be a source of dissention and erosion of team cohesiveness. A good way to keep team meetings successful is to reflect and evaluate how productively the time was spent:

◆ Did the meeting start and end on time?
◆ Did everyone participate?
◆ Were important issues discussed and decisions made?
◆ Did the team reach a consensus solution?

- ◆ Did all team members engage and add to the discussion?
- ◆ Did the team allow new ideas to come forward?
- ◆ Did the team explore these and include them?
- ◆ Was any negative issue brought up?
- ◆ Was the team able to resolve the situation?
- ◆ Was the meeting efficient and effective?
- ◆ Did the team make decisions?
- ◆ Did someone record them?
- ◆ Did the team stay on the agenda?

No one said it would be easy, and often, just as in real-world situations, conflicts, difficulties, and communication issues arise within teams. This should not come as a surprise. Students need to learn to deal with difficulties in the work environment too. Future success depends on being able to deal with adversity and resolve conflict appropriately. Failure to address issues when they first arise will slowly erode team chemistry, impact the quality of the work, and ultimately affect the students' grades. Welcome to the real world.

Key attributes related to effective high-performing teams include:

- ◆ The group has well-defined goals and practices good planning techniques.
- ◆ Each team member has well-defined roles and responsibilities.
- ◆ The group has decided how to enforce accountability.
- ◆ The lines of communication among members remain open at all times.
- ◆ Everyone is encouraged to participate and voice their opinion without reprisal.
- ◆ The group has established a means to manage and resolve conflict.
- ◆ Methods to encourage innovation are in place.
- ◆ There is evidence of strong leadership.

Some early team exercises should be undertaken to build team chemistry. For example, teams should answer the following questions:

- ◆ What are the team goals?
- ◆ What planning is required?
- ◆ What are the appropriate roles and responsibilities?
- ◆ How will the team deal with leadership and responsibility?
- ◆ How will the team resolve or manage conflicts if they occur?
- ◆ How will the team maximize communication among members?
- ◆ How will the team encourage innovation in the process?

Good team members take responsibility for their assigned roles and duties. They deliver on their commitments with professional-quality work on time. They contribute to team discussions in planning meetings, listen effectively, and ask helpful questions. They should give and receive useful feedback and do *more* than their fair share of the work. Bad individual team members can be difficult to deal with, and they come in many varieties that students need to learn to recognize and deal with in a positive manner. Some examples follow:

- ◆ **The know-it-alls**—These people are arrogant and usually have an opinion on every topic, but when proven wrong, they become defensive and aggressive or uncooperative. Students must develop strategies to help them see things from the point of view of others in the group once in a while.

◆ **The passives**—These people never offer any ideas or let the group know where they stand. They just want to get by with exerting a minimum of effort. Students need to find ways to bring them out of their shells so they can contribute in a meaningful way. Otherwise, they can manipulate the rest of the group indirectly.

◆ **The dictators**—These people are bullies. They intimidate and dominate the dialogue. They are constantly demanding and brutally critical without empathy. Students must find ways to tone them down while avoiding direct conflict.

◆ **The "yes" people**—These people agree to any commitment but never follow through. They cannot be trusted to deliver. Students must find ways to get them to come through in a crunch.

◆ **The "no" people**—These people are quick to point out why something cannot be done. They are dogmatic, inflexible, and negative. Students must find strategies to turn that negativity into productivity.

◆ **The complainers**—These people prefer to complain about something rather than try to find a solution or provide useful input. Students must find ways to move them into problem-solving mode.

Unfortunately, it is a distinct possibility that everyone will eventually encounter these kinds of character traits, and it is a good bet that most already have come across examples of these types of personalities or recognize them within. These are the people everyone works with, depends on, lives with, and deals with every day.

In addition to the traits of bad team members, there are basically three personality types that seem to hurt group dynamics and group performance the most (Felps et al. 2006). These are known as *bad apples*, because they can spoil the barrel of good apples. Bad apples can manifest themselves as (a) *jerks*, who attack or insult others in the group and take advantage of weak team members; (b) *slackers*, who do less than they are capable of because they are lazy, exhibit poor time management skills, or simply do not care; and (c) *depressive pessimists*, who hate the project, their team members, and everything else in their lives. If a team includes one of these, there is a good chance that person might spoil the group. Figure out how to engage each one productively.

Research shows that groups with bad apples perform 40% worse (Felps et al. 2006), despite the fact that their teammates may be very talented, very smart, or very likeable. On teams with a bad apple, teammates argue and fight and do not share relevant information. In other words, they communicate less. Even worse, team members may take on the bad apple's personality traits. When the bad apple is a jerk, other team members will act like jerks. When the bad apple is a slacker, other teammates will become slackers too, and so forth. Further, team members will not just act this way in response to the bad apple; they will act this way toward each other in a spillover effect, creating a downward spiral toward poor results. Watch out for individual team members who defer to a "natural leader" and keep their opinions to themselves, as well as people getting frustrated and conflicts erupting. Conflict is inevitable in a team process, and that is why teams should establish the guidelines for high-performing groups discussed earlier.

It all starts with personal accountability, and one way to demand accountability is to create a written agreement or contract called a team performance agreement, which includes some or all of the following items:

- ◆ Agree on the specific roles of each member.
- ◆ Agree on the guidelines for interaction and communication.
- ◆ Agree on where the meetings will be held.
- ◆ Agree on how disagreements will be handled.
- ◆ Agree on how conflict be will resolved.
- ◆ Agree on how final decisions will be made.
- ◆ Agree on the consequences of failing to meet the conditions of the agreement.

Decorum also is important both within and outside of the group. Decorum is correct or proper behavior that shows respect and good manners. Students are expected to respect one another and act in a professional manner at all times. To this end, it is important to state that it is disrespectful to hold a conversation in the audience while others are presenting. Streaming videos, downloading music, watching podcasts, checking email, texting friends, playing video games, updating profiles, doing homework for other classes, etc. are not conducive to learning. Such behavior is disrespectful and unprofessional and could be grounds for dismissal. There is a good reason to observe the other group presentations, which is to learn how to listen and at the same time to learn what presentation strategies are effective and which strategies do not work so well. Students might gain some insight on solutions and challenges with respect to their own projects merely by listening to the other groups. That is actually the point!

The worst offense is for a team member to hold a conversation while another team member is presenting. This shows a complete lack of respect. It also tells the audience that the speaker is not worth listening to. To those in the audience, it is abundantly clear which teams take the assignment seriously versus those who met just before class to try to pull something together at the last minute. Wasting student, faculty, and visitor time is disrespectful to group members, but more importantly, it is disrespectful to the client and the project itself. Ultimately, those who are not fully invested in the success of the team will negatively impact their personal professional reputation as well as the reputation of the engineering firm. Eventually this will affect one's ability to garner work and retain clients. Individuals who take this route do not last long in the profession. Respect is a goal for engineers—to gain the trust and respect of others. But respect is earned, not given, and it starts with decorum.

All engineering students can be successful in class and in their careers. The capstone class allows students to acclimate to the difference between academia and the profession. Through this process, students will develop an understanding of their responsibilities in the workplace and how their contributions can enhance the project for everyone involved. Success breeds more success for the organization, so start with good habits in class—now!

2.6 References

Beer, D.F., and McMurrey, D.A. (2009). *A Guide to Writing as an Engineer*, 3rd Ed., John Wiley & Sons.

Evans, W. (2012). "Eye tracking online metacognition: Cognitive complexity and recruiter decision making." TheLadders, <http://cdn.theladders.net/static/images/basicSite/pdfs/TheLadders-EyeTracking-StudyC2.pdf> (accessed July 2014).

Felps, W., Mitchell, T.R., and Byington, E. (2006). "How, when, and why bad apples spoil the barrel: Negative group members and dysfunctional groups." *Research in Organizational Behavior*, 27, 181–230.

Kaplan, K. (2012). "Job applications: Straight to the top of the pile." *Nature*, 488(7410), 241–243.

Markel, M.H. (1992). *Technical Writing: Situations and Strategies*, 3rd Ed., St. Martin's Press, New York.

McMurrey, D.A., and Buckley, J. (2007). *A Writer's Handbook for Engineers*, Thomson Engineering.

Schawbel, D. (2009). *Me 2.0: Build a Powerful Brand to Achieve Career Success*, Kaplan Publishing.

Sorby, S.A., and Bulleit, W.M. (2006). *An Engineer's Guide to Technical Communication*, Pearson Prentice Hall, Upper Saddle River, NJ.

Sykes, C.J. (1995). *Dumbing Down Our Kids: Why America's Children Feel Good about Themselves But Can't Read, Write, or Add*, St. Martin's Press, New York.

Vesilind, P.A. (2007). *Public Speaking and Technical Writing Skills for Engineering Students*, 2nd Ed., Lakeshore Press, Woodsville, NH.

2.7 ASSIGNMENTS

1. Consultants, entrepreneurs, and job creators are using online profiles to find suitable candidates for employment. This is an emerging way to find a job and market yourself online. Create a profile using an online service such as LinkedIn:
   ◆ Use your full name to create your account.
   ◆ Fill out all of the sections so that the profile is 100% complete. This includes key items such as professional summary, experience, education, recommendations, websites, interests, keywords, professional photo, and groups.
   ◆ Start a network by searching through the site to find personal contacts such as classmates, faculty members, coworkers, employers, etc.
   ◆ Write a strong personal value statement.
   ◆ Print out the complete profile or submit a link to your instructor.

2. Conduct an Internet search of your name and submit a screenshot of the top 10 results. Did you find any of the results in the top 10 to be misleading or damaging to your online reputation? Try the same thing with an images search for your name.

3. Complete a Google profile (http://google.com/profiles). Register your Google account and fill out all fields. Add a suitable avatar (professional photo), and make the profile public. Then link it up to all of the online profiles that you have created.

4. Discuss your favorite rule from Bill Gates' rules about life and describe how it can be useful in managing your career goals as an engineer.

3

The Profession
and Ethical Conduct

After earning an engineering degree, licensure is something that must be obtained to be successful in the long term. As will be discussed later in this chapter, holding a professional engineering (PE) license demonstrates to the public that the holder has the requisite education, experience, and knowledge necessary to make reliable engineering judgments that protect public health, safety, and welfare. The last part about the protection of public health, safety, and welfare is a public trust issue that rivals the expectations of doctors, and the reasons are similar. Most people do not really know what engineers do. In fact, many people, if asked what engineers do, will say that they drive trains (Figure 3.1). Oops. It would be great to alter that understanding!

Figure 3.1 An "engineer"

The public takes for granted such things as the delivery of water used to shower, the sewer lines for drains, the roads and bridges to get to work, the stormwater system that drains the roadways, and the structure of the buildings in which people work. A random person on the street probably does not know how a cellular phone works, how a television set works, or how a car works; people just have an expectation that these things will work. As a result, having a PE license allows professionals to perform engineering consulting, own their own businesses, and bid for public funding. Before talking about licensure, however, let's explore where the concept of licensure came from and the responsibilities that go along with it.

3.1 Engineering Ethics

The concept of licensure stems from responsibility to the public and the expectation of the public that engineers will act to protect their interests. It is an ethical responsibility. Ethics is an issue that comes up in the engineering business on an ongoing basis (some jurisdictions even require an ethics refresher course at regular intervals). But what are ethics? To begin to answer this question, we must turn to philosophy. A cursory review indicates that there are three potential definitions of an ethical person (Popkin and Stroll 1993):

- One who establishes a set of values and lives by them
- One who lives by any set of values that are shared by a group of people
- One who lives by a set of values that are universally accepted

Let's take a look at each one of these. The first definition is a person with a set of values who lives by them. What do we make of this definition? Do we accept it? Do we accept that a person who acts this way is ethical? In reality, few people buy into this first definition of an ethical person because the values can vary and may include individuals with highly personalized sets of ideals (e.g., Robin Hood) or individuals with frequently unaccepted behaviors (e.g., serial killers).

Obviously, then, a person with any set of values that are shared by a group of people who live by them must be better. What do we make of this definition? Do we accept it? Do we accept that a person who acts this way is ethical? These people share many of the same beliefs and conform to an accepted set of "rules" of acceptable behavior. Engineers are among the groups with common values. But alas, so are religious cults and political parties with ethics with which we may not necessarily agree. Worse still, groups with common "values" include terrorists, fascists, racists, and many others not generally associated with ethical behavior.

Obviously, then, a person with a set of values that are universally accepted must be better still. What do we make of this definition? Do we accept it? Do we accept that a person who acts this way is ethical? Find one example of a universally accepted ethical value—just one! There is not one, so that does not work very well either. Thus the philosophical answers are not very helpful.

Another approach is to look at professions. What professions do most people perceive to be unethical (ignoring for the moment whether or not the perception is reality)? Professions that may be perceived to be unethical by the general public on a routine basis include:

- Salespeople of any type
- Lawyers of any type

◆ Politicians of any type
◆ Financial brokers and bankers
◆ Realtors
◆ Mechanics
◆ Contractors

Illegal enterprises also can be included here, but illegality is not necessarily an ethical issue. Organized crime typically has a set of ethical values and core principles which those in it swear by (see groups with the same ethics above). That does not make these syndicates acceptable to society by any stretch of the imagination, but there is a fundamental set of expectations within the organization that fits the working definition. Ethics and legality are different principles. For example, prostitution is considered by many to be an "unethical activity," but if you get what you pay for, what is unethical? The fact is, prostitution is illegal in the United States (except in Nevada) or perhaps immoral, but unethical? The commonality among the professions perceived to be unethical lies in the fact that money is going to those in the profession, and they are those who primarily benefit. Those who are perceived as being unethical all want your money.

Now let's compare professions that generally are perceived as being ethical:

◆ Engineers (at least we hope)
◆ Scientists
◆ Medical personnel
◆ Teachers
◆ Public safety workers
◆ Health care providers
◆ Social workers

Generally speaking, the perception is that people who work in these professions serve the public and protect the public interest. In addition, most have to be licensed and are regulated. If things go wrong with these service providers, they can be brought before professional boards and reprimanded. The public has expectations that those practicing in these professions know what they are doing, even though the public may not understand what they actually do. There is a trust factor for these professions thought to be ethical which presumes competency with accountability.

Another common trait among these professions is that most of their decisions are based on judgment. Doctors, like engineers, have imperfect information, but they make a diagnosis based on their "best educated guess," given the facts. However, the public expects that they will come up with the correct assessment every time. Many situations do not have definitive answers, and furthermore, things are always changing.

Thus, public expectations of competence and application of judgment are considered good, but that leaves us with only a foggy idea of a perception of ethics and no real answers. Perhaps a historical review would help.

3.1.1 Where Do Ethics Come From?

Around 10,000 B.C.E., human civilization was thought to consist of disparate bands, or tribes, of hunter-gatherers that subsisted on wild animals and plants. The people who brought home the food were respected. The elders were respected for their wisdom as well as their leadership. Shelter providers and those loyal to the group also ranked highly. Members who were considered food hoarders, selfish, thieves, or freeloaders were not appreciated. Offenders were, in many cases, cast out and banished from the group, condemned to survive alone in the wild. At that time, few people, if any, were capable of survival alone in the wilderness, so this banishment was somewhat of a quasi-death sentence. The tribe, in order to be successful, figured out how to control its members to ensure the survival of the collective. Legends were passed down to demonstrate ethics and propagate acceptable behavior. Those tribes that were successful figured this out. We know little of those that did not.

Three thousand years later, agriculture began and changed the way people lived and proliferated. Agriculture meant better nutrition, which meant higher birth rates. Larger populations could be supported, which in turn meant a need for more agriculture. With more harvests came the opportunity for a surplus, and efficiency led to a surplus of products to trade with other tribes. Because agriculture is stationary, there needed to be a place to trade goods. Hence, since people settled in one place for agriculture, the tribes established villages and commercial systems for trading purposes. To increase agricultural efficiency with more productive harvests, it became necessary to manage water supplies for irrigation. Armed with a surplus of food, some in the community became free to specialize in other tasks. While the farmers specialized in growing wheat, barley, or grapes, for example, others specialized in raising animals, making tools, processing food, healing the sick, building defenses, and trading with other tribes. There was an expectation that traders would deal fairly with the farmers and merchants. The villages also needed to build infrastructure to maintain order and promote efficiency. This infrastructure included shelters to protect families from the elements, defenses to protect the village water supplies from raiders, roads to transport goods, and storage facilities to protect the harvest.

Successful groups of villages became city-states that traded on a larger scale. They needed roads, bridges, ports, drainage systems, and methods to deal with wastes. They therefore began to create laws to regulate trade among people who were unknown to each other. Laws were later recorded, and trade centers with specialties were created. The best of the products were in high demand, but were dependent on others to provide packaging for shipping. This dependency created the expectation that the packagers knew what they were doing—trust. As demand outstripped a given merchant's ability to supply, more packagers were needed, but the expectation was that the quality would remain the same. Products that failed could be catastrophic for a village, so training guilds and similar associations were created to ensure that the workers provided the level of expertise and workmanship that the public expected of them. Thus, the public expectation that people would do their jobs correctly was beginning to become tradition. This system actually remains much the same today; it is called apprenticeship. In reality, the 4 years that engineers spend in the field working under a licensed engineer after school is their apprenticeship. But none of this quite gets to the definition of ethics. That is where the philosophers weigh in.

3.1.2 The Philosophers Weigh In

Most ancient philosophers lived in challenging times. As the Greeks developed the laws of mathematics, Greek philosophers tried to discover complementary behavioral laws to explain life and explain why life went awry. The belief at the time was that society's failures were caused by the failure to follow some behavioral law. The ancient philosophers, starting with Plato and Aristotle, thought that education and leading a proper life were important to avoiding difficulty, so learning was encouraged. The result was philosophy, which is *the study of consequences for implementation of a series of behavioral principles*. Ethics are those behavioral principles.

Over the last 2,500 years, numerous philosophers have pursued an explanation of this elusive concept. One philosopher whose work directly affects the engineering profession is Immanuel Kant (1724–1804). Kant contributed the concept of "duty" and the need to evaluate whether or not an action was ethical based on its impact to society. He proposed that people have a duty to act in a certain manner and that every action could be evaluated by a universally accepted code of behavior. Society does this today. The question Kant raised was: If everybody did [insert specific activity], would society function (Mantell 1964)? If the answer was no, the behavior was deemed unethical. If the answer was yes, the practice was considered ethical. This relates directly to the engineer's responsibility to protect the health, safety, and welfare of the public and therefore has direct applicability to the engineering profession.

Utilitarianism appeared in the 18th century and espoused that an action is ethical if it pleases the greatest number of people. This scenario is how most democratic governments function today. The majority rules because that makes the most people "happy." Engineers use this to do alternative analyses by selecting the preferred option using an objective set of selection criteria established *a priori* (see Chapter 6). However, making the majority happy does not equate to fairness or a good society. Therefore, in many democratic governments, a court system has been established that permits the "unhappy" to challenge a law or policy that disadvantages them.

3.1.3 Creeds, Codes, and Canons

As the engineering profession developed formally in the late 19th century, societies of engineers were created. For example, relevant to civil engineering is the American Society of Civil Engineers (ASCE), which established a set of codes, creeds, and canons to guide engineers to the proper actions to take, given certain circumstances. This was put forth before licensing boards were created, when self-regulation was the rule.

As the engineering profession developed its rules, it also created consequences in the form of disciplinary actions for failure to abide by them. There is a theme that underlies the rules established by the ASCE: the top priority under *all circumstances is* the health, safety, and welfare of the public. Legal issues are second, and third is the engineering profession itself. This is because the profession must be perceived to be ethical, because damage to that perception will damage all engineers and the profession itself. The client is fourth on the priority list, and the engineer is last on the hierarchy.

There are a series of other provisions to observe. Engineers must perform work within their area of specialty and not in an area where they may lack experience or competence. For example, a chemical engineer may not be qualified to design structures. Therefore, an engineer must not accept a project if he or she does not have the skills to design it. In other words, engineers must recognize their competencies as well as their personal limitations. This is directly related to the protection of the public health, safety, and welfare. Engineers are required to issue objective and truthful comments and avoid conflicts of interest. They must build their personal reputations on *merit*, uphold the dignity of the profession, and continue professional development throughout their careers. Most states require a specific number of hours of continuing education for each license renewal cycle. This is how engineers keep up with new technology and changes to relevant design codes and standards.

Only another professional engineer is allowed to review the plans of a professional engineer. From time to time, engineers may be paid by someone to review another engineer's plans, but the original engineer must be notified in writing first. One reason for doing this is that all the information is probably not on the plans and specifications. Typically, that additional information is needed to provide an appropriate review. Engineers need to do their due diligence when reviewing someone else's work. *Due diligence* is a legal concept that involves ensuring that all information required to make a decision is properly considered. Performing due diligence is necessary in order to write an accurate report for whomever is paying for the review.

Engineers are paid by their clients. The client's fee is to be the *only source of income* on a job. Accepting compensation from someone other than the client on a project is a conflict of interest, which engineers must avoid. It is understood that engineers cannot accept money from suppliers, competing clients, manufacturers, and salespeople, as this compromises the ability to enforce the requirements in the plans and specifications. It is paramount in the contract between the engineer and the client that the engineer protect the client's best interests as long as doing so does not conflict with protecting the public safety and welfare.

If things go wrong, engineers are required to freely and openly admit errors and then offer a solution (of course, the insurance carrier may need to be called). Engineers who have no involvement in an issue should not comment (a legal concept called *standing*).

The National Society of Professional Engineers® (NSPE) also has a code of ethics. It covers all engineering disciplines. Familiar highlights include:

◆ Give the utmost performance.
◆ Participate in honest enterprises.
◆ Live in accordance with the highest standard of professional conduct.
◆ Service before profit.
◆ Honor and standing above personal advantage.
◆ Health and welfare above all else.
◆ Notify the client when judgment is overruled. This is a potential conflict point.
◆ Sign and seal only in areas of competency.
◆ Report violations in the code of conduct.
◆ Avoid conflicts of interest.
◆ Do not accept outside compensation.
◆ Review plans only when requested to do so and advise the standing engineer of the review and why.

Reviewing the history of the engineering profession, licensure comes from codes, which came from guilds, which came from the collective expectations of the public. The engineer has a civic duty, as Kant notes, to protect the public. However, as mentioned before, the public does not need to know exactly what an engineer does; the public need only have confidence that the engineer is competent to do it correctly. Licensure is that notice to the public that the engineer is indeed competent.

3.2 Ethical Issues in Engineering

From a legal perspective, there are three basic areas of failure associated with engineering errors:

- ◆ Negligence
- ◆ Incompetence
- ◆ Misconduct

Negligence is defined as the failure to exercise due care in the performance of the work or something which an ordinarily prudent person would foresee as a risk of harm to others if not corrected. Negligence is one of the most common violations associated with engineering and can constitute grounds for disciplinary action by the Board of Professional Engineers, but not criminal prosecution in most states. The following is a case that deals with negligence (real name not used). Bob allowed his license to be used by someone else to seal a drawing. This is negligence. Bob received a fine, a reprimand, and a year of probation for not securing his license.

Incompetence is defined as a lack of ability to perform a function, a lack of qualification to perform a function, or a lack of physical or mental ability to perform. Like negligence, incompetence can constitute grounds for disciplinary action by the Board of Professional Engineers, but not criminal prosecution. If a person tries to design an electrical system for a building but was neither trained nor had experience in doing so, he or she would be deemed an incompetent electrical engineer and would incur significant personal vulnerability if the system fails.

Misconduct is another matter entirely and is defined as a transgression of some established rule of action where there is no discretion. An example is any violation of a statute, rule, ordinance, or other recognized authority. Misconduct assumes that an engineer ignored a rule or law that he or she should know about, like signing and sealing a drawing that was not the engineer's work and not prepared under his or her supervision or using another engineer's plans and specifications for a project. Willful neglect of a licensing or engineering statute would be construed as misconduct. Some examples of ethical violations include soliciting or accepting gratuities or bribes, practicing engineering without qualifications, revealing facts without the client's consent, filing fraudulent data to obtain a permit, expressing an opinion publicly without the facts, failing to disclose a conflict of interest, and stamping plans without performing due diligence. Let's take a look at some ethical issues that involve engineers.

Case Study 1. Licensure in Multiple States

James has a PE license in Florida and Idaho. His license is being acted upon by the licensing agency in the state of Idaho. The agency fined him $1,000 and would suspend his license if the fine was not paid in 30 days. Does James have an issue with his Florida license? James definitely has an issue with his license in Idaho, and because of this, he now has an issue with his license in Florida. Many states now have reciprocal agreements whereby if there is a problem in one state, other states where that person is licensed will apply the same penalty. In this case, James would be fined $1,000 in Florida for having action taken on his license in Idaho. Having a license in multiple states may be required for a job, but action and requirements (such as continuing education) in one state may translate to others as well.

Case Study 2. Practicing without a License

Ernest worked for many years in construction, and although he did not have an engineering degree, he called himself an engineer. He applied for and got a job with an agency where a job requirement is a PE license. Ernest said he had one, and it took some time for anyone to check if this was true. When someone did check, it turned out that Ernest did not have a license. What could happen? In this case, Ernest was charged with one count of practicing without a license and one count of presenting someone else's license as his own. The Board of Professional Engineers ordered him to cease and desist practicing without a license, which is against the law in all states and is deemed to be misconduct. The profession takes this very seriously. Ernest ignored the Board of Professional Engineers in his state and continued to work. This is a criminal issue, and the Board of Professional Engineers not only fined him $10,000 but also referred his case to the attorney general for criminal prosecution. In this case, Ernest did not have a degree of any type, let alone an engineering degree. Big trouble for Ernest. Also, keep in mind that in many states, without a license, the title "engineer" cannot be used.

Case Study 3. Design Defect

Joe designed a stormwater drainage system but had to deviate from the permit conditions to make the design work in the field. He did not submit the revisions to the permitting agency, and construction moved forward. Does Joe have any issues with his license? Yes. Permitting agencies must be notified of major changes, even during construction. In this case, Joe did not notify the agency, so the Board of Professional Engineers fined him $2,000, plus administrative costs of $3,129, as a result of negligence/defective design. He also got a 1-year suspension of his license and 2 years of probation. Permits are legal requirements; they are documents with specific requirements that are not allowed to be revised without permission from the permitting agency.

Case Study 4. Failing to Seal the Documents

Bill is a civil engineer by training and mostly does subdivision work. He filed several documents to be recorded in the public records department, which typically is at the county courthouse. The clerk of courts notified the Board of Professional Engineers that Bill had not properly signed, sealed, and dated the documents. Does Bill have any issues with his license? Yes. He did not properly seal the documents, so he was fined $1,000 and threatened with suspension of his license if the fine was not paid within 30 days. Again, this is a rule requirement, so it is misconduct, not negligence.

Case Study 5. Sealing Documents That Are Not Final

Mary is a professional engineer. She signed, dated, and sealed some preliminary plans submitted to a permitting agency for permit purposes without indicating the plans were preliminary. Does Mary have an issue? Actually, this example is based on a real case that created a lot of discussion and altered the rules in Florida. The seal means that the plans meet all codes, engineering standards, and rules. It also means final plans. Plans submitted in an effort to acquire permits for a project are not final because the permit requirements may necessitate some changes. In this Florida case, although the rules for sealing were confusing, there was no option but to levy a penalty; however, it was acknowledged that the interpretation of the rule failed to reflect the fact that permit submittals are not final drawings because they may be changed. Drawings that are incomplete should not be sealed. The Florida Board of Professional Engineers altered language in its rule to note that permit drawings are not considered final. If permitting agencies require a seal, then the drawings must be clearly denoted as "for permitting purposes only" or a similar caveat. The rule applies to submittals to clients as well. Partial submission drawings are not final and therefore should not be signed and sealed under any circumstance.

Case Study 6. Misleading Testimony about a Design

Lester made a series of statements before a governing board regarding a specific design created by one of his subconsultants. However, Lester had not done any due diligence and generally was not informed about the project. The project did not work as designed, and a prudent review would have revealed errors. As a result, Lester was found to have given misleading testimony regarding the project and was fined for it.

3.3 Licensure

Every engineering licensing board has its own laws regarding licensure within its jurisdiction (state, province, country), but in general a candidate must complete the following: (1) earn a degree from an engineering program accredited by the Accreditation Board for Engineering and Technology (ABET), (2) pass the Fundamentals of Engineering (FE) exam, (3) gain 4 years of progressively responsible engineering work experience under the supervision of a professional engineer after earning an undergraduate engineering degree, and (4) pass the Principles and Practice of Engineering (PE) exam in the appropriate discipline.

As state engineering boards were being organized in the early 20th century, there was a need to coordinate the effort nationally. The ABET was established in 1932 as a means to create a structured program for engineering guidance, training, education, and recognition. ABET accreditation is an assurance that a college or university program meets the minimum quality standards established by the profession and enables prospective employers to recruit graduates who are well prepared. Because engineers must graduate from an ABET-accredited school to be eligible for a PE license in all states, universities zealously protect and maintain their accreditation. Most of the coursework in an accredited undergraduate engineering degree is dictated by ABET. However, students, faculty, and members of industry play an important role in ensuring a school retains its ABET accreditation.

Students enrolled in an ABET-accredited engineering program are eligible to sit for the licensing exams. These exams are administered nationwide by the National Council of Examiners for Engineering and Surveying® (NCEES). For examinees, the review/approval process can take up to 30 days and will not begin until NCEES has received all required documentation. Authorization to schedule an appointment for testing cannot be granted until this process has been completed. The FE exam is the first step in obtaining the PE license. The FE exam evaluates content knowledge related to the subjects that ABET requires for accredited undergraduate engineering degree programs. Historically, the FE exam was an 8-hour session. From 1980 to 2010, it migrated from individual, hand-calculated problems to an exam that contained 180 multiple-choice questions and was split evenly into a 4-hour morning session (120 questions) and a 4-hour afternoon session (60 questions).

As of 2014, the test became a 6-hour computer-based exam that is administered year-round in four testing windows. The computer-based exam consists of 110 multiple-choice questions, and the 6-hour time frame also includes a tutorial, a break, and a brief survey at the conclusion. For the first half of the test, all examinees take the same general exam common to all engineering disciplines. Essentially, this covers information from the general education requirements and includes subjects such as mathematics, probability, computational tools, engineering mechanics, ethics, and engineering economics. During the second half, examinees can elect to take a discipline-specific (chemical, civil, electrical and computer, environmental, industrial, mechanical, etc.) exam or a more general exam labeled "other disciplines." If a candidate's degree is in one of the six major disciplines, taking the discipline-specific exam is recommended. Examinees must participate in both sessions on the same day.

The FE exam is closed book, but an electronic version of the *FE Reference Handbook* is provided on exam day. Students should be familiar with it before taking the test. The most up-to-date version of the reference handbook is available for download or purchase at www.NCEES.org, along with practice exams, study materials, and other references.

Candidates must bring their own calculators to the exam; however, only models of calculators as specified by NCEES can be used. The NCEES calculator policy is constantly revised because of concern about the security of examination content. Available calculator technology has been used for exam subversion, and calculators that can store and communicate text are considered a security risk. It is highly recommended that candidates check the NCEES calculator policy for compliance and become familiar with the allowable calculators, so that it will not be an issue on exam day.

A passing score on an NCEES exam is the number of correct answers or points required to indicate a knowledge level necessary to meet the minimum performance standard for a discipline, which is determined by an appointed committee of licensed subject matter experts. Beginning with the October 2005 administration, results are issued as "pass" or "fail" only with no numerical score. There is no "curve," and NCEES scores each exam based on its own merits with no regard for a predetermined percentage of examinees that should pass or fail. All exams are scored the same way.

If a candidate fails the test, a diagnostic report (see Figure 3.2 for an example) is generated, which lists the percentage of correctly answered questions in each knowledge area of the exam. This is the best guide for determining strengths and weaknesses with regard to specific subject areas. The diagnostic report summarizes the scores as follows:

◆ If performance in a subject area is significantly below that of the passing examinees, this indicates that substantial study of that content area is recommended prior to retaking the exam.

◆ If performance in a content area is near or just below that of the passing examinees, this indicates that understanding may be improved by further study, thus improving the chances of passing the examination.

Those who fail are allowed to retake the test up to two more times. After the third attempt without success, candidates are required to enroll in and pass 12 credit hours of senior-level and above college classes to be eligible to retake the exam for the fourth time.

After passing the FE exam, the next step is to gain acceptable work experience, under the supervision of a licensed professional engineer as a mentor. After gaining the appropriate experience, candidates are eligible to take the PE exam, which is an open-book exam designed to test the engineering experience gained over the 4 years since the FE exam. As a result, the PE exam tests the candidate's ability to practice competently in a particular engineering discipline. The PE exam is typically the last step in the process of becoming a licensed professional engineer. The specific disciplines as of 2014 are:

NCEES

FE Electrical and Computer

	Knowledge Area	Number of Items	Your Performance (on a scale of 0 - 15)	Your Performance Compared to the Average Performance of Passing Examinees Average of Passing Examinees = Your Performance =
1	Mathematics	11	7.8	
2	Probability and Statistics	4	8.2	
3	Ethics and Professional Practice	3	7.1	
4	Engineering Economics	3	9.4	
5	Properties of Electrical Materials	4	7.1	
6	Engineering Sciences	6	8.2	
7	Circuit Analysis	10	8.3	
8	Linear Systems	5	8.1	
9	Signal Processing	5	8.7	
10	Electronics	7	8.9	
11	Power	8	7.4	
12	Electromagnetics	5	5.3	
13	Control Systems	6	9.1	
14	Communications	5	8.8	
15	Computer Networks	3	7.8	
16	Digital Systems	7	11.0	
17	Computer Systems	4	8.7	
18	Software Development	4	15.0	

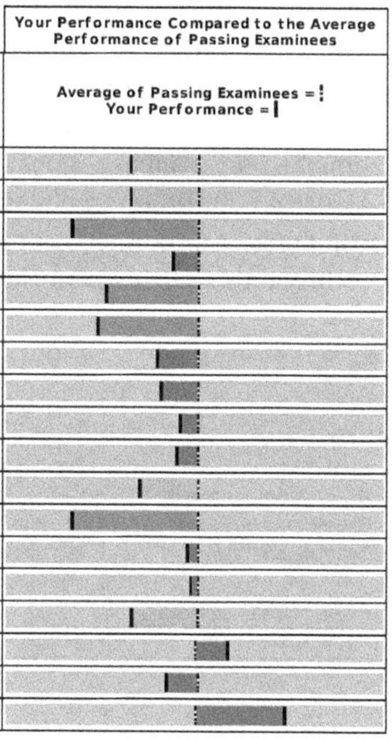

Figure 3.2 Sample diagnostic report for student who failed the FE exam (used with permission from NCEES)

- ◆ Agricultural
- ◆ Architectural
- ◆ Chemical
- ◆ Civil: Construction
- ◆ Civil: Geotechnical
- ◆ Civil: Structural
- ◆ Civil: Transportation
- ◆ Civil: Water Resources and Environmental
- ◆ Control Systems
- ◆ Electrical and Computer: Computer
- ◆ Electrical and Computer: Electrical and Electronics
- ◆ Electrical and Computer: Power
- ◆ Environmental
- ◆ Fire Protection
- ◆ Industrial
- ◆ Mechanical: HVAC and Refrigeration
- ◆ Mechanical: Mechanical Systems and Materials
- ◆ Mechanical: Thermal and Fluids Systems
- ◆ Metallurgical and Materials
- ◆ Mining and Mineral Processing
- ◆ Naval Architecture and Marine
- ◆ Nuclear
- ◆ Petroleum

- ◆ Structural I
- ◆ Structural II

The list continues to grow as the field expands. The goal of every engineering student should be to obtain a PE license.

While students who fail the test receive a report, the university also receives an institutional report as an assessment tool to help improve its curriculum. An example is shown in Figure 3.3. In this example, the university needs to improve its curriculum with respect to computational tools, ethics and professional practice, and fluid mechanics, areas where the institution's students performed below the national average.

NCEES

Examination:	Fundamentals of Engineering (FE)
Report title:	Subject Matter Report by Major and Examination
Exams administered:	January 1–May 31, 2014
Examinees included:	First-Time Examinees in **EAC/ABET**-Accredited Engineering Programs
Graduation date:	Examinees Testing Within 9 Months of Graduation Date

Name of Institution:		**Example**
Major: **Civil**	FE Examination:	**Civil**

	Institution	ABET Comparator (*2)			Uncertainty Range for Scaled Score (*4) +/- 0.30
No. Examinees Taking (*1)	11	1632			
No. Examinees Passing	10	1247			
Percent Examinees Passing	91%	76%			

	Number of Exam Questions	Institution Average Performance Index (*3)	ABET Comparator Average Performance Index	ABET Comparator Standard Deviation	Ratio Score (*4)	Scaled Score (*4)
Mathematics	7	10.5	10.5	3.2	1.01	0.03
Probability and Statistics	4	10.8	10.9	3.8	0.99	-0.02
Computational Tools	4	9.3	10.5	3.6	0.89	-0.32
Ethics and Professional Practice	4	9.3	11.2	3.7	0.83	-0.50
Engineering Economics	4	11.9	10.4	3.7	1.15	0.42
Statics	7	10.3	10.1	2.9	1.02	0.08
Dynamics	4	12.9	10.3	3.5	1.25	0.73
Mechanics of Materials	7	10.7	9.9	2.6	1.08	0.30
Materials	4	9.8	9.4	3.0	1.05	0.15
Fluid Mechanics	4	10.7	10.9	3.5	0.98	-0.08
Hydraulics and Hydrologic Systems	8	11.0	9.4	2.2	1.17	0.75
Structural Analysis	6	10.0	9.2	2.5	1.09	0.34
Structural Design	6	9.5	8.9	2.5	1.07	0.24
Geotechnical Engineering	9	10.6	9.2	1.9	1.14	0.69
Transportation Engineering	8	9.9	9.0	2.1	1.10	0.41
Environmental Engineering	6	10.4	9.0	2.6	1.16	0.55
Construction	4	10.5	9.8	3.7	1.08	0.21
Surveying	4	10.0	8.6	3.6	1.16	0.39

Footnotes:
(*1) 0 examinees have been removed from this data because they were flagged as a random guesser.
(*2) Comparator includes all examinees from programs accredited by the ABET commission noted.
(*3) Performance Index is based on a 0–15 scale.
(*4) These scores are made available for assessment purposes. See the NCEES publication entitled *Using the FE as an Outcomes Assessment Tool* at http://ncees.org/licensure/educator-resources/.

Figure 3.3 Example of a university diagnostic report (used with permission from NCEES)

3.4 References

Kant, I. (1781). *Critique of Pure Reason.*
Mantell, M.L. (1964). *Ethics and Professionalism in Engineering,* Collier-MacMillan, London.
Popkin, R.H., and Stroll, A. (1993). *Philosophy Made Simple,* Broadway Books, New York.

3.5 Assignments

1. Identify and describe a "nearly" universally accepted value. Why is it not universal?
2. Explain the ethical issues involved in the following situations as they relate to the engineering profession. Identify laws, rules, canons, or other references that support your position. Provide a recommended course of action for the engineer involved in each scenario.

   a. An engineering college professor, who also has a PE license, is asked to peer review a bridge design in the context of the bridge design process for the state department of transportation (DOT). After a 5-month investigation that includes interviews with DOT staff, observation of the design process, and review of project documentation, the professor issues a report that is critical of DOT's quality control process. As an example of the quality control process, he chooses a bridge designed by ABC Consulting where the design is unique and local materials were used, which is in contrast to the recommendations of a prior consultant with XYZ Engineering and DOT standards. He further notes that little information was provided by ABC Consulting to the DOT, consisting basically of only design correspondence and the final plans and specifications. ABC Consulting was not forthcoming about added information when asked for it by the professor. The DOT makes the report public, and the regional press is quick to criticize the DOT and outline the issues involving the example project. ABC Consulting is unhappy and threatens legal action against the professor. Define the issues and guidelines involved. What could happen?

   b. A professional engineer is at a conference and decides he wants to play in the golf tournament. He is looking for a foursome. One of his college buddies, who works for a major vendor that supplies water and wastewater parts to the industry, offers a spot in his foursome at no charge since he is short one person anyway. The engineer is not sure if he should accept the invitation to play for free. The slot is worth $100.

   c. A professional engineer at a local utility is in charge of the capital improvement program for the county. One of his engineering consultants proposes to install a different type of piping instead of the C900 PVC the county and most other utilities use as a standard. The consultant claims that this is a major cost savings. The pipe has never been used in the United States. The engineer says no because he has found insufficient data to support use of the pipe. The consultant goes to the engineer's boss (not a professional engineer) and makes the same pitch. The boss says he wants to use it. None of the parties research the piping online, where they would have found a series of failures in Europe due to longitudinal cracking. Of course, 18 months after this pipe is installed, it starts to fail, just as it did in Europe.

   d. A professional engineer conducts asset management for a city government. Included is an evaluation of the sidewalks. The engineer documents the cracks and breaks in the sidewalks throughout the city. However, the city attorney instructs the engineer not to provide the information to the city because to do so would subject the city to lawsuits for negligence if people are hurt before the city can fix the cracks.

   e. A local engineering firm is competing for engineering design projects against some other engineering firms that have substantial political clout with local politicians who award engineering contracts. Three different people approach the firm and offer to provide assistance. One says he has some negative information about a couple of the competitors and would be happy to get the documents to the county commission for a price. Another offers to lobby the commission on the firm's behalf for a portion of the contract amount. The third employs a woman who is an acquaintance of a couple of the commissioners and claims she can help get the job for a price. What should they do?

3. Professional engineer A is asked to review a project where the client believes the cracks in the structural tank walls were caused by a deficiency in the design created by engineer B. Engineer A is asked to tell the client if engineer B is at fault. What is the proper course of action for engineer A?

4. Review an NSPE or ASCE publication from the past 3 years, and pick an issue that included an ethics discussion. Discuss the case and what the ethics conclusion was.

5. Among the common answers to the question "what do engineers do" is:
   a. Drive trains
   b. Get into politics
   c. Go to the moon
   d. Act like Sheldon in *The Big Bang Theory*

6. Professions that often are deemed to be ethical include:
   a. Doctors
   b. Lawyers
   c. Politicians
   d. The mafia

7. Obtain a copy of the rules for professional engineering for your state. Compare them to the rules for another state. List five similarities and three differences.

8. Are all illegal activities always unethical?

9. The difference between what we perceive as ethical professions and unethical ones is:
   a. Money
   b. Influence
   c. Something to sell
   d. All of the above

10. The priority for engineers is:
   a. The public
   b. The client
   c. The firm's shareholders
   d. The mayor

11. An ethical person is a person with a set of values who lives by them.
   a. True
   b. False

12. Engineers are included in a group that lives by a set of values that are shared by a group of people.
   a. True
   b. False

13. The expectation of the public is that engineers will:
    a. Design buildings to protect the public
    b. Report potential failures to their clients only
    c. Tackle any problem handed to them
    d. Find definitive answers
14. Judgment is required by engineers because:
    a. There is often imperfect information
    b. It is better than guessing
    c. It allows them to find the perfect answer to a problem
    d. There is often imperfect education
15. The commonality among engineers, social workers, teachers, public safety personnel, and doctors is:
    a. They are all regulated by state agencies
    b. They all are government employees
    c. They all have an obligation to protect the public first
    d. They are all highly ethical
16. The commonality among engineers, lawyers, teachers, and doctors is:
    a. They are all regulated by state agencies
    b. They all are government employees
    c. They all have an obligation to protect the public first
    d. They are all highly ethical
17. Ethics goes back at least 10,000 years.
    a. True
    b. False
18. The Greeks first came up with the idea of ethics.
    a. True
    b. False
19. Ethics developed from the expectations of members of the tribe, village, and society.
    a. True
    b. False
20. Tribal ethics that would be respected include:
    a. Thievery
    b. Hunting skills
    c. Selfishness
    d. Disrespecting the elders
21. In the ancient world, people were expected to:
    a. Be good at agriculture
    b. Do their share of the work
    c. Be good at hunting and gathering
    d. Be good warriors
22. Explain why lobbying might violate ethical principles for the engineering profession.
23. Apprenticeship requires:
    a. Training
    b. Education
    c. Experience
    d. All of the above

24. Philosophers found that ethics could be explained by the actions of people.
    a. True
    b. False

25. A good starting spot when confronted by an ethical dilemma is to ask: If everybody did _____, would society function?
    a. True
    b. False

26. Utilitarianism is an ancient philosophy that deals with greed and apportionment.
    a. True
    b. False

27. Utilitarianism assumes:
    a. A sense of duty
    b. Education and training
    c. Maximizing the happy people
    d. Being objective

28. There are codes of ethics in many areas of engineering. All of them require:
    a. A sense of duty
    b. Education and training
    c. Being objective
    d. All of the above

29. The highest priority from the engineer's proper perspective according to ASCE is the:
    a. Client
    b. Firm
    c. Public
    d. Law

30. The lowest priority from the engineer's proper perspective according to ASCE is the:
    a. Client
    b. Firm
    c. Public
    d. Law

31. Engineers should not:
    a. Notify the client when judgment is overruled
    b. Report violations in the code of conduct
    c. Accept outside compensation
    d. Put service before profit

32. The concept of protecting the profession suggests that engineers should never:
    a. Market
    b. Notify the client when judgment is overruled
    c. Report violations in the code of conduct
    d. Lobby elected officials for work

33. Ethics for engineers do not come from:
    a. The concept of apprenticeship
    b. Codes
    c. Canons
    d. Due diligence

34. If a consultant is teamed with a contractor on a design/build contract, but has been asked to perform contract administration for another project for which the same contractor has been engaged, what is the appropriate ethical response?
35. If an engineering firm engages a lobbyist to secure a contract and the lobbyist's plans involve trashing the other bidders, as well as overstating the firm's qualifications, what ethical issues arise?
36. Review the laws and rules governing professional engineers in your state and one other state. Compare and contrast the rules in a matrix.
37. Review 5 years of data on cases associated with discipline by the Board of Professional Engineers in your state and one other state. What are the most common issues for which engineers are disciplined? Make a pie chart that compares them.
38. Call your state Board of Professional Engineers and ask what they perceive to be the most interesting case they have had recently.

4

Getting the Design Contract

Society continuously undertakes the construction of capital projects to erect buildings or install the supporting infrastructure such as water mains, sewer lines, stormwater drainage, transportation, parks, power lines, pumps and motors (Figure 4.1), and the like, all of which means *engineering jobs*! For efficient operation, these newly constructed facilities must be developed in accordance with the latest technical and professional standards to protect the health, safety, and welfare of the public served now and in the future, which requires engineers to keep pace with the latest technological advancements. It is the responsibility of engineers to ensure their designs meet or exceed all requirements and to identify the state-of-the-art standards and best practices. Keep in mind that virtually all major roadways (Figure 4.2), bridges, railroads (Figure 4.3), water and wastewater treatment plants (Figures 4.4 and 4.5), schools, pipelines (Figure 4.6), drainage systems, dams, dikes, reservoirs, parks, airports, and large buildings will be constructed by governments (public sector projects). These are projects that sometimes involve enormous investment, which translates to engineering jobs. In most cases, those government projects will be designed and constructed by the private sector and then paid for by funding borrowed by governments. With the declining condition of much of the infrastructure built in the period from 1920 to 1945, and the generally poor ratings by the American Society of Civil Engineers of the current condition of the infrastructure in the United States, there is much design and construction work to be done by engineers in the near future just to update the existing infrastructure, not to mention the infrastructure needed to keep up with population growth and economic growth.

Figure 4.1 Pumps and motors will always be a staple of engineers

Figure 4.2 Roadways and bridges will continue to need improvements and designs by engineers

Figure 4.3 Railroads are both old and ongoing technology being designed by engineers

Figure 4.4 An ion-exchange water treatment plant is an example of an advanced technology designed by engineers to make water safe to drink

Figure 4.5 A reverse osmosis/nanofiltration water treatment plant is another example of an advanced technology designed by engineers

Figure 4.6 Excavation for pipelines requires problem solving by engineers

4.1 Building a Capital Project

The primary issues of importance to owners (public or private) when implementing a capital construction project or upgrade are cost, construction quality, and completion schedule. Each of these issues is separate, but they are interrelated; therefore, owners must choose which objective is most important, as that choice may adversely affect the other two. For example, increasing quality requirements may delay the project completion date and inflate costs, while accelerating the schedule and keeping costs fixed will diminish the finished quality. An owner can theoretically maximize two of the three objectives, but not likely all three. Figure 4.7 represents the conundrum. The triangle can be turned to maximize any two of the three variables, but at the expense of sacrificing the third. Thus an owner can get fast work with high quality but not at a low cost, or the owner can get fast work at low cost but at the expense of quality. The key is to determine the most critical variable within the context of the project scope. Achieving the ideal scenario for all three (cost, quality, and time) typically is difficult because of the exclusivity among them, but the final scope of a project must represent the optimization agreed to by the engineer and the client.

Figure 4.7 Three main factors that govern project completion

While cost, quality, and time frequently are the determining issues, they are not the only concerns of owners. Control over the final product also is important. Governments typically believe that in controlling the project, they can achieve schedule, cost, and quality goals. The owner's personnel also often have specific ideas on the details of the project. Retaining control means more involvement by the owner during design and construction. However, productive oversight can be lacking because public agencies frequently build a limited number of new facilities during any given employee's tenure, whereas consulting engineering firms may have specialists on staff who have built similar facilities many times. The same often is true of private owners. By assuming a degree of control, a private owner will reduce the responsibility of the engineer and contractor because certain decisions will have been made by the owner, sometimes against the best judgment of the engineer or contractor. In such cases, the owner takes the risk. Engineers should be careful with defining exactly what methods the contractor must use to construct the project, as the means and methods are the contractor's responsibility. Costs may increase because the contractor assumes risk as a result of the contract conditions (Bloetscher 2011).

The time commitments of the client's staff can be an important issue on projects. Staff can spend large amounts of time dealing with inconsequential details if the engineer and contractor are not given the freedom to make minor decisions. In addition to the key factors of schedule, quality, cost, and control are the following (Bloetscher 2011):

◆ **Whether or not operational expectations are met**—If the project serves its intended purpose and meets the intent of the owner, then operational success likely will be achieved.

◆ **Whether or not the new facility can be integrated within the framework of the existing system easily and quickly**—If the new project is an expansion or a retrofit of an existing facility, the interface of new and old must achieve continuous service without significant interruption to be successful.

◆ **Responsibility of the parties to one another**—Every project will be different, but change orders, the ability to make changes as needed, on-site staffing, and additional effort by the engineer (shop drawings, etc.) may result in increased delivery time and additional cost.

◆ **Likelihood of change orders from contractor**—Change orders are an issue that can increase cost as well as be a battleground for control. Most projects will have minor change orders to deal with uncertain conditions in the contract documents. Change orders initiated after the contract award for items outside the scope of the project are controllable and should be discouraged.

◆ **Ease of transition to operation**—The transition to operation by the owner generally is improved by having the contractor, or one of the subcontractors, operate the system for a period of time to "get the bugs out." Start-up of a new system is always difficult because the owner's operations staff must address new and unfamiliar equipment, operating procedures, or protocols.

◆ **Operational capability issues must be addressed for long-term facility success**—The owner/operator of the facility must be capable of understanding the operations needs early and clearly for the project to be successful.

The goals of the owner in relation to each of these factors must be considered when making decisions on pursuing a construction delivery alternative.

4.2 What Owners (Should) Look for from Consulting Engineers

What owners look for when selecting a consultant is not always what consultants expect. Most owners look for honesty, reliability, forthrightness, sound judgment, and commitment to quality and customer service, which are subjective and difficult to quantify. Owners need consultants who will be truthful, indicate what the problems are, and be candid about issues rather than furtive (Bloetscher 1999a, 1999b). Because owners generally hire the consultant who is best qualified to provide the type of work required, trust is of great importance in the selection process (Bloetscher 1999a, 1999b).

It is often clear that not all consultants are equally qualified, despite their self-perceptions. For instance, if an owner wants to build a membrane water treatment plant, a consultant who has already designed a similarly sized membrane facility should be preferred. Yet, despite this obvious fact, many consultants will apply for the work and some will actually lobby elected officials to obtain that work, despite the fact that they have no prior experience (Bloetscher 1999a, 1999b). This violates the honesty and forthrightness criterion (and is probably an ethical violation as well). It is far better for consultants who do not have that type of experience to team up with consultants who do; thereby, their lobbying efforts are maintained within more appropriate parameters (Bloetscher 1999a, 1999b).

In addition to being honest, owners look for the right skills to do the job. This includes not only the known capabilities of the consulting firm, but also the specific talents of the personnel who are actually selected to perform the work. A common problem is that a consultant will submit a proposal that includes work done by people who will never work on the proposed project in an attempt to show that the firm has the expertise to do the job. This tactic is misleading and inherently dishonest. It also is an indication that the consultant is not putting sufficient thought into the proposal process, which is likely to end in a poor response from the people making the selection. Accessibility of the consultant also is important in performing any task; however, this does not necessarily mean that the consultant needs to have an office in the community in which the job is located (Bloetscher 1999a, 1999b).

A favorable reputation among clients and within the consulting field is extremely important for the success of an engineering consulting firm; therefore, references should include other consultants as well as clients. The engineering field is surprisingly small, and issues that arose with engineers (and contractors) during previous projects will tend to spread quickly. Therefore, a consultant must have a reputation for maintaining strong working relationships with clients and other consulting firms and their employees to be successful. Another important criterion is performance, particularly in meeting time and budgetary constraints. This means that the consultant will assign the appropriate staff and resources to complete the work in a timely manner. It also indicates the consultant's ability to be flexible in meeting the needs of the project while maintaining progress toward completion on time and within budget.

Finally, the owner should determine whether or not the people assigned to the job as project managers are comfortable dealing with the consultant. This goes back to honesty and forthrightness. If the owner is not comfortable, or does not trust the consultant's personnel, the relationship will be strained.

4.3 What Consultants Do Not Need

There also are a number of issues that consultants often think they need to address but actually do not from an owner's perspective. A consulting firm does not need to be a large national or international consulting organization. Of course, national firms may be able to access the appropriate experts to provide information, but they cannot be expected to focus on one particular agency, especially a smaller one. As a result, a local consultant who has access to or can collaborate with small, local providers matched with a larger consulting firm may be a more effective consulting team than either the small, local or large firm alone. At the same time, the consultant need not always be local, just accessible. If an important member of the consulting team can be to a job site or attend a meeting within 2 to 4 hours or be available by phone at a moment's notice, this usually will meet the "accessible" criterion. The owner should not expect the consultant to have local staff capable of completing all of the design work for a complex water treatment plant, for example.

A single consultant does not need to have all of the skills to do the job, just the important ones. This is where the abilities of subconsultants with which the consultant works are important, as is past experience in working together or with the client. If the consultant can bring in experts from other offices or experts from among the team members, this strengthens the quality of the proposal. Owners look for the firm with the people who are most capable of doing the job and are wary of using public monies or the public trust to be the "first on the block" to try something new.

4.4 The Typical Public Sector Proposal Process

Public sector agencies normally are required to follow stringent consultant selection laws and formal bidding procedures, which can create significant time delays. Such time constraints sometimes limit the entity's ability to have the facilities designed and built quickly enough to meet its needs without significant costs or interruptions in service. The time constraints result from the selection process used by public agencies, which, in most cases, requires consultants to be chosen based on qualifications, not price. There are exceptions, but they are not the norm for good reason. Price and competence are not always related. Complex projects cost more to design. Assuming a qualifications-based approach, owners/clients will request or solicit information on qualifications from interested parties to help select the most qualified applicant to do the work. The terms for these solicitations include *request for qualifications (RFQ), statement of interest and qualifications (SIQ)*, and *request for proposal (RFP)*. In each case, interested parties submit packages outlining their qualifications. Costs are included only with an RFP, and normally this is done as a secondary consideration for a design-build project. An important aspect of the solicitation process is articulating the appropriate design approach and management of the project. The following paragraphs outline how such a solicitation might proceed (based on an actual template).

4.4.1 Scope of the Project

In the original solicitation package, the scope of the project may or may not be well defined. It is up to the consultants to refine the scope of the project because the original authors of the request may not have a clear understanding of what might be feasible to meet their needs from an engineering perspective. In an effective response, the consultant explains the objectives of the project in detail to demonstrate a complete understanding of the project to the client. To illustrate this concept, the following is an example of a poorly defined scope:

> The City desires to retain a design-build project team to provide the scope of services described as follows:
>
> 1. Land surveying
> 2. Site planning
> 3. Construction of the improvements designed
> 4. Various studies, reports, etc. as necessary to accomplish the work
> 5. Decommissioning and removal of existing fuel station within 6 months
>
> This scope of services may be expanded or reduced at the discretion of the City to either include or remove any one or more service listed.

The following is an example of a more well-defined scope:

> The City has created a Transit-Oriented Development (TOD) zone in its downtown as a part of a regional activity center. At present, the downtown area is blighted as a result of economic conditions, with older, mostly vacant buildings that are not conducive to an area that is likely to use alternative modes of transportation. The City desires to redevelop the downtown commercial district to create employment opportunities and an increased tax base. Ultimately the City wants 650,000 ft^2 of commercial space downtown and will add 4,000 residential mixed-use housing units. The City is looking to expand its ocean access and boating industry and desires that the downtown concept be in keeping with this expansion. The theme is that the City is a "Marine Town," and therefore its redevelopment and architecture should be in keeping with that theme.
>
> At the present time the City is constructing a LEED®-certified branch of the County Public Library at its current City Hall site. The City also is evaluating a 400-car parking garage on the site just south of City Hall. The City also owns the three lots directly north of the City Hall site and desires to convert them to a mixed-use development with all parking on-site. It is desired to have this location incorporate 30,000 ft^2 of commercial space and 20 two-bedroom/two bath (minimum) condo units. The building will need to be LEED Gold, meet all TOD zone criteria, and hold all of its stormwater on-site.
>
> To use this site, the City will need to have the existing fire station moved to a better location in the vicinity of the TOD zone. This site has not been identified, but it would be useful to put both the police station and the fire station into one public safety complex. The second task is to identify the needs of the fire station and police offices and designate an appropriate site for same within the downtown area providing appropriate access. The station will need to be LEED Gold, meet all TOD zone criteria, and hold all of its stormwater on-site.

4.4.2 Requirements of Proposers

Proposers interested in providing the services listed in the solicitation must demonstrate considerable relevant experience with the type of work and should emphasize both the experience and capabilities of particular personnel who will actually perform the work. To improve the quality of the team, the proposer may include subconsultants, or subcontractors, who have specific expertise. In all cases, interested parties, both corporate and individuals, must be fully licensed in the type of work to be performed at the time the response to the solicitation is submitted. As will be discussed later, the written content of the response is of paramount importance in getting short-listed for the project. The ability to write a clear, concise, easy-to-understand, well-formatted, mistake-free response is critical to the success of any solicitation response.

In order to ensure that a uniform review process will be conducted and to obtain the best evaluation, owners/clients require that proposal packages be organized in a prescribed order. A typical outline for actual solicitations used by a number of public sector clients includes the following:

A. **Title page**—This section includes the project name, solicitation number, due date, consulting firm name, and any other information specifically requested in the solicitation.
B. **Table of contents**
C. **Letter of interest**—This letter should include the name of the person(s) authorized to make representations on behalf of the respondent, as well as other pertinent contact information such as job title, address, telephone number, and email.
D. **Description of the proposer's team**—This section should identify the design professionals and subconsultants and provide a project and staffing plan, organizational chart, and resumes.
E. **Questionnaire**—Each respondent should address the required information solicited by the client, which typically includes a management plan, technical plan, information about the team members, and a quality assurance plan.
F. **Firm experience**—Each respondent should include recent (within the last 5 years or so) direct firm experience in the design and construction of capital improvement projects, particularly those that involved directing a similar or more complex scope of services than listed in the solicitation, under a parallel or higher responsibility level.
G. **Approach to scope of services**—This section outlines how well the responder understands the issues associated with the project. The proposal should include a statement of project understanding, a planned project approach, and a tentative timeline of performance, with key milestones for the scope of services. This section requires some groundwork to address important issues like zoning, codes, construction costs, potential permitting issues, environmental issues, and any other identified constraints, along with proposed solutions. The idea is to identify items that might impact the owner's schedule or cost. If the scope is not completely understood, the project will be doomed from the start.

It is typical to include a project management plan that shows the client that the proposer understands the scope and timeline of the project. An effective management plan briefly describes any anticipated major challenges and how they will be dealt with, the problem-solving approach, and the communication plan for those doing the proposed work. It is important to describe the organizational chart and how work tasks will be completed, including identification of all key staff members and their duties. The team organizational chart should include those individuals who will be most directly involved with the project and how they will interact with the project and each other (see Figure 4.8).

The proposer should identify the anticipated level of participation for each key individual in comparison to his or her daily workload activities and also should identify a project manager who will be directly responsible for day-to-day communication and coordination with the owner. This person must be able to represent the design team and be capable of making commitments on behalf of the design team. The designated project manager must be duly licensed in the state where the project is located. There may be requirements for experience in the solicitation (e.g., 10 years of experience in the management of projects equal to or more intricate in technical scope than the services required).

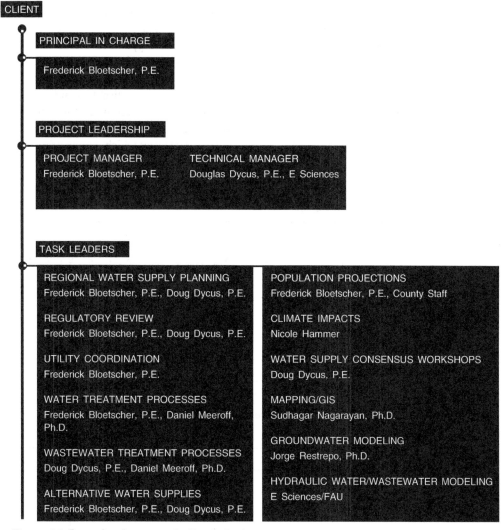

Figure 4.8 Example of an organizational chart

Another item to include is a description of how subconsultants and subcontractors will be integrated into the organization of the project. It helps to identify an actual previous project and describe the technical and managerial methodology used to keep the project on schedule and on budget. Internal and external channels of communication employed to achieve successful completion should be identified in this response. Finally, it is critical to make sure that the owner understands the project approach and timeline by providing a (Gantt-style) schedule (see Table 4.1).

Communication among the owner, proposer, and subcontractors is crucial to the overall success of any project. Technical and managerial interaction at all levels is necessary if a project is to move forward smoothly. The response should outline how the team is accustomed to working under this type of communication environment. Specific details of previous experience and applications should be provided, along with a description of how the organizational chart will be integrated with the owner's staff. Names of personnel to be assigned and their duties should be included, along with comments on how subconsultants are able to adapt to this type of working environment. A summary of how the team plans to overcome communication difficulties should be included. Details on how this has worked well in the past should be provided, and the strategies that were applied to overcome such situations and how these strategies had an impact upon the success of the project should be described briefly.

The respondent also should include a description of the team's applicable quality assurance/quality control program or protocol and indicate specific steps conducted for technical review of any type of deliverable prior to submission to a client. The respondent should identify any standard processes used and describe any success stories which demonstrate that such policies are effective.

Table 4.1 Example of a Gantt-style schedule

Task	Lead agency	Month											
		1	2	3	4	5	6	7	8	9	10	11	12
1. Project kickoff	PUMPS	▓											
2. Potable water treatment/source	PUMPS	▓											
3. Wastewater treatment/disposal	PUMPS		▓	▓									
4. Operational goals	PUMPS		▓	▓									
5. Existing water distribution/piping	PUMPS		▓	▓									
6. Growth forecasts	E Sciences			▓	▓								
7. Potable water modeling	E Sciences					▓	▓	▓					
8. Wastewater modeling	E Sciences						▓	▓	▓				
9. Recommended improvements	PUMPS									▓	▓		
10. Publish master plan	PUMPS											▓	▓
11. Presentation to County Comm.	PUMPS												▓

When discussing direct project experience, the respondent should submit previous directly related project experience in the following form:

1. Annotated list of projects for which services were rendered under the same name as the firm submitting the proposal. Each project listed should include the following information, when applicable:
   ◆ Project location
   ◆ Brief project description or title
   ◆ Name of the client
   ◆ Date of completion
   ◆ Estimated cost and actual cost after completion
   ◆ Whether or not the firm was the principal firm in charge of the project, and if not, the name of the principal firm
   ◆ Specific service(s) performed by the firm submitting the proposal and principal in charge (in the case of a team submittal or joint venture submittal, identify which individual firm was responsible for the project or portions of the project listed)
   ◆ Name, address, and phone number of references familiar with the project (or portions of the project) and the services performed by the firm
2. Identification of key subconsultants who participated in the project task, including the respective tasks they performed and evidence of their qualifications.
3. Description of litigation or other dispute resolution proceedings (e.g., arbitration, mediation) in which the firm or staff of the firm is currently involved or has been involved over the past 5 years, stating points of contention and results if available. This information should be summarized with appropriate appendices that include details.

4.4.3 Evaluation of Proposals

The typical selection process to identify the preferred provider is as follows:

1. The request is issued.
2. There is a period for questions and answers from potential proposers.
3. Responses are received.
4. A selection committee scores each submittal in accordance with the rating guidelines and may ask the top-ranked firms to schedule an interview, presentation, or both. The committee will rank qualified firms in order of preference and present recommendations to the decision makers. A short list of three or four firms is usually generated.
5. The decision makers enter into negotiations with the preferred firm. If negotiations are unsuccessful, then the next top-ranked firm may be invited to negotiate or the entire process will be restarted.
6. Upon the successful completion of negotiations, an agreement will be signed by both parties.

It is typical for the evaluation criteria and selection procedure to be specified in the RFP, local codes or statutes, or otherwise detailed in writing for the respondents. The following are six common categories used by most agencies for scoring proposals:

1. **Company's expertise**—Ratings typically are based on information provided with respect to the type of work described in the solicitation. A firm with directly related experience likely will get a higher rating than a firm that has had only limited experience in other types of projects. Level of difficulty and successfully overcoming strategic challenges in similar projects also may earn higher ratings. Firms with previous successful work locally, or with a well-defined team structure, and functional quality control policies may receive higher ratings.

2. **Previous staff experience**—Ratings typically are based on the experience of the key staff members who will be involved in the day-to-day design and construction that is related to the type of work described in the solicitation. Significant individual experience in performing similar projects may receive more points. Limited staff experience likely will receive fewer points.

3. **Current and projected workload**—Ratings typically are based on the number of other current and projected projects a firm is undertaking, staff assigned to those projects, and the percentage availability of the staff members assigned. This is why current projects that are ongoing in the same office should be noted in the response document.

4. **Office location**—Ratings reflect the ability of a team to execute any level of the contract work and provide subsequent responsiveness in a timely fashion based on geographical location of the office in relation to the job site.

5. **Demonstrated prior ability to complete projects on time**—Respondents should provide a tentative project schedule in a Gantt-style format (Table 4.1). The schedule should include both design and construction phases of the project. Respondents will be evaluated on the logic applied to each timeline, interrelationships between project timelines, and predicted impacts to scheduled projects, as well as subsequent proficiency in establishing a streamlined and successful delivery process. Respondents will be evaluated based on previous experience in the successful completion of and steadfast conformance to similar project time frames. Specific attention will be given to successful strategic and managerial approaches utilized to accomplish timely project completion, as well as the ability of the respondent to provide full dedicated attention to each workload priority. Respondents who have demonstrated an inability to complete projects on time will receive a lower rating.

6. **Demonstrated prior ability to complete projects on budget**—Respondents will be evaluated on their capacity to establish competitive and technically responsive projects, as well as the ability to adhere to initial budgets. Comparisons will be made between initial negotiated task costs and final completion costs. A table that shows initial budget/award and final cost is the best means to present this. Respondents will be given the opportunity to explain budgetary overruns, and consideration will be given to scope modifications as a result of expansions or reductions in original scope. Unjustified budgetary overruns will receive fewer points.

As noted earlier, price is not permitted to be a consideration in many governmental jurisdictions. Among the reasons why price is excluded is to allow local entities to secure the most qualified professional, which may facilitate accomplishing the project more efficiently as a result of a shorter learning curve.

4.5 The Typical Private Sector Proposal Process

The private sector can engage consultants and contractors easily, without a solicitation process, although frequently there is a bidding process that involves cost to secure services. Once proposals are received, negotiations begin. The private sector often will negotiate to reduce the cost while maintaining the anticipated scope. If the private sector asks for proposals, all of the items in Section 4.4.2 are probably required. The process does not include a shortlist and in many cases may not include a solicitation. Direct contracts are common. In both sectors, with proper planning and anticipation of the needs for construction, new capital projects can be completed with minimum impact to the operation of the built environment.

4.6 Stages in the Design Process

Once the design professional has been selected by the owner, the contract documents must be executed so that work can begin. The contract will state whether payment for services will be in the form of a lump sum or time and expenses. In a lump sum contract, payment is based on the scope of work agreed upon, not the amount of time or money it takes to actually accomplish the work. If the consultant takes less time than was estimated, the firm makes a profit; however, if the consultant needs more time than was originally estimated, the firm loses money. In a time-and-expense contract, the consultant gets paid an hourly rate based on the amount of time estimated to accomplish the task. This type of contract generally has a limit in the form of a "not to exceed" dollar amount. Time-and-expense contracts are useful for permitting and other tasks that depend on things outside of the consultant's control. The potential for profit is lower with this type of contract, although the risk is substantially less.

Once the contract documents have been executed, the design phase can begin. There are four stages in the design process:

1. **Conceptual design**—The description of what is needed versus what exists is also known as the *basis of design*. The client may have only partial understanding of the needs or may not understand the impact of all of the goals. The conceptual design helps to ensure the engineer and the client align their goals for the project so that future work will be accomplished more efficiently.
2. **Predesign**—This step frames the design project within the constraints of engineering standards and code issues to align the expectations of the client with the deliverables of the engineer.
3. **Preliminary design**—Once the predesign concepts are accepted, it is time to further develop the engineering calculations, design drawings, and specifications in order to both meet the goals of the client and protect the public. This is where the engineer ensures that all code requirements are met, initiates permitting, and completes the basis of design reports including calculations, materials, equipment, etc. Preliminary design is the bulk of the work effort and is the last step before the final details are completed for bidding purposes.
4. **Final design**—The final design documents are basically the final plans, signed and sealed by the professional engineer and used to obtain bids for construction.

4.6.1 Conceptual Design

The conceptual design is predicated on a clear and focused understanding of the scope of the project. Potential solutions that are acceptable to the owner can be explored once the scope is clear. The design professional may be required to create a revised scope to best meet the intention of the original scope in the context of feasibility, cost, and time constraints. To accomplish this task, it is critical to get answers to the following questions:

◆ What does the client actually need?
◆ What is the function of the project?
◆ What facilities are to be included?
◆ How many people will be served?
◆ What codes and regulations must be met?
◆ What environmental considerations must be met?

As noted previously, many clients do not have a good grasp of the answers to these questions, which makes the consultant's job more challenging (and introduces risk due to changes in direction by the client). The result of these discussions should be a conceptual design or basis of design report. It is important for the users (including operations and maintenance staff) to be involved and provide input during the conceptual design stage as they will occupy and operate the facility for the next 50 years or more. Conceptual design reports outline the general principles upon which the design of the project will be based. Such reports include:

◆ Introduction to the project and scope of services
◆ Background information about the existing conditions
◆ An evaluation of engineering alternatives
◆ Details about the recommended alternative
◆ Site program that details the functions and square footage requirements for the recommended alternative

Conceptual site design also may include drawings that outline the project site location, its surroundings, and current site usage. Drawings should include the existing site plan showing the current facilities on the site and the proposed site plan identifying what the site will look like once the improvements are designed and constructed. Data on the size of buildings, location of roadways and parking, landscaping, and identification of utility needs (water, sewer, etc.) should be provided.

The owner should provide written acceptance (not approval) of the conceptual design report and drawings to move forward on the project. The owner should not "approve" the report or plans, as this may cause the owner to assume some degree of responsibility for the design.

4.6.2 Predesign

Once acceptance of the conceptual design has been issued, the design professional will develop the predesign plans and specifications. This may require an update of the previous proposed site plan to show the accepted improvements, including water, sewer, lighting, parking, roads, power, etc., as appropriate. At a minimum, the predesign report will include the following:

- Applicable codes and engineering standards
- Calculations of facility utilization rates (persons per day, etc.)
- Calculations of facility sizing
- Calculations of pre- and postconstruction impervious area to determine stormwater quantity for drainage
- Calculations of the parking requirements and appropriate site locations for parking
- Calculations of the water and sewer demands
- Appropriate corridor locations
- Appropriate access and egress requirements per code
- Off-site roadway and turn lane improvements
- Architectural layout/models, if desired

4.6.3 Preliminary Design

Acceptance of the predesign will allow the design professional to proceed to the preliminary design phase and produce the preliminary plans, with design specifications, which include the following items:

- Building materials
- Construction methods
- Equipment
- Utilities
- Heating, ventilation, and air conditioning (HVAC)
- Structural components
- Foundation
- Roof
- Roadways and sidewalks
- Stormwater drainage

Upon receipt of the preliminary plans and specifications, the owner should provide comments, with particular attention given to equipment and material preferences. Prior to acceptance, the owner should provide the necessary legal documents for the contract package.

4.6.4 Final Design

After acceptance of the preliminary design, the design professional will begin work toward the final plans and specifications. The final design plan set typically includes the following types of documents (as appropriate for the project), in addition to the specifications:

- Cover sheet
- List of drawings and engineers who created them
- Site plan: existing
- Site plan: proposed
- Site plan: changes highlighted
- Water and sewer plan
- Stormwater plan
- Stormwater management details
- Parking plan

- ◆ Landscaping plan
- ◆ Floor plans
- ◆ Elevation plan
- ◆ Roof plan
- ◆ Structural plan
- ◆ Structural details
- ◆ Plumbing plan
- ◆ Lighting plan
- ◆ Electrical plan
- ◆ HVAC plan
- ◆ Details

Because not everything is described in the drawing sets, specifications that outline the type and quality of materials, equipment, etc. are required. For example, the specifications will provide the type of concrete mix that is needed (e.g., 3,000 psi) and grades of steel that can be used for rebar (e.g., grade 60). The accumulation of all of these documents together, along with the contract, general conditions, and bidding conditions, often is referred to as the *construction documents.*

4.7 Construction Documents

The development of appropriate construction specifications and contract documents will allow the construction project to progress as smoothly as possible. The specifications, like the plans, are the responsibility of the design engineer. The contract documents are the basis of a legal arrangement between the contractor and the owner. All too often, however, preparation of the contract documents is delegated to the engineer, who is not qualified in the legal aspects of writing contracts; therefore, appropriate legal help should be solicited for contract document preparation. Recall from Chapter 3 the discussion of working outside of one's area of expertise. Unless the engineer is also a licensed attorney, he or she should not be developing legal documents.

4.8 Scheduling and Project Delivery

The schedule defines how the engineer during design, or the contractor during construction, expects the project to proceed. Most clients will pay monthly, so a pay request usually is submitted monthly. The pay request should include a progress report that documents the work tasks completed or in progress and should update the schedule for the client. Any revisions to the schedule should be made in comparison to the original schedule and should reflect the changes made from any previous report. Explanations should be provided for any changes to the schedule. A variety of software packages and techniques are available to help inspectors and engineers determine if contractors are keeping up with the schedule. Figures 4.9 and 4.10 show examples of construction project management schedules engineers should be familiar with preparing.

Activity planning	Time to complete	Completion deadline	Month																						
			1	2	3	4	5	6	7	8	9	10	11	12	13	14	15	16	17	18	19	20	21	22	23	24
Approve facilities plan contract		02/07/2012	▓																							
Develop plan	60 days	05/07/2012		▓	▓																					
Public hearing process	30 days	06/07/2012				▓																				
Regulatory review and comments	30 days	07/07/2012					▓																			
Clearinghouse review	90 days	09/07/2012						▓	▓	▓																
FONSI finding	15 days	10/15/2012									▓															
FDEP program approval		11/15/2012										▓														
Approval of design contracts		06/07/2012				▓																				
Design	120 days	09/07/2012					▓	▓	▓	▓																
Permits	90 days	11/15/2012										▓														
FDEP program approval		11/15/2012										▓														
Notice to incur costs		12/15/2012											▓													
Construction		02/07/2013												▓	▓	▓	▓	▓	▓	▓	▓	▓	▓	▓	▓	▓

Figure 4.9 Project schedule timeline used by an engineer to secure funding for a project; this is not sufficient for a contractor during construction however (FONSI = Finding of No Significant Impact, FDEP = Florida Department of Environmental Protection)

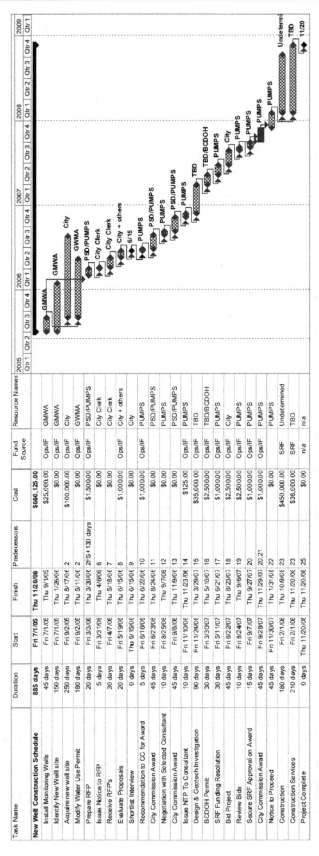

Figure 4.10 Detailed example of a schedule prepared to monitor project progress that shows dates for start and completion of each milestone, responsible party, funds required, and funding mechanism

4.9 References

Bloetscher, F. (1999a). "What you should expect from your consulting professionals (and how to evaluate them to get it)." *Water Engineering and Management,* October, 24–27.

Bloetscher, F. (1999b). "Looking for quality in an engineering consultant." *American City and County,* December, 28.

Bloetscher, F. (2011). *Utility Management for Water and Wastewater Operations,* American Water Works Association, Denver, CO.

Bloetscher, F., Fries, G., Nemeth, J., and Hurlburt, V. (2000). "Membrane installation solves utility's water supply crisis." *2000 Florida Water Resource Conference Proceedings, Tampa, FL,* April.

Halpin, D.W., and Woodhead, R.W. (1997). *Construction Management,* 2nd Ed., John Wiley & Sons, New York.

Nuechterlein, M.F., Watson, L.M., Jr., McGowan, W.F., Jr., Icard, T.F., Jr., Hemke, D.E., Cacciabeve, C.J., Marlowe, S.D., and Campbell, R.B. (1991). *Florida Construction Law: What Do You Mean…? A Practical Guide for the Construction Industry,* National Business Institute, Eau Claire, WI.

4.10 ASSIGNMENTS

1. Identify four characteristics of a successful design firm.
2. Create an organizational chart for your design firm.
3. Compare and contrast an RFP, SIQ, and RFQ.
4. What are the characteristics of a successful preliminary design report?
5. Choose at least four selection criteria to rank team presentations for a response to an RFQ and describe how a high score is achieved in each criterion.
6. Create a matrix to rank the groups in your class that presented responses to an RFQ (do not include your own group). After presenting the matrix, conduct a sensitivity analysis and summarize why the top-ranked group should be awarded the project.

Communication Skills for Engineers

The ability to communicate effectively, both in writing and public speaking, is essential for a successful career in engineering. In this chapter, a part of the discussion focuses on how to compose professional-quality expository, descriptive, and persuasive writing. Engineers also must know how to critique and revise written documents for effective submittals. Because engineers often are asked to create multimedia presentations in a group environment, this chapter also focuses on developing the necessary verbal and visual communication skills. Presenting engineering concepts and ideas to different audiences is an integral part of the profession, and as noted in Chapter 4, communication also is important to securing work via the request for proposal process. Knowing these facts, the aspiring engineer must become familiar with using visual aids and computer-based tools to convey the important design details, so that clients, regulators, politicians, the public, and even other engineers can envision what the final product will look like and evaluate the engineer's ability to successfully execute the project.

5.1 Overview of the Engineering Writing Style

Technical writing skills are a prerequisite for a successful engineering career. They assist in conveying information, serve as a thought process tool, and arguably are just as essential as excellent analytical or computational abilities. For some, writing well comes naturally; for others, it can be a struggle. The difference can stem from experience, confidence, or proper planning. Planning makes writing easier by helping to map out the structure of the document and assign who will write which sections. A good place to start would be to make an outline of topics to adequately cover the necessary content in the appropriate order that allows the reader to follow along in a logical fashion. An outline will keep the document on track and maintain focus on the important points. It is important to avoid overly long documents with too much technical detail and jargon (or specialized terms), as well as too many distractions and tangents. It may come as a surprise, but nearly every written submittal ever prepared was far from perfect after the initial draft, so expect to revise and rewrite often. Writing well takes time, practice, and patience. It simply cannot be done the night before the deliverable is due because the quality will suffer. The consequences of poor writing clearly justify the amount of time and effort required to write well because the written word in a document is permanent; therefore, sloppy work can lead to a bad impression, which can be extremely damaging to the writer's professional reputation. To help keep writing style appropriate to the audience, the writer should use consistent verb tense, vary the sentence structure so that each sentence does not begin with the same word, and use descriptive verbs, as well as limit unnecessary adjectives and prepositional phrases.

5.2 General Document Development and Outlining

The purpose of preplanning by creating an outline is to organize what needs to be said and how to get it across to the reader. An outline acts as a road map on the journey from introduction to conclusion. An effective outline will aid in successful writing by creating the scaffolding structure of the document that contains the essential content in a logical order. There are a couple of rules to follow when creating a proper outline to organize key thoughts. Make sure that each outline level has more than one entry beneath it, and all entries within the same section should be grammatically parallel in structure. The parallel phrasing tells the reader that these sentences belong together as a group. An example of a general outline is as follows:

I. Introduction
   A. Background
   B. Rationale
   C. Scope
   D. Objectives
II. Main idea #1
   A. Supporting detail #1
      1. Case study #1
      2. Case study #2
   B. Supporting detail #2
      1. Case study #1
      2. Case study #2
III. Main idea #2
   A. Supporting detail #1
      1. Case study #1
      2. Case study #2
   B. Supporting detail #2
      1. Case study #1
      2. Case study #2
IV. Conclusion
   A. Summary of findings

B. Schedule
C. Recommendations
V. References
VI. Appendices

Even though the content of the written submittal may be properly outlined, the material presented still must be organized so that the reader understands the main idea, grasps the key supporting details, and remembers the take-home message. So that the reader can easily follow the logical presentation of ideas, there are several strategies that can be used to help organize a document:

- ♦ **Order of importance**—This strategy follows a hierarchy from least to most important or from most to least important. This works well for ranking alternatives.
- ♦ **Chronological order**—This strategy follows events in the order in which they occur or have already occurred. This works well for instructions and progress reports.
- ♦ **Problem/solution, cause/effect, or question/answer**—This strategy begins with a problem, a case, or a question and then is followed by supporting sentences that provide solutions, explain cause-effect relationships, or answer the question posed in the first sentence. This works well for troubleshooting.
- ♦ **Simple to complex or general to specific**—This strategy can begin with a general topic or simple statement and then the writer drills down from the broader topic to a more specific case by progressing incrementally to more complex thoughts supported by specific details or examples. This works well for literature review.

Once a good, well-structured outline is in place, it is time to develop and generate the text of the written document. Armed with a logical strategy for coherence from the outline, each topic must be presented succinctly in blocks of text divided into paragraphs or sets of paragraphs. This will allow the reader to follow the relationship among the key points in the document that lead to a mutually understandable conclusion. Another way to help tie sentences, paragraphs, and sections together is through the use of transitions. It is very common for documents to contain sections written separately (or by separate authors) that are cobbled together using cut-and-paste methods. These individual pieces may not flow well together and may seem to the reader to come out of nowhere unless effective transitions are used. Transitions can be in the form of repeated phrases that link the main concept together with the details (e.g., Dr. Martin Luther King's famous "I Have a Dream" speech), conjunctions, transitional phrases, and parallel structure. Another way to transition cleanly from one paragraph to another is to echo the ending concept from the previous paragraph or section in the first sentence of the next one.

To further help the reader navigate the ideas in the text and follow along with the logical structure of the outline, it is helpful to employ the use of headings and subheadings to signal to the reader what is coming and also to help locate sections of particular interest in written material that is longer than just a few pages. Different levels of headings indicate logical divisions and groupings of topics within the body of the text. The level of a heading can be indicated by placement (indentation), font size, font appearance (bold, italic, underline, or change of font), or numbering system. An example of an effective system of headings is shown in Table 5.1. To be practical, the number of heading types should be limited to less than four to five levels, but it is more important to make sure that headings are consistent throughout the document. Effective headings are informative, specific, and brief.

Table 5.1 Example of heading styles

Level	Heading style
1	CENTERED ALL CAPS, BOLD TYPE
2	Flush Left, Initial Caps, Bold Type
3	Indented, Initial Caps, Bold Type
4	*Same Line as Text, Initial Caps, Bold Italic Type*

Writers should try to make headings parallel in structure. This means that all headings within the same level have the same grammatical structure, as shown in Table 5.2. The poor example column mixes nouns with verbs. The better example column uses a parallel structure that emphasizes the links between each of the subheadings in the form of the type of process.

With the skeletal structure of the document in place, an important question to ask even before embarking on writing a technical report is: Who is the audience? For instance, if the readers will be a panel of experienced engineering professors, it can be assumed that much of the basic terminology would already be known, so defining each technical term will bore or annoy the readers unnecessarily. Conversely, if the document is written for a panel of elected officials who are not engineers, the use of technical jargon and complex explanations will be frustrating. In many of the commercially available word-processing software programs, the user can generate readability statistics (i.e., Gunning Fog Readability Index, Flesch-Kincaid Grade Level, percent of passive sentences, average number of words per sentence, etc.) for a document. These tools can be very helpful to warn the writer when the writing has strayed from too simple to too complex. The Flesch-Kincaid Grade Level readability formula is explained in Figure 5.1 (Flesch 1948; Glinsky 1948; Gunning 2003).

Being able to write at the appropriate level for the target audience is one thing, but expressing ideas clearly, succinctly, and logically is the hallmark of good writing. Professor Hayman from UNC-Chapel Hill (1983) once said that "clear, concise, and free from ambiguity" is the goal. Writers should endeavor to keep the writing style simple and consistent. Awkwardly constructed phrases must be avoided. For engineers, the writing style is factual and descriptive, including specific details and concrete language that cannot be misinterpreted. For example, using numbers without units is unacceptable. Examples of poor specificity and better specificity follow:

Poor specificity: The water must be cold and not too high off of the floor.

What does "cold" mean? What is "too high"? These kinds of uncertainties are unacceptable.

Better specificity: The water temperature should be 5 to 10°C, and the handle for the drinking water fountain must be 29 to 36 inches from the floor.

Table 5.2 Examples of the use of parallelism in headings

Poor example	Better example
Project development	Project development process
Bids	Bidding process
Selecting	Selection process
Negotiating	Negotiation process

Flesch-Kincaid Reading Age = (0.39 × ASL) + (11.8 × ASW) − 15.59

Step 1   Calculate the average sentence length (ASL), which is the number of words divided by the number of sentences.

Step 2   Calculate the average number of syllables per word (ASW), which is the number of syllables divided by the number of words.

Step 3   Multiply the average number of words by 0.39, add that to the average number of syllables per word multiplied by 11.8, and subtract 15.59.

Analysis   Analyzing the results is straightforward. For example, a score of 9.3 means that a ninth grader should be able to comprehend the document.

Theoretically, the lowest grade level score could be −3.4, but since there are no real passages where every sentence consists of one-syllable words, it is a highly improbable result in practice. It is important to note that word-processing programs may not score above grade 12. In this case, any score above 12 will be reported as grade 12.

Figure 5.1 Flesch-Kincaid Grade Level readability formula

5.3 Proper Grammar

Employing specific language to avoid misinterpretation aids clarity, as does following the rules of grammar. Anyone who writes a passage first must choose the words that best express the key thoughts and ideas (*diction*) and then select the most appropriate arrangement of those words (*syntax*) in relation to the main ideas. Within this framework, there are a set of rules that govern the collection of words into sentences, collection of sentences into paragraphs, collection of paragraphs into chapters, and so on and so forth. These are the rules of grammar and punctuation. There are several excellent references on grammar and punctuation, such as Strunk and White (1979), Hodges et al. (1990), McCaskill (1998), Sorby and Bulleit (2006), Vesilind (2007), and Beer and McMurray (2009). Students should refer to these texts to clarify and complement any of the rules briefly discussed here. For a quick reference, the appendix at the end of this chapter provides some tips and rules.

5.4 Reference Citations

Most written documents will require the gathering, synthesis, and use of work created by others. This requires the writer to give proper credit to the original author. Failure to do so constitutes an act of *plagiarism*. Fortunately, with a few basic rules, citing references correctly in a consistent format will avoid plagiarism. There are a number of accepted styles for formatting and citing references, such as APA, Chicago, Harvard, MLA, Turabian, etc. The style selected should be used consistently throughout the text. Two examples of how to cite a reference correctly in the text follow:

> The growth rate of *Escherichia coli* is significantly impacted by chloramine disinfection (Jones 2014).

> Jones (2014) has found that the growth rate of *Escherichia coli* is significantly impacted by chloramine disinfection.

Be careful to cite all references used in developing the document. Avoid copying and pasting from the Internet or someone else's work. When using reference material in the text, try to limit the use of direct quotes by paraphrasing and translating the source's thoughts into your own words. However, if using another author's text directly, make sure the statements are properly quoted and attributed.

The following sections provide examples of an acceptable reference citation format used by the *Journal of Civil Engineering*, published by the American Society of Civil Engineers (http://pubs.asce.org).

5.4.1 Journal Citations

Include author(s), year, article title, journal title, volume, issue, and page numbers:

> Meeroff, D.E., Bloetscher, F., Reddy, D.V., Gasnier, F., Jain, S., McBarnette, A., and Hamaguchi, H. (2012). "Application of photochemical technologies for treatment of landfill leachate." *Journal of Hazardous Materials*, 209, 299–307.

5.4.2 Conference Proceedings and Symposium Citations

Include the sponsor of the conference or publisher of the proceedings and that entity's location (city and state or city and country):

> Bloetscher, F. (2012). "Comparing costs and benefits of wastewater aquifer recharge as alternative water supply to reclaimed water and brackish water." *2012 Sustainable Water Management Proceedings—Portland, Oregon*, American Water Works Association, Denver, CO.

5.4.3 Book Citations

Include author(s), chapter title (if applicable), book title, publisher, the publisher's location, and inclusive page numbers (if applicable):

> Bloetscher, F. (2011). *Utility Management for Water and Wastewater Operators*, American Water Works Association, Denver, CO, 69–129.

5.4.4 Report Citations

Reports use the same format as books. For reports authored by institutions, spell out the institution name at the first use and follow with the acronym in parentheses, if applicable. If subsequent references were authored by that same institution, use only the acronym. For reports authored by people, include the full institution name (no acronym) and its location:

> Carsey, T., Amornthammarong, N., Bishop, J., Bloetscher, F., Brown, C., Craynock, J., Cummings, S., Dammann, P., Davis, J., Featherstone, C., Fischer, C., Goodwin, K., Meeroff, D.E., Proni, J.R., Sinigalliano, C., Stamates, J., Swart, P., and Zhang, J.Z. (2007). *National Oceanographic and Atmospheric Administration (NOAA) Data Report AOML-40, Boynton Inlet 48-Hour Sampling Intensives: June and September 2007*, Atlantic Oceanographic and Meteorological Laboratory, Miami, FL.

5.4.5 Unpublished Material Citations

Unpublished material is not included in the references but may be cited in the text as follows:

(John R. Proni, personal communication, May 16, 2014)

5.4.6 Web Page Citations

Include author(s), copyright date, title of "page," web address, and the date on which the material was downloaded:

Burka, L.P. (2013). "A hypertext history of multi-user dimensions." *MUD history*, <http://www.ccs.neu.edu> (Dec. 5, 2013).

5.4.7 Thesis and Dissertation Citations

Include author, date, title, and the name and location of the institution where the research was conducted. Note that some institutions use specific terminology (for example, "doctoral dissertation" rather than "Ph.D. thesis"):

Romah, T. (2012). "Advanced methods in sea level rise vulnerability assessment." MS thesis, Florida Atlantic University, Boca Raton, FL.

5.5 Persuasive Writing

As pointed out earlier, the process of securing engineering work relies heavily on technical writing that is persuasive enough to convince the reader or the client of a preferred option or solution. This requires the successful use of justification strategies, such as providing evidence, data analysis, statistics, factual statements, logical arguments, and cause-and-effect relationships. In many engineering applications, the solution may not be open to interpretation, such as in the case of a code requirement. There is simply no choice but to comply. However, there are many more instances where an engineering project can have many different feasible solutions. Hence, it is the task of the engineer to convince the client (or supervisor or regulators) that the proposed solution is the most appropriate or the preferred approach given the circumstances. Two techniques can be used effectively in the persuasive style of communication:

1. **Primacy**—The conclusion is stated first (to capture the reader's attention), and then the supporting detailed arguments are presented.
2. **Recency**—The detailed arguments are provided in a logical order, followed by the conclusion.

When employing recency, the reader's short-term memory is targeted. It is important to build a logical argument from simple concepts to complex ones, in a stepwise fashion, gaining reader buy-in at each step. If the reader is in agreement with each argument along the way, then he or she inevitably will agree with the conclusion statement. The following is an example of a recency argument:

- ◆ Our company needs more business revenue.
- ◆ Our competition is getting more business by offering fuel-efficient hybrid vehicles.
- ◆ But the competition's product has a range of only 100 miles.
- ◆ Customers want greater range from their hybrid vehicles.
- ◆ Our company can make a hybrid vehicle with a greater range than our competition (300 miles).
- ◆ We should increase our production of our longer range hybrid vehicles, so we can get more business.

When employing primacy, the reader's first impression is being targeted. Now that the reader's attention has been grabbed, a logical argument to support the stated conclusion must be constructed to reinforce his or her initial buy-in. The following is an example of a primacy argument:

- ◆ We should increase our production of our longer range hybrid vehicles, so we can get more business.
- ◆ Our company needs more business revenue.
- ◆ Our competition is getting more business by offering fuel-efficient hybrid vehicles.
- ◆ But the competition's product has a range of only 100 miles.
- ◆ Customers want greater range from their hybrid vehicles.
- ◆ Our company can make a hybrid vehicle with a greater range than our competition (300 miles).

In either case, the goal is to consider the readers' perspective so that the arguments will appeal to their priorities and resonate with their decision-making sensibilities. The audience is more likely to be convinced by three to five justified arguments that speak to their collective set of priorities than by a laundry list of advantages, many of which are irrelevant to the project. When employing primacy, go with the strongest argument first. Readers want the most appealing argument up front. Otherwise, they may not finish reading the rest of the document. When employing recency, make sure to finish with a strong conclusion.

In developing logical arguments and supporting data, it is important to provide peer-reviewed information or conclusions from multiple sources, as justification to strengthen claims made in the text. Never assume facts without proof. Care also must be exercised to avoid using implied assumptions (e.g., the earth is flat), stating conclusions without proof (e.g., solar panels save money), using unsupported causal relationships (e.g., plenty of non-smokers get lung cancer, so cigarette smoking must not cause lung cancer), employing *non sequiturs* (e.g., you will do what I say because I am older), and misusing statistics (e.g., 50% of textbooks are over $200, because one costs $231 and the other costs $195).

5.6 Engineering Graphics

The text may contain convincing arguments, but engineering communication often requires effective displays of quantitative information that are informative, complete, accurate, and memorable. Long blocks of text rarely convey the point of discussion better than integrating tables and figures as visual aids to the reader. There are many ways to present technical data to help support the arguments in a document. These include:

◆ **Numerical tables**—Used when showing specific values in a data set.
◆ **Line/scatter plots**—Used to demonstrate trends between continuous variables.
◆ **Bar graphs/histograms**—Used when categories are not continuous (or numerical).
◆ **Pie charts**—Used to show parts of a whole with percentages.

5.6.1 Numerical Tables

It is difficult to incorporate sets of data into text in paragraph form. Instead, they should be presented in tabular form to better emphasize some sort of numerical trend. Because tables are read from the top down, the column headings and labels should be at the top. Each table should have an informative title caption that can stand alone without further explanation. However, it is standard practice to introduce a table (or figure) in the text and refer to the major conclusions drawn from the table in the text immediately preceding its appearance. It is important to include a brief summary of the relevant trend that is demonstrated in the table so as to avoid any reader confusion or misinterpretation of the data results. Tables should indicate the proper units in each column heading, and the independent variable should be placed in the left column. The data should be arranged clearly and logically so the reader can arrive at the intended conclusion. This can be accomplished by arranging the dependent variables in decreasing order of importance from left to right, for example. Tables should be formatted to assist the reader in arriving at the trend without unnecessary distractions. Decimal places and/or significant figures should be consistent throughout each table (e.g., 0.00), so the reader's eyes do not wander and miss the point. Commas should be used appropriately in numerical values (e.g., 2,342). Numbers should be aligned so that the decimals and commas in a column line up vertically. Finally, there should not be any blank cells in a table (use N/A, 0, or —). Blank cells automatically draw the eye of the reader and give the impression that the data set is incomplete, reducing the strength of the conclusion or trend presented.

Each table should be located immediately or shortly after it is referenced in the text, so the reader does not have to turn the page or hunt to find it. All rows in a table should be kept on one page, if possible. If not, the column headings should be repeated on each subsequent page. Tables provide useful information to the reader, but if a table is 15 pages long, the reader may get frustrated, and the connection with the accompanying argument in the text will inevitably get lost. An appendix should be used for cumbersome data sets that are very long. Showing data in both tables and figures that convey the exact same information should be avoided, as it is redundant. Finally, the source of the data should be indicated with a correct citation.

Tables should conform to the same margins as the text. If a table is too wide to be accommodated within the required margins, even when displayed in landscape mode, then the type size should be reduced to fit the margins; if that does not work, it may be necessary to reduce the number of columns or split up the data into multiple tables.

To better show data trends, items in the same column should be placed in ascending or descending order of rank or some other logical organization (e.g., most current date or increasing cost). Making the data trends easier to visualize by aligning all columns of number values by decimal point or by using right alignment makes them more understandable to the reader. Scientific notation should be used for very large or very small numbers, but if one value requires scientific notation, then all values must be converted to scientific notation so it is easy for the reader to compare the data. Units should never be mixed. All abbreviations

Table 5.3 Table of data

Concentration (mg/L)	Peak height (mm)
0	0
1	6
2	7
4	9
8	13
16	21
32	37

Table 5.4 Example of bid tabulation

Vendor	Attended mandatory meeting	Provided original +5 copies	Provided bid deposit ($500.00)	Provided Exhibit A	Provided Exhibit B	Provided Exhibit C	Total base bid
A Corp.	Yes	Yes	Yes	Yes	Yes	Yes	$194,198.00
GCT	Yes	Yes	Yes	Yes	Yes	Yes	$212,885.00
SEG	Yes	Yes	Yes	Yes	Yes	Yes	$195,042.00

and symbols should be explained in the footnotes located immediately below the table to alleviate any possible reader confusion.

The following show some good and bad examples of tables. Table 5.3 contains properly labeled columns with units, and the data in the left column seems to be arranged in increasing order, but the table title provides no insight to allow the reader to understand the nature of the increasing trend of peak height with concentration.

In Table 5.4, the vendors seem to be arranged in alphabetical order, but the base bid amounts are not arranged logically. It may be more beneficial to the reader to arrange the rows in increasing order of base bid. The six columns in the middle seem to confirm that the bidding requirements were met by all three vendors listed. Therefore, these columns may not be necessary to highlight the trend. The fact that all three vendors met the requirements could be mentioned in the table title instead, allowing those six columns to be eliminated. If the intent is to highlight the lowest bid, then the title could be rewritten as follows:

Table 5.4 Bidding vendor checklist arranged by increasing base bid

5.6.2 Figures

Graphics that are not considered tables are referred to as figures. As the cliché goes, "a picture is worth a thousand words," and figures provide visual support for readers to understand a concept. Figures are read from the bottom up; therefore, the caption is placed below each figure. Just like table titles, figure captions should be informative and able to stand alone in the text without further explanation. However, it is standard practice to introduce a figure in the text and describe it in more detail immediately preceding its appearance. Figures should have a purpose, be simple and uncluttered, and present a manageable amount of information. Line graphs and scatter plots, bar graphs and histograms, and pie charts will be discussed in the following paragraphs.

Line graphs and scatter plots can be used to determine or visually convey a trend in a data set. The dependent variable typically is displayed on the y-axis (ordinate), and the independent variable usually is displayed on the x-axis (abscissa). An exception to this rule is for altitude or depth readings because they are more easily visualized on a vertical scale. Both axes should be properly labeled with units. Axes typically start at zero and use simple multiples on the scales. Grid lines or tick marks should be used appropriately so as not to overly clutter the figure with unnecessary lines that might obscure the trend. Once the data is plotted, points should be connected only if the data is continuous. Otherwise, broken lines should be used to help the reader visualize the correlation between variables. Every figure must include a legend, which clearly labels and differentiates each different curve or trend line.

An example of good use of a scatter plot graph would be to see if a general pattern exists in data sets of a measured variable, such as temperature, company profits, rainfall, or costs, to name just a few, over a period of time. On the other hand, a line graph assumes that the numerical values on the x-axis are categories and will be shown the same discrete distance apart, much like a bar graph. Figure 5.2 shows a scatter plot on the top and a line graph of the same data on the bottom. The first point of data is (0,0), which is shown at the origin in the scatter plot (Figure 5.2 top), but in the line graph (Figure 5.2 bottom), this point is actually shown as (1,0) because it is falsely interpreted as a measurement category instead of a numerical value on the x-axis of a line graph. In Figure 5.2, the values plotted on the x-axis in the scatter plot (top) depict the concentration data, but in the line graph (bottom) the concentrations are shown as the number of the data in the series and not the value; for example, the third data point is actually 2 mg/L in the scatter plot but is shown as 3 (for the third point in the series) in the line graph. Also notice that the trend lines are different because the line graph plots each point in the series as a discrete data point (like a color or a grade) and not as the actual value of the x-axis measurement.

Figure 5.3 shows an example of an effective scatter plot, where the x-axis (labeled as elapsed time in hours) is the independent variable and the y-axis (labeled as permeate flux in gallons per day per square foot) is the dependent variable, which is measured. An effective graph tells an obvious story. Two trends are evident here. One is for the 10-HFM-251-FNO membrane, which was a used membrane, and the other is for the 10-HFP-276-FNO membrane, which was new. The graph clearly shows that after about 4 hours, both membranes, regardless if new or used, operate with essentially the same permeate flux. The other observation here is that for the first 4 hours, the new membrane has a higher permeate flux. It is important to note that when plotting a graph for the purpose of determining a trend line and correlation coefficient, it is preferable to use a scatter plot instead of a line graph because a line graph assumes that all x-axis values are the same magnitude apart, which may not always be the case.

A bar chart or histogram is designed to show the frequency of a parameter when the independent variable is not a continuous variable. In other words, the independent variable is a category, such as letter grades, colors, months, or equipment manufacturers. An example of a grade distribution bar chart is shown in Figure 5.4. All bars should use the same solid pattern, unless there is a difference that needs to be pointed out for emphasis. If different colors are used, they should be of similar saturation and brightness. Notice that error bars

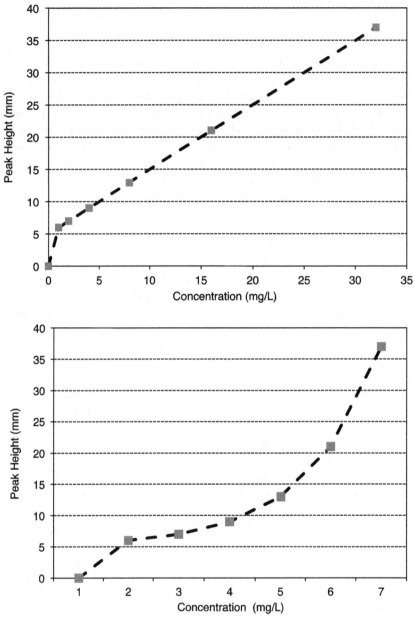

Figure 5.2 The graph on the top is a scatter plot and the graph on the bottom is a line graph of the same data

are plotted in Figure 5.4. These are meant to show the variability in the data by showing the value of the standard deviation of the average scores over five different semesters for the same class.

A pie chart is designed to display the importance of different categories with respect to the percentage of the whole. Usually it is best to begin with the largest slice and arrange the categories clockwise in order of decreasing size. The number of slices should be limited to less than seven if possible. A miscellaneous slice should be included for small categories;

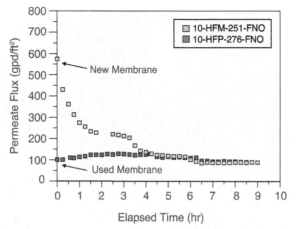

FIGURE xx. Comparison of permeate flux for a 2.5% seawater dilution of Orimulsion® with 0.1:1 clay:solids addition ratio using both membranes.

Figure 5.3 Example of a scatter plot (Meeroff 1997)

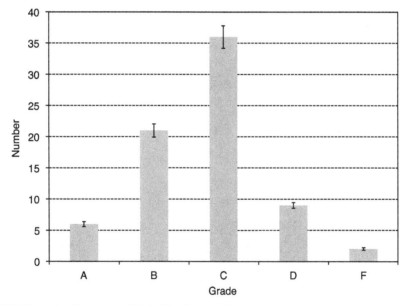

Figure 5.4 Example of a grade distribution bar chart

otherwise the labels will be too small or too cluttered to see. Slices should be labeled horizontally (not radially), and the inside of each slice should be labeled, if space permits; if not, call-outs can be used. Before finalizing, the percentages in the pie wedges should be added up to make sure they total 100%. An example of an effective pie chart is shown in Figure 5.5.

If the figure is a diagram (Figure 5.6), it should be carefully labeled so as to be understandable to the reader. If the figure is a photograph (Figure 5.7), the caption should explain what the photograph is meant to show, which should be pertinent to the project. A technical

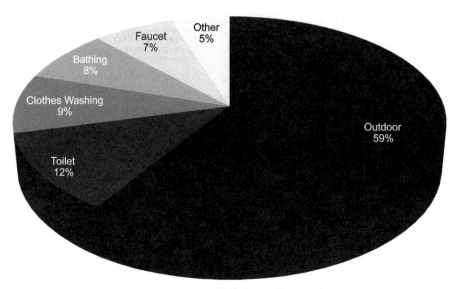

Figure 5.5 Pie chart showing water use in a community

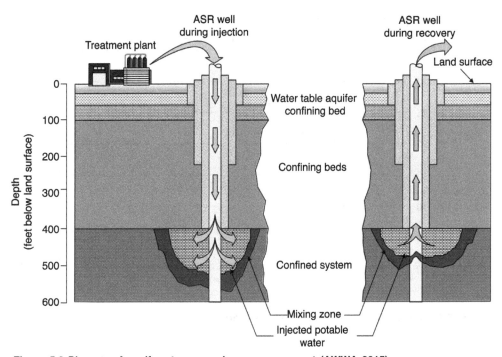

Figure 5.6 Diagram of aquifer storage and recovery concept (AWWA 2015)

report should use high-resolution photographs, not clip art or other icons that detract from the formality of the document. Photograph captions should be labeled as figures and should be located below each photograph. As with all tables and figures, each graphic should be referenced in the text prior to its appearance, and proper credit must be given to the photographer or original source.

Figure 5.7 Aquifer storage and recovery well in South Carolina

5.7 Proofreading Strategies

Now that all of the text and graphics have been incorporated into the rough draft and organized logically, it is time to edit the document with an eye toward revising its readability and effectiveness. Submitting a document plagued with annoying errors suggests to the reader that the writer is careless, incompetent, or lazy; therefore, proofreading is necessary to ensure the quality of the written submittal. When proofreading one's own work or someone else's writing, it is important to read carefully, without skimming or skipping over words. Writers should methodically go through the document to correct grammar, diction, punctuation, spelling, capitalization, missing words/letters, sentence-level errors, transitions, and syntax. It also is important to address higher order issues such as clarity, organization, and proper citation of references. If groups of multiple authors and editors are involved, it may be helpful to use the "track changes" option in the word-processing software and the "comment" tool to mark major issues such as organizational changes, formatting (fonts, margins, line spacing, alignment), missing items, voice, tone, structure, and recurring sentence-level errors.

After proofreading has been completed, revisions will be required. It is a good idea to become familiar with common shorthand and abbreviations typically used for editing, such as *sp.* for spelling error, *RW* for rewrite sentence, *CS* for comma splice, *frag* for sentence fragment, and *ref* for missing reference. Dealing with minor corrections is relatively straightforward, but often the comments involve more than just mechanical errors; in these cases, it will be necessary to rewrite or reorganize entire sections of text. In some cases, missing content will need to be added. Revision is always more involved than simply adding a comma here and there. After the revision process is complete, another spelling and grammar check should be run and the entire document reread to catch any last-minute stylistic errors before accepting the revised version. One more piece of advice is to save the file at regular intervals to avoid losing work if the computer crashes.

5.8 Fonts

The author should choose a standard font and size and try to avoid mixing fonts throughout a document. Font sizes that work best for engineering writing include the following:

- 10 point
- 11 point
- 12 point

Fonts are classified as *serif* or *sans serif*. Serif fonts have leading and trailing strokes on each letter to help make the text easier to read. Classic serif fonts include Times New Roman, Garamond, and Cambria. Sans serif fonts do not have these strokes and are used primarily for headings and graphics. Examples of sans serif fonts include Arial, Calibri, Tahoma, and Verdana. There are some unusual fonts that are designed for emphasis, but they are difficult to read for long sections of text and should be avoided. These include Haettenschwiler, Brush Script, Algerian, and Stencil, among others.

5.9 Margins

Report formats may include specific rules for margins (top, bottom, right, and left) and line spacing (single, double, 1.5 spacing). With regard to line spacing, if using a double space between paragraphs, it is not necessary to also indent. However, an indent should be used when there is only a single space between paragraphs; otherwise, the paragraph structure will be lost. Alignment of text can follow several different formats:

- Left justification (or ragged right)
- Right justification (or ragged left)
- Centered
- Full justification (both right and left justified), which looks like a block of text that is aligned at the left margin as well as the right margin (this type of justification will tend to have some squeezed words on some lines and some uncharacteristically large gaps between words on other lines, but the effect is that there will be a clean margin on the left and right—perfect for editorial comments).

5.10 Typical Engineering Documents

Now that general writing and editing have been discussed, it is time to learn how to create specific engineering documents, such as meeting minutes, emails, memos, letters, proposals, progress reports, technical memoranda, and technical reports. Knowing how to craft these kinds of documents properly can enhance the engineer's credibility and professional reputation. Keep in mind that engineering documents may be different than the deliverables that a client expects, but documents like memos, emails, and letters can be inserted into other documents, such as technical reports, final reports, etc., which have specific due dates.

5.10.1 Meeting Minutes

An agenda should be created for each meeting. Every time a design team gathers together for a group meeting, someone should be designated to record the minutes of the meeting, in writing. Otherwise, the important decisions made will get lost or there will be confusion about them, and progress will be slowed. The important components of effective meeting minutes include the name and title of each attendee, the date/time and location of the meeting, and brief notes of items discussed, important decisions reached, roles/tasks assigned, action items, and due dates. Before adjourning and distributing the draft version to the team, attendees should review the minutes for completeness and accuracy. Ideally, draft copies can be made available at the end of the meeting; otherwise, they should be sent out electronically. Engineers should always aim for 100% accuracy and should not be afraid to ask for clarity during the meeting itself. At the end of the minutes document, the time the meeting ended should be recorded, and sometimes it is beneficial to record the approval signatures or the results of the vote to accept the minutes. An example of meeting minutes from a construction meeting is shown in Figure 5.8.

Monthly Construction Meeting
November 20, 2013 10:00 AM
Public Works Conference Room
1201 Stirling Road, Dania Beach, FL
Lime Softening Upgrades—Accelerator Contract (13-002)

Present for the meeting were: Sheryl Wells (by phone) and Mickey Bartlett of Close Construction, Dominic F. Orlando and Phil Skidmore of the City of Dania Beach, and Frederick Bloetscher with Public Utility Management and Planning Services, Inc.

The following items were discussed:
1. Pay request for October 2013 has been processed.
2. Catwalks are installed.
3. Staircases are installed—will check kick plates.
4. Simpson is still blasting, but they anticipate completion this week for the east accelerator. Dr. Bloetscher noted that the paint has not been inspected.
5. A solution for the aeration splashing is needed.
6. The lime pumps, mixer, and new mixing tank have been installed. Some minor piping needs completion on the pipe manifold.
7. Chemical piping will have the color coding and painting installed by the end of December.
8. Project completion is waiting on the manufacturer to install the west accelerator. The drive and spray system is on back order.

Dr. Bloetscher noted that the piping paint change order #3 must be approved. Dominic Orlando agreed to approve after the correct form is submitted. There is a $50,000 allowance in the contract to cover this contingency.

Ms. Wells noted that change order #2 needs to be resolved. The City must respond to the information request by Close Construction, noting the reasons that this item was included in the original scope of work.

Phil Skidmore ordered the knife gate with a smooth shaft needed to repair the east accelerator before it can go back into service.

Given that there was no other business, the meeting was concluded at 10:20 AM.

Figure 5.8 Example of meeting minutes

5.10.2 Emails and Informal Notes

Although emails, text messages, tweets, blog posts, and other electronic forms of communication are pervasive in society, it is easy to forget that these informal notes are not private and can remain public in cyberspace until they resurface and potentially become incriminating. Anyone can forward these items to someone else without the writer's knowledge or consent. In emails, the *bcc* (blind carbon copy) function may be used to send a message to another person without the original recipient's knowledge.

Figure 5.9 is an example of a poorly written email from a student to his teacher. There are numerous problems with this email. Imagine being the recipient (as one of the authors was). First of all, the opening, *Hey Mirhof,* is not appropriate. The email should be addressed *Dear Dr. Meeroff.* Spelling the recipient's name correctly and using the person's formal title gives a much better first impression. Next, who is Jason? The email handle does not provide any indication as to the last name of this individual, and no last name was provided in the body. Does the author think he is the only Jason in the entire world? Take a look at the subject line: *Class stuff.* Does Jason think that Dr. Meeroff only teaches one class? To which class is he referring? Overlooking the obvious grammar and spelling mistakes, the tone is way too informal for a student-teacher email. Finally, what day is tomorrow? The email was sent on Sunday, but may not actually be read until Monday, so the reader might think tomorrow is Tuesday, when the author meant Monday.

Before sending an electronic message, give the reader a concrete reason to open it. Some people get hundreds of emails every day and are much too busy to waste time reading every one of them. They need a compelling reason to open a message while sifting through the daily email spam. The first things someone typically notices when scanning the inbox are the name of the sender and the subject line. By including a clear, meaningful subject line in the header, an email has a better chance of being read, particularly if the author's email handle is not immediately recognizable. This is another example of primacy. Because email is not the same as a text message, a rapid informal back-and-forth exchange should not be expected. Email is an important part of most people's daily work routine, so most professionals do not really want to engage in a leisurely back-and-forth exchange; they want to clear out the inbox and move on to the next task. If the subject line is vague or blank, the first opportunity to inform or persuade the reader was missed, and the message implies that the sender's name in the *From* line is all the recipient should need in order to make it a top priority. That could come across as arrogant or inconsiderate.

As with any engineering writing, the sender should remember the rule of primacy and get to the point immediately. Email requires keeping the text as brief as possible. Avoid large

```
From:      maniac@out.com
Sent:      Sunday, June 2, 2014, 12:19 PM
To:        dm@xyz.com
Cc:        mom@out.com
Subject:   Class stuff

Hey Mirhof, this is Jason and i was just wondering what materails i need for class tomorrow?

Thankyou
```

Figure 5.9 Example of a poorly written email

blocks of text by breaking up the writing into short paragraphs or bullet items. Again, never assume privacy; because anyone can forward an email, the highest level of professionalism should be maintained at all times.

Figure 5.10 is an example of a more effective email. In this example, the writer introduces herself, explains the situation, and asks for an accommodation to be made for her absence due to official university business at an athletic event. Furthermore, the student is contacting the instructor over a week in advance to arrange to take the makeup quiz and also has arranged for someone to take notes for her during her absence. This is much more professional.

Emails always should include the author's name, title, company, and any other important identification information in the first few sentences or at least in the signature line, if the recipient is a close colleague. If following up on a face-to-face contact, it is likely that the recipient has no memory of the encounter, so it is courteous to provide casual or tactful hints, such as: *I really enjoyed talking with you about LEED features at the USGBC conference last week in Atlanta, GA.* If contacting someone for the first time (cold calling), then provide a more formal introduction, such as *My name is...and I am an engineering associate at....My reason for contacting you is....*

There is no subtlety in email, so humor, sarcasm, irony, etc. do not always go over well. Some writers use emoticons such as a smiley face (☺) to convey emotions, but in professional emails it is more appropriate to use clear, formal language to minimize misinterpretation. Also try to avoid using familiar text speech, such as *FYI* (for your information), *lol* (laughing out loud), *IMHO* (in my humble opinion), and *btw* (by the way). Do not use all caps to indicate emphasis because it is the email equivalent of shouting at someone. On the receiving end, if a rude or angry email is read, do not reply right away. When in doubt, err on the side of caution. Negative emotions can escalate all too quickly in emails. Not everyone knows the rules.

From: mylilponi27@mail.com
Sent: Wednesday, October 1, 2014, 11:31 AM
To: professorx@ABCU.edu
Cc: JenJones1991@mail.com
Subject: CGN4803 Absence, October 10, 2014

Dear Prof. X:

This is Jane Down, from CGN4803 class that meets on Thursday nights. I am a student athlete, and this morning our university swimming team qualified for the conference finals in Orlando, FL. Since I will be out of town on October 10, 2014 for the race, I request to be excused from class that day. According to the syllabus, I will miss a lecture and a quiz. I have asked my classmate Jennifer Jones to take notes for me, and I would like to schedule a time to take the quiz early. May I come to your office hours at 2 PM on October 8, 2014 to take the quiz early?

Thank you very much,

Jane Down
Senior Undergraduate Civil Engineering Student
ABC University

Figure 5.10 Example of an effective email

Another common problem with emails involves the use of attachments. Some attachments are automatically deleted by virus protection software or automatically filed in the junk folder. Make sure that attachments are virus-free; otherwise, the recipient may never open any email from you again. As a recipient, be aware that emails with attachments from unknown senders may contain viruses, malware, worms, or worse. Hopefully by now, everyone knows that the winning lottery notification email from overseas is a scam and should not be opened. All too often, a writer mistakenly sends an email before adding the attachment and then has to send an apology email with the file. To avoid this common misstep, get in the habit of proofreading all messages and taking a deep breath before pressing the "send" button.

Also avoid using the "reply to all" feature unless absolutely necessary. Figure 5.11 is one unfortunate example. Any guesses as to why no one replied to tf121? When the "reply to all" function is used, the original sender of the bad news about the holiday bonuses (Boss@ABCU.com), who happens to be the president of the company, is copied on the message. Talk about awkward.

From: tf121@mail.com
Sent: Monday, December 3, 2013, 12:35 PM
To: Boss@ABCU.com
Cc: ept1@ABCU.com, ftv21@ABCU.com, jcm@ABCU.com
Subject: Regarding holiday bonuses

Dear Colleagues:

That last email sent by our boss (see below) about our holiday bonuses is pure nonsense. We all know that our boss is hoarding the money for himself so he can go buy himself a new yacht. He doesn't care about us, but I would never say that to his face. What should we do? Boycott the Christmas party or plan a practical joke in his office so that he gets egg on his face?

—— Forwarded Message ——
From: Boss@ABCU.com
Sent: Monday, December 3, 2013, 10:51 AM
To: tf121@mail.com, ept1@ABCU.com, ftv21@ABCU.com,
           jcm@ABCU.com
Cc: Boss@ABCU.com
Subject: Regarding holiday bonuses

Dear Design Team:

I regret to inform you that this year our company recorded a loss and will not be able to provide the standard holiday bonus to any of its employees including the upper level management. The company has also canceled the annual Christmas party. I am so very disappointed, as I know you were all counting on the annual bonus and our holiday party, so I am going to host a Christmas party downstairs in the lobby after hours. I will provide the food and refreshments because you have all worked so hard this year and deserve to celebrate. All you have to do is come and have a good time.

President, ABCU Consulting

Figure 5.11 Example of an unfortunate mistake in writing an email

5.10.3 Memoranda

A memorandum (memo) is a document that is internal to an organization. Memos are used to disseminate information to a large group of employees. Information in a memo can document work accomplished, establish internal policies, or record technical information. Classic memos are brief and to the point. They state the purpose immediately, then provide context, and finally end with the decision or ask the reader for more information. A special type of memo, the technical memorandum, will be discussed in more detail later.

The format of a memo starts with five standard lines in the header:

1. TO: Jane Doe, P.E., Project Manager
2. FROM: Steve Smitty, P.E., Project Engineer
3. DATE: June 1, 2014
4. RE: Field sampling activities
5. CC: C. Lions, C. Gary, P. Berg

The tab function is used to line up the information provided in each field. The first line provides the names and titles of the recipients. The writer's name and title go on the second line (multiple authors should be included). The third line is for the date. The fourth line is the subject line, and the fifth line should include the names of everyone copied on the memo. If the distribution list is very long, then "employees" or "distribution list" can be used. The memo is initialed near the name in the *FROM* line (line 2), as shown in Figure 5.12.

5.10.4 Standard Business Letters

A letter written on company letterhead is an external document that represents the agency to the outside world. If a letter is poorly written, it will reflect negatively on the whole organization, not just the writer; therefore, it is imperative that every step possible be taken to make a positive impression. Starting at the top, the *company letterhead* should contain the important contact information, such as the company name, address, telephone number, web address, logo, etc. The next item should be the *date line*. The full name of the month is followed by the day and then the year:

TO: Jane Doe, P.E., Project Manager
FROM: Steve Smitty, P.E., Project Engineer *SS*
DATE: June 1, 2013
RE: Field sampling activities
CC: C. Lions, C. Gary, P. Berg

As per our conversation on May 31, 2013, C. Lions, C. Gary, and P. Berg have been assigned to conduct the remaining sampling activities at the XYZ job site in West Palm Beach, FL while I am away on vacation. These three team members have already been to the site with me on several occasions and are very familiar with our sampling procedures. I am confident that they will perform their duties to the highest standards without my direct supervision until I return. The two sampling events have been scheduled for Tuesday, June 13, 2013 and Thursday, June 15, 2013. Should any issues arise, I will be available via telephone and email, but Miss Lions will be in charge until I return.

Figure 5.12 Example of a classic email

August 14, 2014

The next element is the *address line*, which is the contact information of the recipient. It should include the full name, job title, company name, and address of the recipient:

John Adams, P.E., President
United States Consulting Firm
777 Main Street
Washington, DC 20001

Immediately following the address line is the *salutation line* to formally address the recipient. It is customary to use *Dear [addressee's name]* followed by a colon:

Dear Mr. Adams:

It is appropriate to use *Sir, Madam,* or *To whom it may concern* if the name of the recipient is not known, but every effort should be made to find out the contact person's name, so that the letter can be addressed directly.

With the formalities out of the way, the *opening statement* should contain the purpose of the letter (primacy). The reader should not have to wonder what the letter is about. Remember that a letter is basically a formal request for something from the reader. One good way to start the process of writing a letter is to articulate the purpose of the letter out loud and then begin writing formal and convincing supporting statements for that purpose. The following is an example of a succinct opening statement:

Enclosed please find the first interim progress report for the project entitled *Zone 4 Neighborhood Site Plan.*

Immediately following the statement of purpose is the *body*, which concludes with a specific request for the recipient to accomplish after reading the letter. The writer should be positive, considerate, well mannered, clear, specific, and concise. The letter should end using the principle of *recency* by clearly stating what the reader of the letter must do (action item).

After the *conclusion* comes the *closing line*, which starts with a phrase such as sincerely, regards, truly yours, etc. This is followed by three blank lines for the *signature,* with the writer's name, title, and address typed below for legibility. The last element is to document if anyone is copied on the letter or an enclosure is included:

Regards,
Tony Flags

Tony Flags, P.E., Project Manager
ABC Consulting
555 S. Main Street
Anywhere, NY 12345

cc: H. Johnston, B. Jones
encl: "Zone 4 Neighborhood Site Plan Interim Progress Report"

Do not forget to sign the letter!

In engineering applications, there is a special type of *cover letter* that is attached to most engineering reports and submittals: the *letter of transmittal*. This communicates to the client that the agreed upon deliverable is attached so that payment can be processed. The same guidelines should be followed as for a standard business letter, plus the following:

◆ The first paragraph is focused on the reason for the report and, invoking primacy, it references the name of the report (in italics):

> Please find enclosed a progress report entitled *Phase I Environmental Site Assessment for the Main Street City Hall Expansion Project.*

◆ The next paragraph focuses on the reason for the report and provides a brief synopsis of the content:

> This report outlines all of the recognized environmental conditions associated with the property as a result of the site reconnaissance, records review, and interviews conducted by our team. It is our conclusion that...

◆ The next paragraph acknowledges any support and mentions any limitations or omissions:

> We would like to thank the staff of Washington City for their help in obtaining the environmental records. Unfortunately, no aerial photographs prior to 1973 could be located.

◆ The final paragraph is the closing and asks the reader to reply with feedback. Employing recency, the last paragraph should contain anything that the writer wants the recipient to do after reading the letter. This might include questions that need to be answered, a reminder for billing/payment, a request for a follow-up meeting, or other important items:

> I hope this report meets your needs, and if you have any questions, comments, or concerns, please do not hesitate to contact me at 123-456-7890 or at JSpring@ABCD.com.

5.10.5 Progress Reports

A major part of an engineer's job is the development of technical reports and in particular progress reports that document work toward completion of a project. In view of the amount of time and effort it can take to develop them, why do engineers need progress reports? There are several important reasons. Engineering managers price the cost for work based on time and drawings to be developed. They are rewarded based on the profitability of their office via billable hours. All engineers must keep track of time, so the project manager can better organize the tasks and also bill the client for the work completed. Billable time does not include vacations, volunteer work, marketing, or other work/courses/jobs/projects. Critical to this time management is ensuring that the hours worked on the project are productive. The client pays based on work completed, not hours wasted or tasks promised but not delivered. The progress report is the document used to update the client on the project timeline and work accomplished.

The audience for an engineering progress report can include any of the following groups:

◆ **Owner** (who is paying for the engineering services)—The progress report may be used as a basis for periodic billing.

◆ **Engineering internal staff**—This includes managers and accountants who track progress and define or adjust the projected internal workload based on tasks accomplished and the projected timeline toward completion.

◆ **Design team**—To determine what percentage of the work has been completed to date, what potential challenges remain going forward, and what options are available to overcome those challenges.

Each of these groups may require different data or at least different levels of data. This is why a report should be organized into multiple sections, make use of appendices, and contain a concise executive summary. Also keep in mind that when new employees, the owner, or internal personnel see a report, they can use it to get up to speed on the project.

Progress reports are needed for other reasons as well. Inevitably there will be decisions or preferences that need to be addressed in a timely manner (typically by the owner or client). A progress report sets a framework to seek this input. The goal of a progress report is to review the status of the project overall (from a macro-scale). As a result, the report must include an outline of the scope of work of each task and provide justification for key design decisions as they relate to the agreed upon goals and objectives of the project. In this manner, information in a progress report serves to summarize the status of the project at present, provide details about the work accomplished to date, document future work necessary to complete the project, and justify the design approach. Concurrently, a progress report helps to outline potential (current and future) challenges and includes an updated schedule so that everyone understands the progress to date and the implications for the overall project timeline.

It is not uncommon for a project to require a series of progress reports. In that case, the major discussion in each progress report should focus on what has happened since the last one was issued. This means in detail because, after all, this is also a justification for getting paid. Depending on the audience (more so internal than external), it may be necessary to include calculations and diagrams. The schedule should outline work completed and by whom, how this affects (or limits) other parts of the work, and whether the project is on schedule or delayed (which must be explained). Also, issues that arose prior to or during the reporting period that have been resolved should be discussed, along with how they were solved. Note that much of this information can be summarized for the next report. The client also wants to know what will happen in the coming progress report period (in detail), which means outlining what work will be completed in the next period and by whom. Any potential barriers that may delay progress and how they may affect future periods must be identified, along with any issues that require owner input. The owner needs to know about anticipated problems with the plan/project and how they will be resolved. Owners and managers cannot provide useful input or feedback if they do not know about the progress and issues associated with the project. Suggestions to resolve challenges must be put forth in the document. There is nothing worse than outlining problems and offering no solutions. Note that this section is expected to be summarized in the next report as the progress made (hopefully) during the previous period. Finally, the schedule needs to be updated, including plans to recover from delays or get back on schedule, if necessary. Every progress report should update the schedule, and a comparison should be made against the original schedule, with explanations provided for any major changes.

The format of a progress report may vary by organization and client, but generally should include the following key items:

◆ Transmittal memo or letter of transmittal
◆ Review of project scope and overall status of the project
◆ Review of what has happened since the last progress report
◆ Review of what will happen in the coming progress report period
◆ Updated schedule
◆ Issues that need owner input
◆ Appendices (e.g., technical design memoranda)

If progress includes draft site plans, floor plans, and elevations, these may change as the next design phase begins, so the subsequent progress report should include any changes made since the prior progress report. This could include items such as a revised site plan, revised floor plans, new building elevations that match up with the floor plans and roof plan, a revised schedule with who will do the work and when, thoughts on how to approach the design pieces, and so on, depending on the type of project. Once the floor plans and site plans are approved, the exact parameters for the structural design concept can begin to take shape. This can include steel, concrete, other combinations, etc. A discussion of the structural concepts and foundation options should be included in the report; likewise, concepts for other important aspects of the project, including transportation, water, sewer, drainage, and heating, ventilation, and air conditioning (HVAC), should be addressed.

Depending on the nature of the project, engineers may find that:

◆ Additional space may be needed in the ceiling for utilities, ductwork, and other building amenities such as sensors.
◆ The structural system must not conflict with HVAC, utility corridors, or roof drains.
◆ The stormwater and utility details need to be sorted out as the utilities are being designed.
◆ There will be detail sheets that must be provided in the design documents and/or plans.
◆ Completed floor plans and elevations are needed to do the roof design.

The client's key interest will be the drawings and specifications. The drawings should be in the correct form and include informative title blocks. Examples of drawings and specifications are provided in Chapters 12 and 13. During the crafting of a progress report, engineers must begin collecting the design calculations, which will be summarized and included in technical memos in the final project report, also known as the *basis of design report*. However, the design team should be writing these reports as the project develops, as opposed to waiting until the project ends.

For a capstone design course, the project will involve a series of progress reports to provide scaffolding. In a capstone course with emphasis on construction projects, five progress reports are suggested, with the content of each report as follows:

Progress report #1
◆ Final site plan, floor plans, elevations, and surveying in AutoCAD®

◆ Discussion of structural/foundation concepts and options
◆ Discussion of transportation concepts and options
◆ LEED® checklists and "green" features
◆ Preliminary cost estimates

Progress report #2
◆ Final structural concept and justification
◆ Soil borings, geotechnical engineering analysis, and grading plan
◆ Horizontal/vertical curves, pavement design, and cross section in AutoCAD
◆ Preliminary drainage plan calculations
◆ Number of fixture units, meters, and line sizing for utilities
◆ Updated LEED checklists and "green" features
◆ Updated cost estimates

Progress report #3
◆ Structural design (technical memorandum)
◆ Water/sewer plumbing plan and profile and EPANET simulation output with lift station design and details in AutoCAD (technical memorandum)
◆ Transportation design (technical memorandum)
◆ HVAC design (technical memorandum)
◆ Updated LEED checklists and "green" features
◆ Updated cost estimates

Progress report #4
◆ Foundation design (technical memorandum)
◆ Drainage design (technical memorandum) with drawing set with details
◆ Final landscaping design (technical memorandum)
◆ Updated LEED checklists, documentation, and "green" features (technical memoranda)
◆ Preliminary cost analysis

Final progress report
◆ Final structural details in AutoCAD
◆ Final foundation details in AutoCAD
◆ Final transportation details in AutoCAD
◆ Final water/sewer details in AutoCAD
◆ Final HVAC details in AutoCAD
◆ Final drainage details in AutoCAD
◆ Final landscaping details in AutoCAD
◆ Final LEED checklists, documentation, and "green" features
◆ Final cost analysis

5.10.6 The Basis of Design Report

The ultimate goal of the design process is to create three sets of documents that are all related: plans/drawings, specifications, and technical memoranda. All three are specific to a given project, and the plans and specifications are used by the contractor to eventually construct the project. The exact reasons why certain specifications, equipment, or layouts were used, from the myriad of potential options, are justified within the technical memoranda that make up the *basis of design (BOD) report,* which is among the most important documents created by the engineers and *only* the engineers on a project. The BOD report is where the alternative analyses, design calculations, codes, material options and decisions, and construction assumptions are outlined, typically in a series of technical memoranda that explain the entire design process from top to bottom. It is the BOD report that helps, or hurts, the engineer if something goes wrong on a job. If anything goes wrong during any phase of the design or construction, from the beginning to the end of a project, that event can be called a *failure.* In the event of a failure, one of three parties will be held responsible: (1) the owner (normally the owner is the plaintiff in a lawsuit when a project experiences some form of failure), (2) the contractor (who normally blames the engineer's plans or specifications), or (3) the engineer. The BOD report often will outline why certain decisions were made. If the contractor did not construct the project as designed, the blame will shift to the contractor. If the owner made a decision to override the engineer's judgment, that must be documented in the BOD report to clear the engineer of wrongdoing. However, if there were errors in assumptions or calculations or a failure to perform due diligence, the engineer will be held responsible.

An example will help illustrate this concept. On July 17, 1981 there was an incident at the Hyatt Regency Hotel in downtown Kansas City, MO. The structure had two connected walkways suspended from the ceiling over the lobby. The fourth-floor bridge was suspended directly over the second-floor bridge, with the third-floor walkway offset several meters from the others. During a dance competition in the lobby, the walkways collapsed and plunged into the lobby, killing 114 people and injuring 216 others. Upon review of the contractor's and engineer's records (the BOD report in the case of the engineer), it was noted that a field change, as a result of difficulty in spinning nuts up a long threaded rod, caused the original engineering design of the hangers to be modified in the field (see Figure 5.13).

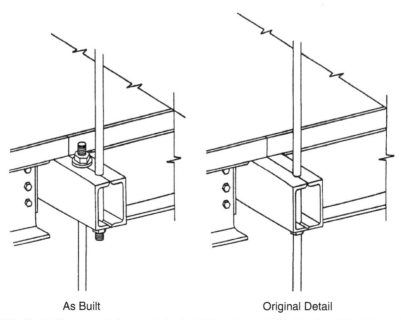

As Built   Original Detail

Figure 5.13 Hyatt Regency failure: original detail and as built drawing (http://www.nist.gov/ customcf/get_pdf.cfm?pub_id=908286)

This subtle change had the effect of doubling the load on the connection between the fourth-floor walkway support beams and the rods by creating a previously absent moment between the rods and the beams. Calculations verified that the original design in the BOD report was adequate to support the anticipated loads. However, the field change was sufficient only to support the dead load of the structure, not any of the live loads, which is why the failure occurred. The original design engineer could, as a result of the BOD report, demonstrate that the design was not in error and was cleared of any wrongdoing. The construction manager who supervised the installation was not so fortunate.

The goal of the BOD report is to document the client's needs, describe the function of the project, detail the facilities to be included in the design, state the capacity of the project in terms of how many people it will serve, and identify how all codes, regulations, and environmental considerations will be met. All engineers should get used to developing BOD reports. Admittedly, far too many professionals neglect to create these documents, and when lawsuits happen (and they do), those engineers will be unprepared, which increases the potential for paying out large claims resulting from an apparent, but perhaps not actual, design defect. As a wise construction attorney once said, "Juries have no technical expertise to litigate construction, so they follow the rule of Solomon; they figure everyone probably has some blame, so they split the baby" (M. Nuechterlein to one of the authors, 1990). However, such a resolution could destroy the career of the engineer involved.

Be methodical about the design process. Meticulously cite references, use accepted best design practices, list assumptions, refer to the appropriate code sections and standards, record any change orders and requests for information, and document important phone conversations and meetings.

5.10.7 Technical Memoranda

A professional-quality BOD report includes a technical memorandum (technical memo, tech memo, or TM) or a series of TMs that outline how the project was designed. Because different engineers will be involved in different aspects of a project, there should be discipline-specific TMs. The benefit is that some time later, should a problem arise, the engineer involved has a written document that can be used in court as a defense for the calculations and assumptions developed for the project. Without it, the engineer is in a vulnerable position as opposed to being able to answer design questions immediately. The goal, then, is to have a TM that will stand the test of time and allow someone to understand the design and recreate the calculations. Such reports should include, as a minimum, the following sections:

- ◆ Scope of work
- ◆ Background information and current existing facilities
- ◆ Alternatives considered to resolve the problem
- ◆ Evaluation of alternatives
- ◆ Description of the recommended alternative
- ◆ Supporting documentation
- ◆ Calculations
- ◆ Drawings
- ◆ Specifications
- ◆ Relevant codes and engineering standards

As with any technical report, primacy dictates that the scope of work section include a brief introduction to the project, the background goals and objectives that outline why the project is needed or a selection was made, some review of prior work and the results (make sure this is properly cited), and the alternatives that meet the needs of the project.

Once the goals and objectives of the project are clearly laid out, the engineer can specify some concept or methods to meet them. References to literature, how data was used, and why the methods chosen are appropriate to the goals are required. The reader must understand the reasons for selections and assumptions used. Some of these may be technical issues, but some may not. The TM should contain a section that discusses the details of the recommended alternative, complete with calculations to illustrate how the issue will be addressed with the proposed design. If there are unresolved issues, solutions should be developed. Extensive appendices that include codes sections, calculations, assumptions, data relied upon, and literature referenced should be attached. Examples of TMs include the structural TM, the geotechnical TM, the water and sewer service TM, the drainage TM, the transportation TM, the HVAC TM, the landscaping TM, and the cost TM.

Analyzing any one of these TMs will be useful to understanding how one is put together, and the structural TM makes a decent entry point. The structure of any building will be based on several considerations, such as type and availability of materials, the exterior skin, the placement of columns, cost, reasonable spans, codes, and local labor expertise. Concrete and steel are commonly used construction materials, as discussed in detail in Chapter 13. Whether a building is to be constructed of steel, reinforced concrete, masonry, glass, timber, or some other structural system will be outlined in the TM. A statement justifying why a certain material was chosen must be included. For example, codes may drive the use of concrete block with reinforcement for a one-story housing project, a curtain wall skin for a multistory office building, or tilt-up panels for a school building. The exterior makes a huge difference in the structural design. Concrete block walls are load bearing, and the others are not. Thus, connections, column locations, and moments become significant issues to address and calculate. A prefabricated steel structure may be an excellent choice for a commercial warehouse, but what if the owner wants a different skin (concrete block, masonry, or glass)? If the expertise to install block or brick is locally available but the expertise to install steel is not, the engineer would be foolish to specify steel. Costs and skill sets may adjust the owner's preferences, but the report should clearly state all of the reasons for design decisions and the sources used to derive those recommendations.

Once decisions regarding the type of construction are made, the structural planning, location of columns, spans, etc., as outlined in Chapter 13, can be undertaken. All of the critical elements will require calculations, which can be done either by hand or by computer (the latter should be verified by hand). Many state engineering boards hold professional engineers responsible for results generated by computer programs. All calculations, assumptions, relevant codes sections, and computer printouts, including input data, should be included in TMs for future reference.

TMs (and plans and drawing sets) created by an engineer should undergo a quality assurance/quality control review by another engineer who is assumed to have as much or greater expertise than the design engineer of record. The idea is to determine that there are no missing or improper assumptions, omissions, misreads of codes, or errors in calculations that lead to improperly designed components. The quality assurance/quality control process also could include a value engineering component, which evaluates the economic priorities of the project and makes suggestions on potential cost-saving opportunities. The TM assumptions are critical for supporting good decision making and for providing arguments to refute value engineering suggestions, if appropriate.

Students often have little experience in developing an outline of their thought processes, and too often, professionals ignore the need to fully develop them as well. This is a mistake. Writing down critical thinking components within a TM is required for a successful career free from failure. If an issue arises, a TM should stand alone in a manner such that an engineer can quickly review it and recreate the prior work to verify its correctness. If a review of a TM by another person does not permit replication of the work, then the TM is not adequate and needs to be revised until someone can read and recreate the work without the need for discussion.

5.10.8 Interim and Final Technical Reports

Writing technical reports requires the same skills and attention to detail as have been discussed throughout this chapter. It is customary for a final technical report or a series of interim reports culminating in a final technical report to be required as the deliverables for a project. An interim report could be a summary of progress to date or a draft of the final report summarizing the work required by the contract, which must be reviewed by stakeholders before submitting the final report that signals the end of the project. Many organizations will have a set of strict guidelines that govern the format of these kinds of reports. The final technical report for a project typically will have an outline such as the following:

◆ Transmittal memo or letter of transmittal
◆ Cover page
◆ Table of contents
◆ List of figures
◆ List of tables
◆ Executive summary
◆ Introduction (project scope, location, goals, criteria, etc.)
◆ Body of the final progress report (summary of work done including site assessments, site plans, building renderings, and floor plans and summary of design parameters for each of the TMs included as the appendices)
◆ Schedule update
◆ Conclusions (cost estimates, LEED checklists, etc.)
◆ References
◆ Appendices

The letter or memo of transmittal (discussed earlier in this chapter) should follow the proper format, address the purpose of the report in the opening (primacy), and ask the reader for comments or payment (recency). The tone should be professional, with no obvious errors, and the letter should be signed on company letterhead or initialed if using memo format.

The cover page should contain all of the essential elements including the title of the project, the contributors' names, the date of submittal, etc. A table of contents provides the outline structure of the report, so that the reader can easily find the exact location of key items. If the appropriate formatting for the headings and subheadings was used, then the table of contents is merely a list of those. The lists of figures and tables provide the page number locations of the graphics included in the report. If the table titles and figure captions were created appropriately, then these lists simply repeat them with the page number locations.

The executive summary is a stand-alone synopsis of material contained in the report, with all essential elements summarized. This allows the readers to decide if they want to read all of the report or focus on certain sections. The executive summary generally is written for a nontechnical audience and does not have to be limited to one page or three paragraphs in length.

The introduction to the report describes the project goals, the location of the project, and the objectives. It must start strong (primacy), with scope and objectives clearly presented. The introduction should fully express the primary purpose and scope in the context of the project goals at the beginning of the report. The introduction also should present a clear statement that demonstrates how the report will track the fundamental, secondary, and implied problems, questions, and issues described within. The body of the report contains all of the supporting information to address the major items, subdivided into sections that are all related to the goals of the project. These typically include existing conditions, discussion of alternatives, selection of alternatives, recommendations, and issues that require owner input, among others. The text should convey a professional level of knowledge of the subject matter, with no important content left out and no incorrect material presented. The focus of the report must be clear to the reader such that paragraphs logically and coherently build upon each other through the complete and fluent use of transitions toward a logical conclusion supported by the data presented or referenced. The writing should exhibit substantial depth and complexity of thought supported by well-developed ideas, analysis, and evidence that tie back to the original purpose and goals of the project. Facts should be presented in a logical sequence, and sections must transition effectively between topics and different authors. Some other characteristics of a well-written body are as follows:

- Seamlessly incorporates and explains the accuracy and relevance of data/evidence/quotations/visuals
- Offers evidence from a variety of sources, including counterarguments, contrary evidence, and quantitative analysis
- Presents data in graphical, tabular, or sketch format, following all rules for graphics, including proper units and labels
- Spelling and grammar are checked
- Provides sentences that consistently communicate thoughts clearly, are relatively free of sentence-level patterns of error, and use a technically sound sentence structure that is varied, convincing, nuanced, and eloquent, with appropriate tone
- Demonstrates evidence of proofreading/editing

Most importantly, the supporting material must build toward an effective conclusion to finish strong with a reasonable summary and/or recommendations, as justified from the body of the report using recency techniques. A complete reference section is included that cites literature sources accurately and consistently in the proper format.

The last section is a set of appendices that contain information relevant to the project but are too detailed or too cumbersome to integrate into the body text of the report itself. For example, a list of typical appendices for a final report in a capstone design course project would be as follows:

Appendix A. Site Plan Tech Memo
  Appendix A.1. Calculations
  Appendix A.2. Drawings
  Appendix A.3. Technical Documentation and Specifications
  Appendix A.4. Codes
  Appendix A.5. Other Pertinent Information
Appendix B. Structural Engineering Tech Memo
  Appendix B.1. Calculations
  Appendix B.2. Drawings
  Appendix B.3. Technical Documentation and Specifications
  Appendix B.4. Codes
  Appendix B.5. Other Pertinent Information
Appendix C. Geotechnical Engineering Tech Memo
  Appendix C.1. Calculations
  Appendix C.2. Drawings
  Appendix C.3. Technical Documentation and Specifications
  Appendix C.4. Codes
  Appendix C.5. Other Pertinent Information
Appendix D. Water and Sewer Service Engineering Tech Memo
  Appendix D.1. Calculations
  Appendix D.2. Drawings
  Appendix D.3. Technical Documentation and Specifications
  Appendix D.4. Codes
  Appendix D.5. Other Pertinent Information
Appendix E. Drainage Engineering Tech Memo
  Appendix E.1. Calculations
  Appendix E.2. Drawings
  Appendix E.3. Technical Documentation and Specifications
  Appendix E.4. Codes
  Appendix E.5. Other Pertinent Information
Appendix F. Transportation Engineering Tech Memo
  Appendix F.1. Calculations
  Appendix F.2. Drawings
  Appendix F.3. Technical Documentation and Specifications
  Appendix F.4. Codes
  Appendix F.5. Other Pertinent Information
Appendix G. Heating, Ventilation, and Air Conditioning Engineering Tech Memo
  Appendix G.1. Calculations
  Appendix G.2. Drawings
  Appendix G.3. Technical Documentation and Specifications
  Appendix G.4. Codes
  Appendix G.5. Other Pertinent Information
Appendix H. Landscaping Tech Memo
  Appendix H.1. Calculations
  Appendix H.2. Drawings
  Appendix H.3. Technical Documentation and Specifications

Appendix H.4. Codes
Appendix H.5. Other Pertinent Information
Appendix I. Cost Estimate
Appendix J. LEED Templates
Appendix K. Request for Proposal
Appendix L. Phase I Environmental Site Assessment
Appendix M. Progress Report 1
Appendix N. Progress Report 2
Appendix O. Progress Report 3
Appendix P. Progress Report 4
Appendix Q. Meeting Minutes
Appendix R. Time Sheets
Appendix S. Other

Each appendix is designed to be a stand-alone TM with supporting materials that offer a complete summary of the design for a particular aspect of the project. It is expected that a reader will be able to take this information and replicate the work. Hence, all calculations, codes, and drawings must be included, and a written description of methods, assumptions, and application of technical data is required. An example rubric to grade the written assignments for student work is shown in Table 5.5. It is not uncommon to find that certain organizations also will have developed rubrics to evaluate the quality of submittals and to ultimately determine if a firm will be used again for engineering work in the future.

5.11 Public Speaking

So far, this chapter has focused solely on written communication, but public speaking is an integral part of every engineer's career too. From dealing with clients, contractors, and other engineers to presenting the work at proposal defenses, board meetings, and conferences, it is time to get over being nervous. The successful engineer is evaluated on reputation and professional judgment, and many times this evaluation is made during presentations or other opportunities involving public speaking and personal networking. In these situations, the speaker is the center of attention, so it is important to engage the audience and make eye contact. This is the best way to determine if the audience has understood the material presented or is even actually listening. In most cases, the speaker(s) will probably know more about the topic than anyone in the audience, so the key is to demonstrate that knowledge. The best way to accomplish this is to simply talk to the audience about the topic. This is the secret to developing a highly successful, confident speaking style and becoming an expert in planning, preparing, and delivering presentations.

No matter the setting, make sure that the presentation is memorable, and above all else, prepare and practice. The audience can tell right away if a speaker knows what he or she is taking about. If a speaker stumbles with words, makes gross mistakes in presenting concepts, or is surprised by the content of slides, the audience will know that the individual did not do the work. Anything that undermines credibility in a presentation will reflect unfavorably upon the engineer's judgment and experience. There will be many public speaking opportunities in an engineer's career, so presentation skills should improve with time.

Table 5.5 Sample rubric for written submittals by capstone design students

	Excellent	Good	Fair	Poor	Unacceptable
☑ **Letter of transmittal**	Format is correct. Opening and closing provide primacy and recency. Professional tone. No obvious errors. Signed.	Format is correct, but has deficiencies in opening, closing, or tone. Includes obvious errors or not signed.	Format is incorrect or has deficiencies in opening, closing, or tone. Includes obvious errors or not signed.	Format is incorrect and has deficiencies in opening, closing, or tone. Includes obvious errors or not signed.	No letter included.
☑ **Executive summary**	Stand-alone, with all essential elements summarized.	Too long, too short, or missing one of the essential elements.	Too long or too short and missing one of the essential elements.	Too long or too short and missing more than one of the essential elements.	No summary included.
☑ **Opening**	Report starts strong with scope and objectives clearly presented. Fully and completely expresses the primary argument in context at the beginning of the report.	Generally expresses the primary argument in context at the beginning of the report.	Vaguely or partially expresses the primary argument with minimal context in the report.	May not express the primary argument or provide context anywhere in the report.	Not an argument-driven report.
☑ **Organizational structure**	Presents a clear statement located in the beginning of the paper that demonstrates how the argument will track the fundamental, secondary, and implied problems, questions, and issues.	Presents a general statement located in the beginning of the paper that demonstrates how the argument will track the fundamental, secondary, and implied problems, questions, and issues.	Presents a vague or partial statement located somewhere in the paper that demonstrates how the argument will track the fundamental, secondary, and implied problems, questions, and issues.	Presents no organizational statement. Readers are not able to determine how the report will proceed.	Not an argument-driven report.
☑ **Content**	Report displays professional level of knowledge of subject matter, with no important content left out and no incorrect material presented.	Report displays professional level of knowledge of subject matter, with minor amount of subject material left out or minor amount of incorrect material presented.	A substantial amount of the report fails to display professional level of knowledge of subject matter, with substantial amounts of subject material left out or substantial amounts of incorrect material presented.	A substantial amount of the report fails to display professional level of knowledge of subject matter, with substantial amounts of subject material left out and substantial amounts of incorrect material presented.	Not an argument-driven report.

	Excellent	Good	Fair	Poor	Unacceptable
☑ Rhetorical structure	The argument's focus is clear to the reader and paragraphs logically and coherently build upon each other through the complete and fluent use of transitions and/or headings toward a logical conclusion supported by data.	The argument's focus is generally clear to the reader, and the use of transitions lends a sense of progression and coherence toward a logical conclusion with some data support.	The argument's focus is unclear to the reader Some transitions are used, providing little or no sense of direction. Conclusion is unclear or not supported.	Transitions and sense of progression are absent. Conclusion is weak with little or no data support.	Not an argument-driven paper.
◆ Reasoning	Exhibits substantial depth and complexity of thought supported by sophisticated ideas/analytics/evidence that supports the report's argument.				
◆ Continuity	Facts are presented in a logical sequence and transitions effectively between topics and authors. Builds toward an effective conclusion. Considers context, assumptions, data, and evidence.				
◆ Data support	Seamlessly incorporates and explains the accuracy and relevance of data/evidence/quotations/paraphrase/visuals; offers evidence from a variety of sources, including counterarguments, contrary evidence, and quantitative analysis. Presents data in graphical, tabular, or sketch format; follows all rules for table/figure format, including proper units and labels. Raw data goes in appendix. Tables/figures are numbered independently and all are mentioned in the text.				
◆ Conclusion	Finishes strong with a reasonable summary and/or recommendations presented, as justified from the body of the report using primacy and recency.				
☑ Overall impression	Report addresses all important subject matter, demonstrates conceptual understanding of the content, and responds to the purpose; arguments are cohesive, clear, concise, and organized well; report has many strengths; delivery is professional and shows excellent engineering judgment.	Report addresses most of the important subject matter, demonstrates conceptual understanding of the content, and responds to the purpose; majority of the work is cohesive, clear, concise, and organized well; report has strengths; delivery is professional and shows good engineering judgment.	Report addresses some of the important subject matter, demonstrates conceptual understanding of the content, and responds to the purpose; some of the work is cohesive, clear, concise, and organized well; report has few strengths; delivery is professional and shows good engineering judgment.	Report addresses little of the important subject matter, demonstrates conceptual understanding of the content, and responds to the purpose; work is not cohesive, clear, concise, or organized well; requires major revision; delivery is not professional and shows lack of engineering judgment.	Report is completely unprofessional.

Table 5.5 Sample rubric for written submittals by capstone design students (continued)

	Excellent	Good	Fair	Poor	Unacceptable
☑ **References** follow the format in http://pubs. asce.org	Cites and formats sources accurately and consistently and provides appropriate and complete references.	Cites and formats sources consistently and provides appropriate references. Some errors or flaws are present.	Cites some sources but often cites inaccurately. May neglect to cite some sources altogether. References typically present but inaccurate.	Little or no use of citation formats.	No references.
☑ **Appendix**	Raw data/photos correctly arranged and labeled.	Missing one item, except raw data, or unnecessary items in the appendix.	Missing two items, except raw data, and unnecessary items in the appendix.	Missing more than two items and unnecessary items in the appendix.	No appendix.
☑ **Writing format**	Follows all format requirements (e.g., 1-inch margins, 1.5-spaced, 11-point Times/ Arial font, block justification).	Missing one of the format requirements.	Missing two of the format requirements.	Missing three of the format requirements.	Failed to respect any of the format requirements.
☑ **Grammar and syntax**	Spelling/grammar checked; sentences consistently communicate thoughts clearly, while relatively free of sentence-level patterns of error; technically sound sentence structure that is varied, convincing, nuanced, and eloquent, with appropriate tone. Evidence of good editing.	Spelling/grammar checked, but minor sentence-level patterns of error, improper sentence structure, or tone issues. Evidence of decent editing.	Minor spelling or grammar errors, with sentence-level patterns of error, improper sentence structure, or tone issues. Evidence of fair editing.	Spelling or grammar errors throughout, and major sentence-level patterns of error, improper sentence structure, or tone issues. No evidence of editing.	Gross disregard for readability.

Speaking clearly, enunciating words, varying voice inflection, and projecting one's voice so that the words can be heard are all tools that will encourage the audience to pay attention. Speaking with passion and being persuasive also will help engage the audience. If a speaker talks under his or her breath or mutters, the audience will not be able to hear, will quickly lose interest, and will form an unfavorable impression of the speaker's engineering skills and judgment. A common mistake is reading a presentation from note cards or reading from slides while facing the screen instead of the audience. This is misinterpreted as lack of preparation or lack of knowledge of the subject, which also reflects poorly on the presenter. Another way to leave a poor impression is to start sentences that do not finish. At the same time, however, if the search for the perfect word goes on silently in the speaker's head for several awkward and uncomfortably long seconds, it is not acceptable to just give up and move on without finishing the thought. This is where teammates can chime in to help rescue the thought. As mentioned before, the lack of practice preparation and the inability to think quickly are negative characteristics that engineers do not want associated with their presentations or their reputations.

Another means to undermine a presentation is the inclusion of crutch words, such as *ah*, *um*, and *er*. A speaker should try to avoid or cut down on the use of these annoying words because they give the impression that the speaker has not practiced and is making things up on the fly. An engineer also should try to resist the temptation to thank the audience for not falling asleep (do not give them any ideas!) and should not apologize for poor quality slides or poor color contrast on the projector. Again, these things reflect a lack of preparation. If a slide is lousy, then why is it included? Instead, a speaker should start the presentation strong (primacy) with a good first impression and leave the audience with a convincing conclusion (recency).

To make an effective presentation that flows smoothly, first develop an outline and a logical progression, just as for any written form of communication. Then select the appropriate visual aids to make the important points. Begin with an agenda (road map) for the audience (primacy), present the data (body of the presentation), and then finish with a strong conclusion (recency). After planning, creating, reviewing, proofreading, and revising the presentation, it is time to rehearse, after which it may be time for some more editing and revisions to make the presentation flow better. Before delivering the presentation, know the material, find out who the audience is, and then answer the following important questions:

◆ Can some of the introduction be skipped because the audience already is familiar with the work, or will extra time be allocated to describe the background details of the project?

◆ Can engineering jargon be used, or will everything have to be explained in layman's terms?

◆ Will there be someone in the audience who helped with gathering information, and is that person's name mentioned (and spelled correctly) in the acknowledgments at the end of the presentation?

Knowing the level of knowledge and sophistication of the audience can help tailor the presentation appropriately so that it is not overly complex or too boring. If the audience is large and contains members of the general public, the content may have to be toned down to the lowest common denominator, but the presenter still should be prepared to answer detailed technical questions from more expert members of the audience. As shown in Figure

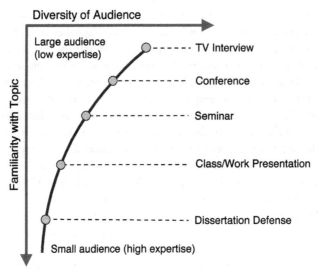

Figure 5.14 Spectrum of audience diversity

5.14, when the size of the audience decreases and the level of sophistication increases, the presentation content needs to be at the highest level of detail. For example, if a presentation is for an audience comprised of the engineering team's three principal engineers, much more detail will be required than if the presentation is for a large diverse audience of taxpayers with no engineering background.

It is extremely important for a speaker to know his or her strengths, weaknesses, and limitations. Practicing in front of a mirror or videotaping a presentation can go a long way toward helping to identify these strengths and weaknesses and help a speaker improve his or her delivery. Some presentations have a time limit that is strictly enforced; therefore, speakers should allow a small cushion of time to account for nervous energy and transitional pauses between team members during a live presentation.

Content and delivery are the hallmarks of an effective presentation, but a poor first impression can doom it to failure, so make sure to look the part of a professional. Do not overdress, but be neat, because, to quote Shakespeare and Tau Beta Pi (national engineering honor society), "although the clothes do not make the man, they proclaim him." Some other helpful tips during a presentation can make all the difference, so try to:

◆ Avoid playing with personal belongings (e.g., keys, cellular phone, loose change, etc.).
◆ Avoid moving around too much without purpose.
◆ Avoid walking in front of the projector and covering up or blocking the audience's view of the slides.
◆ Avoid looking directly into the projector, which will cause momentary blindness and temporary disorientation.
◆ Avoid anything that might distract the audience's focus from the words or content of the presentation.
◆ Avoid slouching or never making eye contact, as the audience infers lack of confidence in the material, which does not bode well for their perception of the speaker's engineering judgment.

The best way to know if any of these issues will creep up during a live presentation is to practice in front of a mirror or record the delivery of the presentation. Once up on the stage, it is not the time to panic.

Unfortunately, Murphy's Law dictates that whatever can go wrong will go wrong, so be prepared for the unexpected. The following is a list of potential issues and solutions:

- Problem: The podium audiovisual equipment is not compatible with the software version of the presentation file.
  ◇ Solution: Bring your own laptop (and cables) with the correct software as a backup.
- Problem: The presentation file gets misplaced.
  ◇ Solution: Bring an extra copy of the file on a USB drive or other suitable media as a backup.
- Problem: The projector screen is missing.
  ◇ Solution: Bring a portable screen or project the presentation onto a blank wall in the room as a backup.
- Problem: The projector bulb goes out.
  ◇ Solution: Bring a backup projector.
- Problem: The project manager or a key team member is late.
  ◇ Solution: Be prepared to start without the person and know his or her part of the presentation.
- Problem: The power goes out in the room or there are other technical difficulties.
  ◇ Solution: Bring hard copies of the presentation for the audience as a backup.

The best suggestion is to be prepared to deliver the presentation without the support of technology if need be. The message here is to be prepared because it reflects well on the evaluation of engineering judgment. Engineers are expected to practice, be professional, speak clearly and coherently, and make their points confidently and correctly. The keys to an effective presentation are content and delivery.

5.11.1 Visual Aids

One way to more clearly convey the content is to use visual aids. It is common to use a presentation software package like Microsoft PowerPoint®, Prezi®, or others to develop multimedia visual aids. Remember, as the saying goes, a picture is worth a thousand words, so the slides that accompany a speaker's presentation can go a long way toward helping the audience make sense of the content.

Slides should be reviewed to make sure they "flow" to make the desired point. To do this properly, it is necessary to create an outline or a storyboard before developing the visuals. An example of an appropriate, logical order for a design presentation is as follows:

1. Introduce the team
2. Introduce the project scope and site location
3. Establish the project rationale
4. Discuss building program requirements
5. Discuss goals and design considerations
6. Review existing conditions or findings of a Phase I Environmental Site Assessment

7. Discuss threshold criteria and alternative selection criteria
8. Present alternatives (advantages and disadvantages)
9. Present alternative analysis selection matrix
10. Discuss the preferred alternative in the context of the project goals
11. Present the proposed site plan and floor plan details
12. Present "green" features and LEED checklists
13. Present preliminary costs
14. Present schedule and timeline of milestones
15. Present strong conclusion
16. Acknowledge those who assisted in the work and entertain questions

Note that while the storyboard contains 16 key points, the final presentation need not be exactly 16 slides. Some concepts can be combined and others need multiple slides to discuss the details. When creating effective slides for a presentation, it is important to keep in mind some key pointers. Slides are supplementary to the speaker's words, so they must provide emphasis and context, not be a distraction. If the audience focuses too much on the slides, the spoken words will not be heard. The speaker should use the slides as a visual cue, not read from the slides. The content of the slides should be BIG (see Figure 5.15). Type should be readable (larger than 20 point) from the back of the room, and illustrations should be high resolution, in color, and BIG. Type in 40 point is great for titles. Minimum size should be 20 point. When the font size gets down below this, slides begin to get harder for the audience in the back of the room to read. Illegible images should be avoided, as should saying "I know you can't read this, but...." Above all else, the slides should be relevant to the key points of the presentation.

To make the wording on the slides more readable, stick with sans serif fonts like Arial, Calibri, Tahoma, or Verdana. Presenters should not mix fonts. To make a point on a slide, use bold type or colors to create emphasis, but limit the use of all caps and avoid using thin

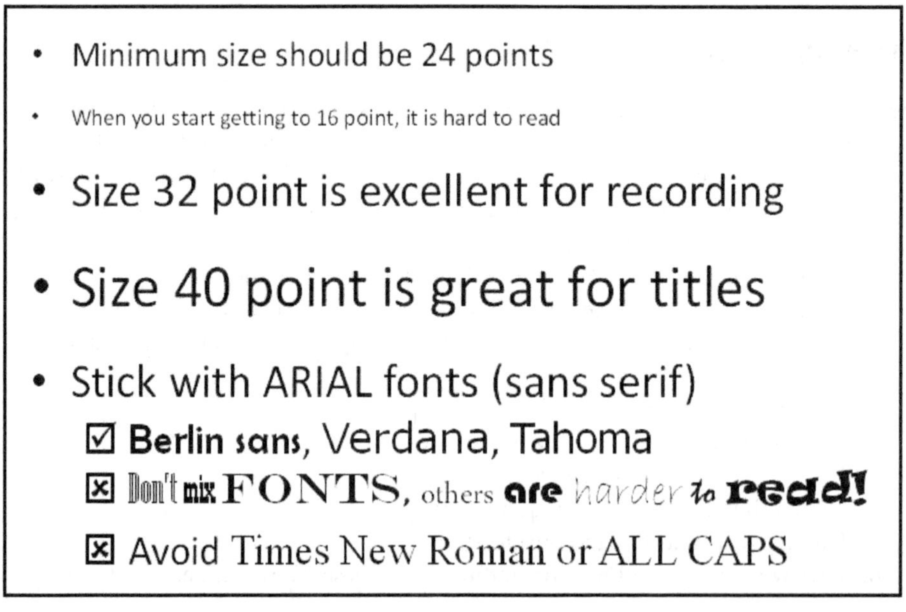

- Minimum size should be 24 points
- When you start getting to 16 point, it is hard to read

- Size 32 point is excellent for recording

- Size 40 point is great for titles

- Stick with ARIAL fonts (sans serif)
  - ☑ **Berlin sans**, Verdana, Tahoma
  - ☒ **Don't mix FONTS**, others **are** harder to **read!**
  - ☒ Avoid Times New Roman or ALL CAPS

Figure 5.15 Examples of type fonts and sizes

lines or underlining because they might appear to be hyperlinks or start wiggling or vibrating if the projector lamp power is too low. Try to adhere to the six-six rule of thumb: no more than six words on a line with no more than six lines on a slide.

Make sure the colors work well. A dark background with light lettering really pops off of the screen (see line 2 in Figure 5.16), particularly for presentations that will be taped. On the other hand, light backgrounds (except white) do not work so well on projector screens, but are great for printing out copies of a presentation. Strive for greater contrast, clarity, and sharpness in color selection. Note that too much contrast can be confusing, overwhelming, and harsh, while too little contrast is boring, bland, and difficult to read. One more tip about color is to avoid importing AutoCAD graphics with a black background (see Figure 5.17) because they are difficult to read as slides.

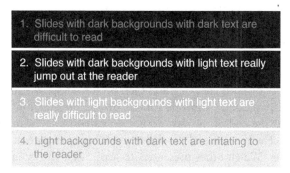

Figure 5.16 Examples of different backgrounds/text schemes for slides

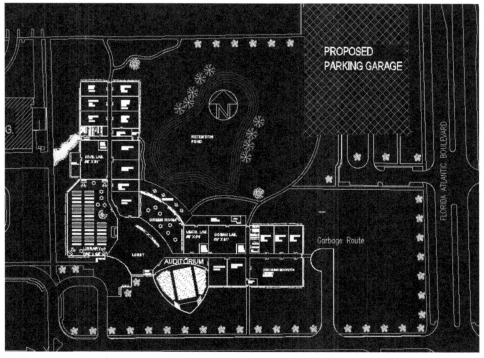

Figure 5.17 Example of a slide with black AutoCAD background

An important lesson with respect to the technology is to keep it simple and avoid annoying audio clips (sounds) and annoying transitions and animation. It may be impressive to proudly show off technological prowess, but remember that someone may ask to go back a few slides, and the audience must then wait while the presenter scrolls through all of the animation and transitions to get there. Also, avoid including useless images ("You didn't need to see that..."), overly complex images ("What this slide is trying to show is..."), or chart-junk and chart-toons that use clip art and logos.

Try not to stay on one slide too long. Visuals can be hidden from view when no longer needed by pressing "B" to black out the screen or "W" to white out the screen (and then pressing the space bar to return to the presentation). At the same time, nothing can be more annoying than someone having trouble clicking to the next slide. Make sure that the wireless mouse is working correctly so that the slides advance without issue. Sometimes if the wrong button is pressed, the slides go backwards or an annoying popup menu appears. If the mouse or pointer is not working correctly or the presenter cannot operate the device, simply hit "N" or the right or down arrow on the keyboard to advance slides or hit "P" or the left or up arrow to go back to the previous slide (the page up/page down keys will do the trick too).

Slides can become a distraction if they have blatant spelling errors. Make sure typos are eliminated in all slides by spell checking and then proofreading. A potentially "fatal" error is to misspell or mispronounce someone's name on the presentation team or during the acknowledgments. At the end of the day, do not get carried away with the technology and neglect the material. The content and the delivery are just as important if not more so than the visuals themselves. Watch other presenters to learn what works and what does not.

5.11.2 Questions and Answers

In any presentation, there should always be time to answer questions. The question and answer period is an excellent opportunity to leave the audience with a strong impression of one's capabilities and level of knowledge (recency). The key to success during this period is to remain poised and think quickly. It is during these pressure-packed moments that the recency of the presentation is at stake. When faced with unexpected questions or controversy, the speaker should calmly compose a thoughtful answer. This means the speaker should never feel rushed to answer; this is a common mistake that normally ends with a poorly constructed response.

When faced with a difficult question, the speaker should not panic, but instead make sure to listen carefully to the question and understand what is being asked before attempting to answer. A common problem with some speakers is that they will cut off the person asking a question and impatiently start to answer too soon. There is no sense in starting to talk until the full question is asked and fully understood, and to do otherwise makes the speaker appear arrogant. When responding to a question, the speaker should look directly at the person asking the question and observe his or her body language while trying to quickly interpret what is being suggested by the question or request. The speaker should not interpret any question as a personal attack or a test, but instead should treat every question as a message of interest in the presentation and a legitimate request for more information or an opportunity to show off one's engineering judgment and content knowledge.

To ensure the entire audience is impressed by the response, the speaker should make sure that everyone in the room (or on the recording) knows what was asked. Repeating the question helps the audience digest what was asked and also gives the speaker some time to gather his or her thoughts. If more time is needed to generate a good response or if the question is unclear, the speaker should ask the person who posed the question to repeat it. Asking for a query to be repeated also will allow the person asking the question to better articulate it.

Speakers should formulate responses using simple sentence construction, as there is no sense in constructing a complex answer that merely confuses the issue. During the question and answer period, speakers should stay focused and be concise with their answers to avoid losing audience interest, coming across as verbose, or saying things that are better left unsaid. There is a tendency among speakers with vast technical knowledge to, under pressure, answer a question with too much information or respond in several different directions. This is called the "dartboard" approach. Only one dart can hit the bull's-eye. All the others will miss. The player who throws one dart and hits the target appears much more polished than the one who throws five darts and misses with all but one.

Speakers should stay above the fray and remain professional by formulating answers in a positive manner, while never appearing to be defensive or combative, even when disagreeing with a comment. Other guidance for the question and answer period includes:

- ◆ Use a strong voice that conveys confidence.
- ◆ Make eye contact with the person asking the question, and scan the rest of the audience as well.
- ◆ Use the appropriate level of formality.
- ◆ Coordinate responses within the project group. If the question was directed to the project manager but others have something important to add, they should be encouraged to chime in without interrupting the speaker. There is likely one person on the team best qualified to answer, but be careful of throwing a teammate under the bus.
- ◆ Answer honestly. Do not fake an answer, as this will ruin all of the credibility that the team worked so hard to gain from the audience during the presentation itself (recency).

There is actually nothing wrong with not knowing the answer to something, but this tactic cannot be used for every question that is asked. Speakers should beware of the possibility of being misinterpreted in the first few words (primacy), so words should be chosen wisely. Also, speakers should be aware of the rhythm of the question and answer period; a long pause is awkward and could be interpreted as negative by the audience.

5.11.3 Evaluating Presentations

Often, the purpose of giving a presentation is to obtain a contract. In these competitive situations, the presentations will be rigorously evaluated to identity the top firms qualified to do the work. For capstone design students, the components of an effective presentation can be evaluated with a simple rubric as shown in Table 5.6. Many organizations will have developed their own rubrics to assess the quality of presentations and help in selecting the most qualified firm to do the work. In presentations of this kind, the essential elements being evaluated are content, organization, delivery, and performance in the question and answer period. In terms of content, each team member must convey a professional level of knowledge of the subject matter, with no important points left out and no incorrect material presented. The presentation must be organized with a strong introduction (primacy) that addresses the scope and objectives; an effective body of material that supports the main content, flows logically, and transitions effectively between speakers; and a strong conclusion

Table 5.6 Sample presentation rubric for capstone design students

	Excellent	Good	Fair	Poor	Unacceptable
☑ Content	All team members display professional level of knowledge of subject matter, with no important content left out and no incorrect material presented.	All team members display professional level of knowledge of subject matter, with minor amount of subject material left out or minor amount of incorrect material presented.	Majority of team members display professional level of knowledge of subject matter, with minor amount of subject material left out or minor amount of incorrect material presented.	Some team members display professional level of knowledge of subject matter, with minor amount of subject material left out or minor amount of incorrect material presented.	No team members display professional level of knowledge of subject matter, with minor amount of subject material left out or minor amount of incorrect material presented.
♦ Subject matter	All important topics are covered during the presentation, with no essential elements missing or misrepresented.				
♦ Knowledge of subject	Each member of the team demonstrates an understanding of the essential topics presented.				
☑ Organization	Presentation has a strong introduction, an effective body of material that supports the conclusions, and a strong ending.	Presentation is deficient in only one of the following: introduction, body, or conclusion.	Presentation is deficient in two of the following: introduction, body, or conclusion.	Presentation is deficient in all of the following: introduction, body, and conclusion.	Presentation is missing introduction, body, or conclusion.
♦ Introduction	Presentation starts strong, with scope and objectives clearly presented.				
♦ Continuity	Facts are presented in a logical sequence and presentation transitions effectively between speakers.				
♦ Conclusion	Finishes strong with reasonable summary and/or recommendations presented, as justified from the body of the presentation.				
☑ Delivery	Presentation is effective in terms of rhythm, visuals, and presenters' body language.	Presentation is deficient in only one of the following: rhythm, visuals, or presenters' body language.	Presentation is deficient in two of the following: rhythm, visuals, or presenters' body language.	Presentation is deficient in all of the following: rhythm, visuals, and presenters' body language.	Presentation clearly is not rehearsed, visuals are unprofessional, and/or presenters' body language is unprofessional.

	Excellent	Good	Fair	Poor	Unacceptable
◆ Rhythm	Presentation demonstrates effective use of time, presenters seem well prepared, and presentation appears rehearsed.				
◆ Visuals	Visuals are effective, free of clutter, related to the discussion, and meaningful.				
◆ Body language	Presenters maintain eye contact with the audience and are free of any distracting or annoying mannerisms.				
☑ Discussion	All questions are fielded professionally, confidently, and correctly while avoiding defensive or argumentative responses.	Majority of questions are fielded professionally, confidently, and correctly while avoiding defensive or argumentative responses.	Some questions are fielded professionally, confidently, and correctly while avoiding defensive or argumentative responses.	Only one question is fielded professionally, confidently, and correctly while avoiding defensive or argumentative responses.	None of the questions are fielded professionally, confidently, and correctly while avoiding defensive or argumentative responses.
◆ Question and answer period	Answers supplied reflect an understanding of the topic.				
☑ Overall impression	Presentation addresses all important subject matter, demonstrates conceptual understanding of the content, and responds to the purpose of the report; slides are cohesive, clear, concise, and organized well; presentation has many strengths; delivery is professional; question and answer period shows excellent engineering judgment.	Presentation addresses most of the important subject matter, demonstrates conceptual understanding of the content, and responds to the purpose of the report; majority of slides are cohesive, clear, concise, and organized well; presentation has strengths; delivery is professional; question and answer period shows good engineering judgment.	Presentation addresses some of the important subject matter, demonstrates conceptual understanding of the content, and responds to the purpose of the report; some of the slides are cohesive, clear, concise, and organized well; presentation has few strengths; delivery is professional; question and answer period shows some engineering judgment.	Presentation addresses little of the important subject matter, demonstrates conceptual understanding of the content, and responds to the purpose of the report; few of the slides are cohesive, clear, concise, or organized well; presentation requires major revision; delivery is not professional; question and answer period shows lack of engineering judgment.	Presentation is completely unprofessional.

at the end (recency). With respect to delivery, the presentation should be effective in terms of rhythm, visuals, and presenters' body language. Rhythm refers to the effective use of time without rushing or slowing down, such that presenters seem well prepared and the performance appears rehearsed. Visuals must be free of clutter, related to the discussion, and meaningful. Presenters should maintain eye contact with the audience and avoid any distracting or annoying mannerisms. During the question and answer period, all questions must be fielded professionally and confidently, but more importantly, the answers must correctly reflect an understanding of the topic, while avoiding defensive, argumentative, or blatantly incorrect responses. The overall impression is evaluated at the end of the presentation and is based on the presenters' ability to address all important subject matter while demonstrating conceptual understanding of the content and responding to the purpose of the presentation. The slides should be cohesive, clear, concise, and well organized. In other words, the presentation should have many strengths, the delivery should be professional, and the question and answer period should demonstrate excellent engineering judgment.

5.12 References

AWWA (2015). *Manual of Water Supply Practices, M63: Aquifer Storage and Recovery,* American Water Works Association, Denver, CO.

Beer, D., and McMurray, D. (2009). *A Guide to Writing as an Engineer,* 3rd Ed., John Wiley & Sons, Hoboken, NJ.

Ebbitt, W.R., and Ebbitt, D.R. (1982). *Writer's Guide and Index to English,* 7th Ed., Scott, Foresman.

Flesch, R. (1948). "A new readability yardstick." *Journal of Applied Psychology,* 32(3), 2211–2233.

Glinsky, A. (1948). "How valid is the Flesch readability formula?" *American Psychologist,* 3, 261.

Gunning, T.G. (2003). "The role of readability in today's classrooms." *Topics in Language Disorders,* 23, 175–190.

Hodges, J.C., Whitten, M.E., Horner, W.B., Webb, S.E., and Miller, R.K. (1990). *Harbrace College Handbook,* 11th Ed., Harcourt Brace Jovanovich, Washington, DC.

McCaskill, M.K. (1998). *Grammar, Punctuation, and Capitalization: A Handbook for Technical Writers and Editors,* NASA SP-7084, Langley Research Center, Hampton, VA.

Meeroff, D.E. (1997). "Development of filtration processes for separation of emulsified bitumen fuels from water." Master's thesis, University of Miami.

Rosen, L.J., and Behrens, L. (1997). *The Allyn & Bacon Handbook,* 3rd Ed., Allyn & Bacon, Boston, MA.

Sorby, S.A., and Bulleit, W.M. (2006). *An Engineer's Guide to Technical Communication,* Pearson Prentice Hall, Upper Saddle River, NJ.

Strunk, W., and White, E.B. (1979). *The Elements of Style,* Macmillan, New York.

Vesilind, P.A. (2007). *Public Speaking and Technical Writing Skill for Engineering Students,* 2nd Ed., Lakeshore Press, Woodsville, NH.

5.13 Assignments

1. Write a brief personal statement in the form of a letter (using complete sentence structure). Make sure to include the following:
   ◆ Your full name.
   ◆ Where you are from and where you currently reside.
   ◆ Describe your specific areas of interest within civil engineering (i.e., environmental, hydraulics, water resources, structures, materials, geotechnical, transportation, construction, etc.).

- Describe your strengths/weaknesses.
- Describe your employment experience over the past 2 to 3 years (include co-ops, internships, and any supervisory experience).
- List the reasons why you chose to go to this particular university for your higher education learning experience.
- Describe your expectations for this course (be sure to discuss this in the context of your short-term academic and career goals).
- Describe your career goals and where you see yourself in 5 years.
- Describe your best and worst experiences in a team setting.
- Describe how you will deal with adversity this semester.
- Describe your hobbies and extracurricular activities.
- Describe any personal obstacles you had to overcome to get to this point in your academic career.

2. Write a draft resume.
3. Rewrite the following sentence passages and justify why they need to be changed:
   a. After the first team presentation and report elaboration I could state that the communication among the members has improved considerably; all members are complement of others.
   b. The report of the first presentation was composed by a group effort from all of the members. Each member had task assigned to them for their part for their part of the report. Dividing the report enable us to have a successful outcome. This is because the report was structure in such a way that the area each team member was in charge of went into specific details.
4. Give two reasons why minutes of project meetings should be kept.
5. Create a table (label it as Table 1.2) that lists examples of green features for each of the LEED categories.
6. Sketch a figure that is appropriate for the following data obtained by measuring the head loss and velocity of a fluid when pumped through a 200-foot-long, 6-inch-diameter PVC pipe, using a Grundfos Mq3 45 (115-volt) 1-horsepower centrifugal booster pump:

Velocity (feet/second)	Head loss (feet)
0	0
1	0.7
2	2.4
3	4.7
4	7.7
5	11.5
6	16.0

7. Write a memo to your professor describing your plans to relax and unwind over the semester break.
8. What is the Fog Index? Choose one of your previous writing assignments and determine the Fog Index score. What strategies can you use to improve the score? What strategies can you use to make the writing more universally readable?
9. Rewrite the following statement: To the immediate north, a SPA/Physical Therapy has been proposed, with emmediate entrance to the poop area to facilitate access tot eh handicap and physical therapy costomrs. To the Northeast section figure 23 the enclosed

poop area is found. The pool will offer various of services ranging from party events to the use of isolated lanes for theray and rehabilitation

10. Write a brief memo that argues the business case for meeting during the summer to continue to make progress on the project.

11. The following sentences were incorrectly submitted as similes (see the discussion on similes in Section 5.14.8 of the appendix at the end of this chapter). Rewrite each sentence as a successful simile:
    a. This cup is as blue as the sky.
    b. Your car is the same size as the boulder.
    c. Though the class consists of 30 students, each writing paper has an introduction.
    d. I'm calm as a duck.
    e. Live your life as if there is no tomorrow.
    f. University is like a buffet where students come to pick up their favorite foods.
    g. The car was smoking so bad, I thought I was driving in smog.
    h. A girl feeling is like a delicate flower.

12. The following sentences were incorrectly submitted as metaphors (see the discussion on metaphors in Section 5.14.8 of the appendix at the end of this chapter). Rewrite each sentence as a successful metaphor:
    a. That Lamborghini is as fast as a bullet.
    b. It's hotter than a teakettle in here.
    c. Visilind is an eye opener for stupidity.
    d. Keep your eyes on the road.
    e. He runs like the wind.
    f. Woman is a rose covering with fume and thorn.
    g. It's raining money.
    h. Life is like a box of crayons.
    i. Dress me slowly when I am in a hurry.
    j. Working long hours is like walking up a steep bridge.
    k. Swimming upstream will get you nowhere fast.
    l. Uncle John is over the hill at 60.
    m. I need a seed of wisdom plant in my head.
    n. My handwriting so bad you need the Rosetta Stone to decipher my note.
    o. I steer away from drama.
    p. Arguing about politics is like talking to the wall.
    q. Happiness is a warm gun.
    r. Watch your back.
    s. Nothing is impossible.
    t. I will be strong when I get older.
    u. His arms are invisible.
    v. Caniggia shocked my soul when he scored against Brazil during the 1990 World Cup in men's soccer.
    w. You're a chicken, be a man confront them if needed.

13. What is wrong with the following sentences? Rewrite them correctly:
    a. United State is like a big melting pot of different ethnic.
    b. In view of the fact that in order to write the paragraph he didn't.

14. Create a map from your home address to the classroom. Also create a set of written directions. Compare and contrast them. Which is easier to follow?
15. Pick a recent newspaper article and determine the Fog Index. Obtain a transcript from a recent political speech and determine the Fog Index. Compare and discuss.
16. Present the following information in an effective bar graph that describes the temporal trends in relative energy production from fuel oil:

Year	Annual oil production (quadrillion BTU)	Annual energy production from other sources (quadrillion BTU)	Total energy production (quadrillion BTU)
1960	14.11	27.39	41.5
1970	24.1569	37.9431	62.1
1980	22.1616	42.6384	64.8

5.14 Appendix

This appendix provides a brief explanation of some of the most common grammatical and sentence-level patterns of error in engineering writing. For more in-depth coverage on these topics, consult the many excellent references listed in this chapter.

5.14.1 Grammar

5.14.1.1 Pronoun Antecedents

Pronouns are words like *it, him, her, them, this*, and *these*. Each pronoun must have an antecedent, which refers to the noun that it replaced. The pronoun must agree with the antecedent in number and gender. When a pronoun can refer to multiple antecedents, the lack of specificity can lead to misinterpretation by the reader. Grammatical errors that involve pronoun antecedents can be avoided by checking every pronoun for a clear, appropriate antecedent and then ensuring agreement between the pronoun and its antecedent:

> **Example 1**: If the baby refuses the milk, boil it.
> *Comment*: Does the word *it* refer to the baby or to the milk?
> *Improved*: If the baby refuses the milk, warm up the bottle and its contents in a pan of boiling water.

> **Example 2**: After listening to group A and group C present, I liked their ideas the best.
> *Comment*: Does the word *their* refer to group A or to group C?
> *Improved*: After listening to both teams, group A presented the best ideas.

5.14.1.2 Subject-Verb Agreement

Verb forms are used to indicate tense or time that an action takes place. The English language has six tenses, as illustrated by the verb to see:

◆ **Present tense**—The verb indicates a situation in the current moment:
  ◇ I *see* stars right now.

- **Present perfect tense**—The verb indicates a situation that happened at an unspecified time before the current moment:
  ◇ I *have seen* those stars many times.
- **Past tense**—The verb indicates a situation that happened previously or prior to the current moment:
  ◇ I *saw* those stars last night.
- **Past perfect tense**—The verb indicates a situation that occurred before another action in the past:
  ◇ I *had seen* those same stars before when I was camping in the forest.
- **Future tense**—The verb indicates a situation that has not happened yet but is expected to happen:
  ◇ I *will see* those stars later tonight.
- **Future perfect tense**—The verb indicates a situation that is expected to happen before a time of reference in the future. It refers to a past event in the future:
  ◇ I *will have seen* those same stars after I get back from my trip to Utah next week.

Sometimes it is important to follow a specified tense in certain forms of writing; however, it is more critical to make sure that the point of view remains consistent and the writing does not switch from one tense to another within the same passage. For example, in writing directions, the reader has not performed the task yet, so the future tense is appropriate. However, when writing a summary or a methodology for a report or writing a progress report of work completed to date, the writer already performed the tasks described, so the past tense is more appropriate. When writing about tasks that will be completed by the next progress report, then the future tense is appropriate.

A verb, no matter its tense, must agree in number with its subject:

- The table *is* long (not the table *are* long).
- The plans *are* ready (not the plans *is* ready).

This is exacerbated when there are many words in between the subject and the verb that confuse the numerical relation.

The matter becomes trickier when subjects are joined by a coordinate conjunction like *and, or, nor,* etc. Subjects joined by the word *and,* no matter if singular or plural, require a plural verb:

- The bolt and the nut *were* fastened together (not *was* fastened together).
- The columns and the beam *were* attached (not *was* attached).

Whenever this situation occurs, simply replace the conjunction with the word *they* temporarily and conjugate the verb; this will help avoid making an error.

Singular subjects when joined by *or* or *nor* require a singular verb:

- Either the boy or the girl *was* yelling (not *were* yelling).

Notice how the verb *was* agrees with the singular noun *boy* or *girl.*

If one of the subjects is singular and the other is plural joined by *or* or *nor,* then the verb agrees with the closer noun:

- Neither the hammer nor the nails *were* found.

Notice how the verb *were* agrees with the nearest word, the plural *nails*. If we flip it around, then it changes to:

◆ Neither the nails nor the hammer *was* found.

If there are intervening phrases in the sentence, they are ignored as if they were not there in the first place for purposes of subject-verb agreement.

5.14.1.3 Plural Nouns

Most nouns in the English language are made plural by adding an "s" to the end of the word; for example, schools, papers, pencils, doors, cars, etc. However, words that end in an "s"-like sound are made plural by adding "es" to the end, such as churches, lunches, kisses, gases, boxes, buses, successes, etc. Some words have special plural forms, such as child-children, man-men, person-people, mouse-mice, goose-geese, fungus-fungi, ox-oxen, index-indices, appendix-appendices, criterion-criteria, datum-data, alumnus-alumni, etc. For words that end in a consonant followed by the letter "y," the "y" is changed to an "i" and "es" is used to create the plural form, such as baby-babies, lady-ladies, reality-realities, etc. Some words that end in the letter "o" are made plural by adding "es," such as potato-potatoes, tomato-tomatoes, and hero-heroes, but some are not (e.g., memo-memos). For words that end in an "f" sound, the "f" is changed to "v" for pluralization, such as knife-knives, leaf-leaves, life-lives, self-selves, etc.

Avoid adding an apostrophe when forming the plural of common nouns. The apostrophe indicates the possessive form of a word, not the plural form:

Original: My *grandparent's* were in the other room.
Correct: My *grandparents* were in the other room.

Do not add an unnecessary apostrophe when forming the plural of numbers or letters as words:

Original: During the *1980's*, the *ABC's* were...
Correct: During the *1980s*, the *ABCs* were...

unless absolutely necessary:

Original: Remember to dot your *is* and cross your *ts*.
Correct: Remember to dot your *i's* and cross your *t's*.

5.14.2 Punctuation

Punctuation marks are placed in the text to clarify the meaning and make reading easier. They separate phrases, connect ideas, group thoughts, and provide nuance. Periods and commas as ending punctuation should be placed inside the ending quotation mark:

Original: The professional engineer said, "I approve the plans".
Correct: The professional engineer said, "I approve the plans."

However, question marks, exclamation marks, colons, and semicolons should be placed after the ending quotation mark unless they are part of the quotation.

Also, punctuation should be placed after a reference citation:

Original: It is standard practice to put a comma after an introductory clause or introductory phrase that contains a verb form. (Ebbitt and Ebbitt 1982)

Correct: It is standard practice to put a comma after an introductory clause or introductory phrase that contains a verb form (Ebbitt and Ebbitt 1982).

5.14.2.1 Comma Use

The comma is probably the least understood punctuation mark in technical writing. Pay attention to the rules of comma use, and apply commas appropriately in the following situations:

◆ Before a coordinating conjunction (such as *and, but, or*) that links main clauses:
  ◇ The team designed the column layout, and the project manager reviewed the plan.
◆ After introductory words, phrases, or clauses in a sentence:
  ◇ Before practicing, the team prepared the first draft of the presentation.
◆ Between items in a series:
  ◇ Engineers need computational skills, communication skills, and leadership skills.
◆ To provide a visual cue for the reader to pause within a sentence for greater clarity, such that if the phrase is removed, the sentence retains its original meaning:
  ◇ The building designed by Joseph Smith, in 2007, was built for $13 million.

A *comma splice* occurs when a comma is used to connect two independent sentences without a coordinating conjunction (such as *and, or, nor, for, so, yet*, etc.):

Original: The project manager organized the meeting, he kept the meeting on schedule.

Correct: The project manager organized the meeting, and he kept the meeting on schedule.

5.14.2.2 Colon Use

The function of a colon in writing is to separate and introduce lists, clauses, and quotations. Basically, it brings the sentence to a halt, so that the reader can begin going through a separate list, element, or thought. A colon is not needed between a verb and its direct object, nor is one needed after an introductory phrase like *that is, for example*, and *such as*:

Original: The items you need to bring for sampling are: sample bottles, data sheet, cooler with ice…

Correct: The items you need to bring for sampling are sample bottles, data sheet, cooler with ice…

Remember to use a colon after the salutation line in a formal business letter:

Original: Dear Dr. Jones,
Correct: Dear Dr. Jones:

5.14.2.3 Semicolon Use

The semicolon is used between main clauses that are closely related but are not linked by a coordinating conjunction:

Original: Some walls are made of concrete block. Others are not. The choice of materials depends on many factors.
Correct: Some walls are made of concrete block; others are not. The choice of materials depends on many factors.

Keep in mind that *however, therefore, for example, on the contrary, nevertheless*, and many other transitional expressions are not coordinating conjunctions:

Original: The long front room in the floor plan ended up with a column in the center. Therefore, the room was divided into two smaller spaces.
Correct: The long front room in the floor plan ended up with a column in the center; therefore, the room was divided into two smaller spaces.

Do not overuse the semicolon. It is more appropriate to revise the passage as a standard compound sentence.

5.14.2.4 Hyphen Use

The hyphen is used to express multiple-word phrases as one unit and avoid ambiguity by linking two or more words that function as a single word:

Original: The house has a three car garage.
Correct: The house has a three-car garage.

Original: The awkward room was in the shape of an L.
Correct: The L-shaped room was awkward.

Use a hyphen to connect a number and unit if they directly precede and modify a noun:

Original: The ceiling was 12 feet high and airy.
Correct: The 12-foot ceiling made the room feel airy.

The trend in modern usage is to limit the use of hyphens. Words that begin with short prefixes, like co-, non-, post-, pre-, and re-, generally are spelled as part of the word they modify, without a hyphen; examples include codisposal, pretreatment, and rewrite. Sometimes it is necessary to use a hyphen to avoid doubling a vowel or tripling a consonant; examples include anti-inflammatory and hall-like. A hyphen should be used where two or more hyphenated compounds have a common basic element that is omitted in one or more of the terms; for example, three- or four-person crews. Prefixes that typically require hyphens include ex-, self-, and quasi-.

5.14.3 Capitalization

Capital letters are used for the first word of a sentence, but not to convey emphasis within the sentence structure. Instead, bold or italic type is used to denote emphasis.

Capitalize when using the official form of a proper noun:

Original: The *engineering department* officially approved the plans.
Correct: The *Department of Engineering* officially approved the plans.

Furthermore, only capitalize an official title if it precedes a name, and do not capitalize words just to give them importance:

Original: The ceremony was attended by *President John Smith of ACL, Inc.*
Correct: *John Smith, president of ACL, Inc.,* attended the ceremony.

Original: The *Civil Engineer* visited the job site to supervise construction.
Correct: The *civil engineer* visited the job site to supervise construction.

Geographic names, holidays, languages, religions, book titles, days of the week, months of the year, and biological names (except the species name) are capitalized, whereas seasons and directions are not:

Original: This occurred in the Fall in South Florida and in the Spring in North Dakota.
Correct: This occurred in the fall in south Florida and in the spring in North Dakota.

When using the genus-species name of an organism, the genus name is capitalized, but the species name is not, and the full genus-species name is italicized:

Original: The sample tested positive for Escherichia Coli.
Correct: The sample tested positive for *Escherichia coli.*

5.14.4 Common Spelling Errors

Always spell check *and* also proofread. Beware of typos that are actual words spelled correctly but significantly change the meaning of a phrase or are nonsensical when used incorrectly, as they will not be caught by the spell-checking function, which can be quite embarrassing:

Original: We hired a civil engineering *form.*
Correct: We hired a civil engineering *firm.*

Original: The professional engineer was *component* and experienced.
Correct: The professional engineer was *competent* and experienced.

Original: Getting my PE license is *soothing* I plan to do.
Correct: Getting my PE license is *something* I plan to do.

Original: The project manager is an *impotent* part of the team.
Correct: The project manager is an *important* part of the team.

Words commonly misspelled:

accommodate	judgment	reconnaissance
adherence	occasion	seize
advertise	occurred	siege
adviser	permissible	toward (no "s")
afterward (no "s")	personnel	vacuum
attendance	privilege	weird
indispensable	recommend	

5.14.5 Misused Words

Always choose words carefully to avoid jargon or unnecessary technical terms and complicated phrasing. The thesaurus is a great tool, but it does not highlight shades of meaning. Many common words are used improperly because they look and sound alike, leading to embarrassing errors in writing. Examples of commonly confused and misused words are as follows:

accept, except	council, counsel	sense, scents, cents
adverse, averse	eminent, imminent	than, then
affect, effect	ensure, insure	there, their, they're
allusion, illusion	farther, further	to, too, two
already, all ready	flier, flyer	weather, whether
a while, awhile	its, it's	where, were
between, among	lie, lay	which, that
capital, capitol	precede, proceed	whose, who's
complement, compliment	principle, principal	your, you're
composed, comprised		

Another common misuse is the phrase "try *and* do something," when the phrase should be "try *to* do something."

5.14.6 Abbreviations

Try to avoid using too many abbreviations in the text. Write out months of the year (e.g., August instead of Aug.). State names can be abbreviated, but with the official postal abbreviation (e.g., MN instead of Minn.). For any word, name, title, or phrase that will be repeatedly shortened throughout the text, spell out in full the first time it is used and include the acronym in parenthesis right next to it. Thereafter, the acronym should be used; for example, U.S. Environmental Protection Agency (USEPA) or Leadership in Energy & Environmental Design (LEED®).

5.14.7 Numbers

Try to describe the difference between a lion and a diamond ring without using numbers, descriptive shapes, or units of measure. A lion has bushy hair, and a ring is made of metal, but that would not give a clear picture of either. Signs along the highway indicate the speed limit. The speedometer on the dashboard shows velocity and helps a driver avoid exceeding the speed limit. The gas gauge tells the driver when the fuel is running low, and the gas station pump specifies how much a gallon of fuel will cost.

Engineers frequently have to use observable quantitative data to generate new information and form conclusions that ultimately lead to decisions. For example, it is a quarter to 12, and the employees are hungry for lunch. The next client meeting begins at 12:30 P.M., but the nearest fast-food restaurant is 3 miles away. If it takes 10 minutes to get to the car, then another 15 minutes to drive to the restaurant, and finally 5 minutes in line to order, is there enough time to get lunch and return before the meeting starts? How long will it take to eat? How long will it take to drive back to the office? How long will it take to find parking? Will the employees be able to eat lunch and also make the meeting on time? Math can help with these answers.

The importance of units is illustrated by the following example:

Which of the following is heavier?
◆ 13 !!
◆ 250 @@
◆ 3,750 ##

For a reader who is unfamiliar with these nonstandard units, it is impossible to tell. What do these symbols represent? If !! is grams, @@ is tons, and ## is pounds, then it is easy to see that the correct answer is 250 tons. The take-home message here is that any form of quantitative information must include numerical values and the proper units of measure in order to be of any value to the reader.

Generally, spell out whole numbers for values less than 10; for example, five steel bars that measure 12 inches in length. Use the number for dates (June 15, 2010); also, do not use June 1st, 1st of June, or June first. For monetary amounts, write out the number for values less than one million; for example, $250,000 is correct and $14.1 billion is correct.

Engineers are good at math and must remember the importance of math in their writing, but many people do not understand the magnitude or scale of numbers. A simple technique to help novice readers understand scale is through the use of analogies and metaphors. These comparison tools are employed to put information in perspective for the reader and help compare it to something more familiar to the reader, particularly with respect to size, shape, color, and texture. For instance, in a document written for professional football executives, where a 90-meter stretch of concession stands leading toward the stadium entrance must be described, that stretch can be compared to the length of a football field (100 yards).

5.14.8 Figurative Language Use

In figurative language, words are used in a nonliteral sense to make a comparison between nonsimilar things. These types of comparisons are particularly valuable when they clearly emphasize a relationship that is difficult to describe in words by evoking an image in the reader's mind. The two most common are similes and metaphors.

A simile is a comparison that uses *like* or *as*, and a metaphor is a comparison that does not use *like* or *as*. However, just because the word *like* or *as* is used does not mean the sentence is a simile:

- ◆ I will be ready in like two minutes. (not a simile)
- ◆ We took the samples as the sun set in the sky. (not a simile)
- ◆ The pickup truck is as American as apple pie. (simile)
- ◆ My little brother eats like a pig. (simile)

By contrast, a metaphor is an implied comparison without the use of *like* or *as*:

- ◆ My little brother is a pig. (metaphor)
- ◆ This classroom is a zoo. (metaphor)
- ◆ Her headache is bad. (not a metaphor)

Be careful to avoid mixed metaphors:

Original: Playing with fire can get you into deep water.
Correct: Playing with fire can get you into hot water.
Correct: Playing with fire can get your fingers burnt.

5.14.9 Voice

The voice of a verb refers to a subject either performing the action (active voice) or receiving the action (passive voice). The active voice is direct, clear, and strong. It is a set of commands. On the other hand, the passive voice is used when the one doing the action is not known or should not be mentioned or if the recipient of the action is emphasized. It generally is used to obscure responsibility:

- ◆ Record the weight of the sand. (active voice, present tense)
- ◆ The weight of the sand was recorded. (passive voice, past tense)

5.14.10 Gender Issues

In today's society, masculine words such as he, his, him, etc. should be avoided, particularly when the person being referred to may be a man or a woman. To combat this problem, rewrite the sentence without the pronoun, replace the pronoun with a genderless word (e.g., person, one, individual, team member, engineer, etc.), or if all else fails go with the awkward "his/her" construction. Avoid adding gender to job titles or descriptions:

Original: The *crewman*...
Correct: The *crew member*...

Original: The *chairman* of the committee...
Correct: The *chair* of the committee...

Original: The *policeman*...
Correct: The *police officer*...

Do not make the mistake of changing to the plural form to avoid the awkward "he/she" construction:

Original: The engineer left *his/her* plan set at the field station.
Worse: The engineer left *their* plan set at the field station.
Correct: The engineer left *the* plan set at the field station.

5.14.11 Writing Pitfalls to Avoid

5.14.11.1 Redundancies

In writing engineering reports, it is important to get to the point and avoid wordiness and needless repetition. Wordiness refers to the use of more words than is necessary to adequately convey the concept, and needless repetition can distract or confuse the reader. This is a common problem with speaking that gets translated to writing. Try to make every word count, and omit phrases that simply fail to add meaning to the content:

Original: We believe the findings are adequate *enough*.
Correct: We believe the findings are *adequate*.

Original: During the late days in the month of April, the final exam looms menacingly for the capstone students.
Correct: The capstone design final exam is in late April.

Original: The bridge was used to *bridge* the ravine.
Correct: A bridge was constructed to *cross* the ravine.

Original: It may be possible to *work* on the capstone project while *working*, but I will not know if this will *work* until I get closer to *working*.
Correct: It may be possible to *work* on the capstone project while *employed*, but I will not know if this *is feasible* until I *start my new job*.

Original: Once *I* have completed this task, *I* want to get a job. *I* want to work in a construction firm because *I* want to work somewhere *I* know *I* will get hands-on experience. Although *I* respect all engineering disciplines, *I* must say *I* dislike the ones that *I* worked at that had no field work.

I, I, I! When writing about one's own accomplishments, it is common to use sentence construction that begins with "I." However, the repetitiveness will begin to annoy the reader, so try to vary the sentences by using introductory phrases and try to limit the use of "I":

Better: Once this task is completed, *I* want to get a job, preferably in a construction firm where *I* will get hands-on experience from field work.

Be careful with using unnecessary or redundant introductory phrases that are already implied. Instead, be more direct and eliminate the extra words that do not add meaning to the sentence:

Original: Because of the fact of time constraints to check the design calculations, the quality control officer will perform a check on behalf of the project manager in order to check for errors.

Correct: If the project manager is unavailable, the quality control officer will check the design calculations for errors.

However, sometimes the opposite is true. Watch out for the omission of necessary words:

Original: They *been* trying to install lighting systems.

Correct: They *have been* trying to install the lighting systems.

5.14.11.2 Informal Language

Informal language is communication used in daily speech. It generally is not appropriate for technical writing in which more formal language construction is expected. The following are just a few examples:

Original: Turn right, and after you *hit* the stop sign, turn left.

Correct: Turn right, and after you *hit* the stop sign, pick it up and put it back!

Better: Turn right, and after you *stop at* the stop sign, turn left.

5.14.11.3 Inventing New Words or Phrases

This also means avoiding contractions, colloquialisms, and slang. It is more of an issue in the legal field, but addressing technical terms is often where engineers get into trouble. Here are some examples:

- ◆ Strategery
- ◆ Irregardless
- ◆ Ain't
- ◆ Refudiate
- ◆ Misunderestimate

Original: After you pass one *light intersection...* (as opposed to a heavy intersection?)

Correct: After you pass one *traffic signal...*

5.14.11.4 Double Negatives

This refers to the use of two negative words to express a single negation, which is grammatically redundant:

Original: We didn't do nothing in class today.

Correct: We did nothing in class today.

Now you should be familiar with all of the rules needed to craft effective and grammatically correct writing for engineers. Remember to visit some of the excellent references cited in this chapter for more information on these and other less common grammar errors and sentence-level patterns of error.

Alternative Analysis

Unlike the homework that students tackle in most undergraduate engineering classes, real-world problems rarely have only one correct answer. In fact, there may be many acceptable solutions. Think about trying to select the right tool at a hardware store for a particular job. There are many things to consider. If the task requires screws to be fastened, then a hammer or a wrench would not be appropriate. What size are the screws? Are they Phillips head? How many screws need to be fastened? What is the budget? Will a drill be required or will a screwdriver suffice? What about quality of the manufacturer? Clearly, this is an important decision that depends on many criteria, and this decision-making process is something everyone does all the time. However, it is the job of the engineer to objectively select the preferred options, or at least optimize the feasible alternatives, to provide the best solution possible, given the constraints. Engineers call this process an *alternative analysis*, the purpose of which is to help systematically sort through the available options to choose the preferred approach. Fortunately, there are guides for this.

Planning any infrastructure project involves nine major components (Dzurik 2003):

- Development of goals and objectives
- Collection and analysis of data
- Definition of scope
- Clarification of scope and ancillary issues
- Identification of reasonable alternatives
- Analysis of alternatives
- Summary of recommended actions
- Development of an implementation program
- Monitoring of results

Often, before a project can be identified and solutions sought, a series of relevant goals and objectives should be developed. Solutions to be considered will be evaluated using these goals and objectives. Therefore, if the goals and objectives do not adequately conform to the client's intent, the solutions proposed may fail to adequately address the project's needs.

While defining the goals and objectives sounds relatively easy, this may be where the most conflict can arise in planning a project due to a myriad of issues, such as economics, preference trade-offs, and controversies over the goals of the project. Goals and objectives are general in nature, but clear in application (i.e., whether or not they are met or optimized in a given situation or used in ranking alternatives). Objectives are defined as more specific subsets of the goals, and several objectives may be used to attain a goal.

The data collection and scope identification steps may vary in order. For example, effluent water quality data from a wastewater treatment plant may indicate the need to upgrade the facility to meet regulatory standards, or water demand data may indicate the need to increase capacity of a drinking water plant. Maintenance data may indicate where pipelines, bridges, roadways, or buildings are in need of replacement, expansion, or repair. Thus, these facts must be obtained up front to better inform decision making and help refine the scope of the problem. However, once the problem is defined, additional data and analysis will be required to identify and characterize the alternatives that should be considered for the engineering solution. Data should include future projections of the appropriate variables in order to meet the future needs of the project. Often this data is not provided by the client, unlike what students have grown to expect in classroom environments, so young engineers need to be resourceful in researching and tracking down the information needed for this part.

An engineering project results from the intersection of the goals and objectives developed initially by the client (owner, governing board, politicians, customers, projected growth demands, regulatory requirements, failures of infrastructure, or some combination of sources) and the defined needs of the system as interpreted by the consultant. The project scope may be defined differently by different constituencies, so it is important to agree on the scope of work before getting too far along in the process. For example, the replacement of septic tanks with sanitary sewers may be required by regulatory agencies to resolve actual or perceived public health issues, but the residents who will use the sewers may view the added expense of sewer service, connections, and assessments as intrusive, unpopular, and unnecessary. Certain projects will require some consensus among the constituents in order to proceed smoothly, while other needs will be obvious. Some consensus is required in the initial stages of the project in order to properly define the goals and gather data for evaluating the viable engineering alternatives.

The next step in the process involves brainstorming to identify reasonable engineering alternatives. Say a goalkeeper must be selected for a soccer team. In addition to being physically talented with quick reflexes, the best candidate also would have some experience playing the position, so those without previous experience will be excluded, meaning that experience is a *threshold criterion*. Thus, the first question to all of the candidates would be whether they have ever played goalkeeper before. If the answer is no, then the candidate is eliminated from consideration, and the pool of potential goalkeepers gets smaller and more manageable. To narrow down the possible options quickly, it is necessary to establish *threshold criteria*, which refer back to the project's goals. For example, a typical set of threshold criteria for selecting a suitable site for a new hydroelectric plant might be that the project must be located within the city limits along the riverbank. Given this threshold criterion, it is inappropriate to suggest locating the hydroelectric plant far away from a water body in another county.

Thresholds eliminate poor alternatives from the analysis right away to avoid clouding the picture. Often, poor alternatives are of high or very low cost but will only partially meet the goals of the project, so they do not belong in the analysis. Bad alternatives should never be presented in a public forum. They are a magnet for elected officials because bad alternatives are often lobbied for by outside entities that see potential opportunities to profit from inadequate planning, policies, or solutions.

Analysis of the remaining viable alternatives should use the project goals and objectives defined earlier as criteria for evaluation. Depending on the specifics of the alternative being evaluated, economic evaluations should be made using a benefit-cost analysis or present worth analysis (see Chapter 11 for more about these engineering economics tools) to determine if the option is economically justified. A major note of caution is that many times the benefits are unknown or difficult to quantify at the time the benefit-cost analysis is conducted. There may be future benefits or the alternative may encourage other action. For example, one reason water/sewer utilities typically are publicly funded is that a benefit-cost analysis of extending pipelines to undeveloped areas will always fail if measured solely for economic reasons, because future development is unknown and there are few current customers. Nevertheless, failure to extend lines guarantees economic development goes elsewhere. The benefit of extending piping may be to expand development or discourage future competition, which might have significant disbenefits to the utility. For the moment, discussion of economic benefits and disbenefits will be deferred until Chapter 11.

The criteria used in evaluating viable alternatives should match the goals and objectives defined for the project and agreed upon by all parties. The comparison of alternatives should be clearly delineated using professional judgment. Trade-offs and impacts to social, political, economic, and environmental factors should be addressed. The preferred solution should be the alternative that best meets the goals and objectives, realizing that no one solution may meet all the objectives better than all others.

6.1 Application to Design Projects

Let's take a look at specific issues associated with design projects and create a simplified means to evaluate engineering alternatives. Keep in mind that usually there is more than one way to do something. Good engineers expand their thought process to include these other ways, but when they do, they may realize that there are too many choices or that the possibilities can be endless. Thus, deciding which of these alternatives to pursue is a challenge. This is the art of alternative selection.

The first step is to recognize the feasible alternatives. There are so many different manufacturers, contractors, construction methods, technologies, and site locations that it can be easy to lose sight of the goal. For example, if someone requests a soda from a vending machine, how does one choose from among Coca-Cola™, Diet Coke™, Pepsi™, and RC Cola™? What about Sprite™, 7-Up™, and Slice™? What if they are all the same price? What about spring water, lemonade, or Gatorade™? Should they be considered, even though they are not sodas?

Before rushing to make an uninformed decision, engineers must brainstorm all the feasible options and include the "do nothing" alternative, but realize that only appropriate alternatives can be analyzed. Therefore, begin by asking the following questions:

◆ **What is the perspective?** Make sure that the understanding of the project is in the right *context* or frame of reference. As an example, in choosing between several options for development of a site, are the costs analyzed from the perspective of the buyer or the seller? The goals would be different. From the buyer's perspective, the price must be minimized. From the seller's perspective, the price must be maximized. Establish the perspective first to eliminate confusion in evaluating alternatives.

◆ **What are the project goals that need to be met?** It is important to clearly define the project goals to make sure that the proposed alternative actually is feasible and makes sense within the context of the project itself. The goals later give rise to the evaluation criteria for selection.

◆ **Does the alternative meet the minimum requirements of the project goals?** This is the threshold criterion. If a particular alternative does not meet the minimum requirements for the project goals, then it should be dropped from further consideration. For example, would a particleboard building, built with glue, that takes over a year to construct seem to be an appropriate option for a classroom building in a hurricane zone? Even a bad alternative will look great compared to poorer alternatives, but bad alternatives generally do not meet the project goals! The threshold criteria typically are in the form of a yes or no response. For example, is the cost of the option below the maximum budget? Any option with a cost above the maximum budget would be removed from further consideration.

◆ **Is the alternative feasible?** If the alternative meets the minimum requirements as stated in the project goals but has a footprint that goes beyond the property guidelines, for example, then the alternative may not be feasible. If the alternative requires the development of new technology that does not exist today or parts that are not commercially available, then it may not be feasible at this time. Only appropriate alternatives should be analyzed.

Once this preliminary filter is applied, what remains are a set of feasible alternatives that meet the goals of the project, some better than others. The benefits or issues associated with each remaining option are evaluated by analyzing the advantages and disadvantages of each alternative. Remember that only equal independent alternatives can be evaluated using appropriate selection criteria derived from the project goals.

6.2 Selection Criteria

A balance must be struck among cost, time, and the goals of the project. Imperfect solutions may be the most cost effective or the fastest to build, but do they pay off in the long term? Going back to the initial process, achievable and relevant project goals already have been set. Note that all goals may not be equally weighted. These goals are representative of the project needs or scope of work. Drawing from these goals, engineers develop a set of selection criteria to evaluate the alternatives against each other with respect to the needs of the project. Examples of typical selection criteria include but are not limited to the following:

◆ Cost criteria
  ◇ Capital

- ◇ Property value
- ◇ Power/electricity
- ◇ Staffing
- ◇ Operations and maintenance
- ◇ Chemicals
- ◇ Disposal
- ♦ Performance criteria
  - ◇ Reliability
  - ◇ Ease of operation
  - ◇ Return on investment
  - ◇ Efficiency
  - ◇ Longevity
- ♦ Physical criteria
  - ◇ Footprint
  - ◇ Appearance
  - ◇ Construction time
- ♦ Manufacturer
  - ◇ Availability of parts
  - ◇ Customer service
  - ◇ Reputation
  - ◇ Proximity/location
- ♦ Future considerations
  - ◇ Expandability
  - ◇ Flexibility

Additional examples of criteria could include accessibility, ease of construction, access to labor pools, lot size, access to utilities, and conformance with the surroundings. Keep in mind that cost is embedded in many criteria. At the end of the day, it is reasonable to select three to seven criteria for evaluation. Appropriate alternative analyses should have multiple criteria that are independent. For each criterion used, there needs to be an explanation as to how each one will be evaluated. For example, if future expansion is a criterion, then it can be evaluated on the basis of maximizing the available lot size such that the option with the largest lot size will receive the highest score. This should be done before assigning scores to the engineering alternatives to avoid confusion.

When writing an alternative analysis report, it is critical to define the selection criteria clearly in the text and explain how each is quantified and scored. Each criterion must be explained and must be quantifiable. For example, in recruiting a goalkeeper for the soccer team, the selection criteria could include wingspan, jumping ability, and years of experience. The wingspan criterion would be measured by height in feet and inches, with the largest number achieving the highest score. The jumping ability criterion could be evaluated by vertical leap height, with the highest value garnering the highest score. Finally, the experience criterion would be measured by years playing competitively, with the highest score achieved by the largest number of years. The highest number does not always translate to the best score. For example, if quickness was a selection criterion, this would be measured by reaction time, and the smallest number of seconds would be the most favorable in this category.

6.3 Scoring System

The next step is to establish a scoring system, which consists of a *rating factor* and perhaps a *weighting factor* for each individual alternative with respect to each selection criterion. The rating factor typically is assigned a numerical score depending on the number of alternatives. For instance, if there are five alternatives, the one that ranks the highest in a particular category should receive a score of 5, and the one that ranks the lowest should receive a score of 1.

For the reader or listener to follow along, the selection criteria must be defined and prioritized by importance using a weighting system. In the simplest scoring system, all of the selection criteria are evaluated equally or assigned a weighting factor of unity (1). This scoring system is unweighted. Although weighting is not necessary, if some of the selection criteria have greater priority than the others, then a weighting factor can be applied. The weighting factor is allocated based on the project goals. This is the level of importance assigned to each criterion, but an explanation is required to indicate why certain criteria are weighted more heavily than others. For example, if cost of operations is weighted 10 times more heavily than any other criterion, then the cost ends up being the most critical criterion and has a major impact on the results of the alternative analysis. That may or may not be appropriate.

One simple weighting approach is to prioritize the selection criteria in order, such that the criterion with the greatest importance receives the highest weight and so on until the criterion with the least importance is assigned the lowest weight of 1, but the weighting factors need to be explained so the audience can understand why certain criteria are weighted more heavily than others.

6.4 Alternative Selection Matrix

Once the selection criteria and the weights have been explained, the alternative selection matrix can be developed. The matrix is a visual representation of the analysis. It is recommended that the selection criteria be placed in the left column of the matrix table, arranged from highest to lowest weight, so that the audience can more easily follow along. For ease of presentation, it makes sense to organize the options so that the one with the lowest score is listed first or to the left, and the remaining options are listed in order by increasing score, until the preferred option is presented last or farthest to the right (recency). The scores for each option and for each criterion are then totaled. Table 6.1 is an example of a simple,

Table 6.1 Example of an unweighted alternative selection matrix

Criteria	Option A	Option B	Option C
Cost (present worth)	1	3	2
Sustainability	2	1	3
Ability to reuse existing	2	1	3
Flexibility	3	1	2
Ease of operation	1	3	2
Longevity	1	3	2
Construction time	1	3	2
Total score	11	15	**16**
Max score: 21			

Table 6.2 Example of a weighted alternative selection matrix

Criteria	Weight	Option A	Option B	Option C
Cost (present worth)	7	1 (7)	3 (21)	2 (14)
Sustainability	6	2 (12)	1 (6)	3 (18)
Ability to reuse existing	5	2 (10)	1 (5)	3 (15)
Flexibility	4	3 (12)	1 (4)	2 (8)
Ease of operation	3	1 (3)	3 (9)	2 (6)
Longevity	2	1 (2)	3 (6)	2 (4)
Construction time	1	1 (1)	3 (3)	2 (2)
Total score		11 (47)	15 (54)	**16 (67)**
Max score: 21 (84)				

unweighted matrix. Notice how the options have been arranged so that the alternatives are listed in order by increasing score, with the preferred option listed on the right. Also note that the maximum score (21) is shown in the matrix, so that the reader can instantly evaluate how well the preferred option meets the project goals (16 out of a possible 21 in this case).

Table 6.2 is an example of a weighted matrix. The value in parentheses is the rating factor multiplied by the weighting factor. Make sure to include a row with the total score, computed as follows:

$$\text{Final result} = \sum_{1}^{i} (\text{Weighting factor})_i \times (\text{Rating factor})_i \tag{6.1}$$

After tallying up the scores, the alternative with the highest numerical value becomes the objectively determined preferred option. In the case of Tables 6.1 and 6.2, this turns out to be option C. Again, it is critical to provide the scoring range to be able to assess if the best alternatives were analyzed. In the example, the highest unweighted score is achieved by option C, with a tally of 16 out of 21, and when the scores are weighted, the preferred alternative is still option C, with 67 out of a possible 84.

Two cautions: if the results are all similar, an additional preferred option probably has not been included in the matrix. If a certain option rates highest every time, the alternatives it is being compared against may be inappropriate. Seeking more options is recommended in both cases.

6.5 Sensitivity Analysis

In the event a different option scores higher when the matrix is weighted, then it is possible that not all of the top alternatives were evaluated or the weighting system is skewed unfairly toward one of the options. In this case, a sensitivity analysis can be performed. A sensitivity analysis can assume the most important criterion, which is cost (present worth) in this example, is removed from consideration. In this case (see Table 6.3), the highest total score is once again option C, with an unweighted tally of 14 and a weighted score of 53 out of a possible 63. This shows that even if cost is not considered, option C is still preferred. A sensitivity analysis can be performed by changing weights, removing a category, or adding a new one.

Table 6.3 Weighted matrix with sensitivity analysis

Criteria	Weight	Option A	Option B	Option C
Cost (present worth)	7	1 (7)	3 (21)	2 (14)
Sustainability	6	2 (12)	1 (6)	3 (18)
Ability to reuse existing	5	2 (10)	1 (5)	3 (15)
Flexibility	4	3 (12)	1 (4)	2 (8)
Ease of operation	3	1 (3)	3 (9)	2 (6)
Longevity	2	1 (2)	3 (6)	2 (4)
Construction time	1	1 (1)	3 (3)	2 (2)
Sensitivity (without cost) **Max score: 18 (63)**		10 (40)	12 (33)	**14 (53)**
Total score **Max score: 21 (84)**		11 (47)	15 (54)	**16 (67)**

6.6 Automobile Purchase Example

Let's consider the purchase of a hybrid automobile; the available options at the dealership are listed in Table 6.4. An alternative analysis is performed by going through the step-by-step process, as follows:

- ◆ **Step 1. Define the perspective**—Assume the point of view of the buyer who wants to purchase a hybrid vehicle.
- ◆ **Step 2. Define the goals and selection criteria**—The buyer is a young single working mother, with a 10-year-old son and a 7-year-old daughter, who wants the best possible fuel economy at the lowest sticker price, with an excellent crash safety rating, maximum storage space, and sporty performance. Using these goals, the selection criteria are defined as follows:
    - ◇ **Fuel economy** (miles per gallon [mpg]) is important because the buyer wants to lower fuel costs at the pump and reduce her carbon footprint. A higher mpg rating will achieve a higher score.
    - ◇ **Price** (manufacturer's suggested retail price [MSRP]) is important in terms of initial cost to the buyer. A lower MSRP will achieve a higher score.
    - ◇ **Safety** (crash rating) is important because the buyer wants to maximize protection for her children when driving around town. A higher crash rating will achieve a higher score.
    - ◇ **Storage space** (cargo capacity, cubic feet) is important because the kids are involved in sports and there is lots of equipment to cart around. A larger cargo space will achieve a higher score.
    - ◇ **Performance** (0 to 60 acceleration, seconds) is important in order to be able to speed up to get out of difficult or life-threatening situations. A shorter acceleration time will achieve a higher score.
- ◆ **Step 3. Perform research on the available alternatives to determine the value for each quantifiable criterion**—In this step, the appropriate numerical values for each selection criterion are determined so as to be able to make an informed ranking for each category for each available option (refer to Table 6.4).
- ◆ **Step 4. Establish any threshold criteria**—In this case, the vehicle must be a hybrid, so the Hummer H3, which is not a hybrid, cannot be considered. This threshold criterion leaves only three options left to evaluate (see Table 6.5).

◆ **Step 5. Select a weighting scale for the selection criteria**—If the buyer is most interested in maximizing fuel economy with a low sticker price, then a weighting scale could be set as follows:

Fuel economy	5
Price	4
Safety	3
Cargo capacity	2
Acceleration	1

Note that some of the categories could be weighted equally (e.g., safety and cargo capacity), which is perfectly acceptable if this is the buyer's preference.

◆ **Step 6. Construct the matrix and perform a sensitivity analysis**—The preferred option from this perspective turns out to be the Toyota Prius because the buyer is most interested in maximizing fuel economy with a low sticker price. As shown in Table 6.6, the Prius scores highest in these two categories. If another buyer had different goals or prioritized differently, the Ford might score higher. A word of caution: avoid

Table 6.4 Summary of automotive purchase options and quantifiable selection criteria as reported by the manufacturer

Parameter	Unit	Hummer H3	Ford Escape	Honda Civic	Toyota Prius
Type		SUV	Hybrid SUV	Four-door sedan	Hatchback sedan
Fuel economy	mpg	12	32	42	46
Price	MSRP	$53,000	$29,300	$22,600	$22,000
Safety	crash rating	5	5	5	4
Cargo capacity	cubic feet	198	100	10	16
Acceleration	seconds	13	9	11	10

Table 6.5 Summary of remaining automotive purchase options and quantifiable selection criteria as reported by the manufacturer

Parameter	Unit	Hummer H3	Ford Escape	Honda Civic	Toyota Prius
Type			Hybrid SUV	Four-door sedan	Hatchback sedan
Fuel economy	mpg		32	42	46
Price	MSRP		$29,300	$22,600	$22,000
Safety	crash rating		5	5	4
Cargo capacity	cubic feet		100	10	16
Acceleration	seconds		9	11	10

Table 6.6 Alternative selection matrix for a hybrid car purchase showing the preferred option based on the criteria selected (weighted and unweighted)

Parameter	Weight	Honda Civic	Ford Escape	Toyota Prius
Fuel economy	5	2 (10)	1 (5)	3 (15)
Price	4	2 (8)	1 (4)	3 (12)
Safety	3	3 (9)	3 (9)	2 (6)
Cargo capacity	2	1 (2)	3 (6)	2 (4)
Acceleration	1	1 (1)	3 (3)	2 (2)
Final score		9 (30)	11 (27)	12 (39)
Max score: 15 (45)				

saying that the Prius (or the Escape) is the winner! The matrix provides the *preferred option* based on the criteria used.

The scores for the Prius are 12 out of 15 for unweighted and 39 out of 45 for weighted. But what if the weights were different? A sensitivity analysis will increase the level of confidence in the purchasing decision. For example, what if safety is more important and the sticker price is not so important? Then the weights can be rearranged. What if annual maintenance costs are also considered? Then an additional category can be added to the analysis. What if buying American-made is also considered? Then the Ford would be the only option to consider because the other cars are Japanese-made and do not meet the "buy American" threshold criterion.

6.7 Reference

Dzurik, A.A. (2003). *Water Resources Planning*, Rowman & Littlefield.

6.8 Assignments

1. Select three different vacation destinations that you would like to visit. Identify four factors that you would consider as a part of the decision. Research these destinations and create a matrix that allows you to reach a conclusion.
2. Assume you and your partner are deciding on options for your new place to live. You have looked at apartments, condos, and single-family homes. Identify four factors that you would consider as a part of the decision. Research these properties, specify the perspective, list the threshold criteria, and create a weighted matrix that allows you to reach a conclusion by performing a sensitivity analysis.
3. Assume you are buying a townhouse and need to decide where in the community you want to live. Identify four factors that you would consider as a part of the decision. Research these options, specify the perspective, list the threshold criteria, and create a weighted matrix that allows you to reach a conclusion by performing a sensitivity analysis.
4. You are constructing a roadway. You have the option to use asphalt or concrete for the roadway surface. Identify four factors that you would consider as a part of the decision. Research these options, specify the perspective, list the threshold criteria, and create a weighted matrix that allows you to reach a conclusion by performing a sensitivity analysis.
5. You are constructing a pipeline. You have identified several options for the route. Identify four factors that you would consider as a part of the decision. Research these options, specify the perspective, list the threshold criteria, and create a weighted matrix that allows you to reach a conclusion by performing a sensitivity analysis.
6. You are constructing a new commercial building. Identify four factors that you would consider as a part of the decision on where to locate this building. Research these options, specify the perspective, list the threshold criteria, and create a weighted matrix that allows you to reach a conclusion by performing a sensitivity analysis.
7. You are constructing a new school and need to determine where to locate it in your community. Identify four factors that you would consider as a part of the decision on where to locate this building. Research these options, specify the perspective, list the

threshold criteria, and create a weighted matrix that allows you to reach a conclusion by performing a sensitivity analysis.

8. You are constructing a high-density residential development with 500 units. Identify four factors that you would consider as a part of the decision on where to locate this project. Research these options, specify the perspective, list the threshold criteria, and create a weighted matrix that allows you to reach a conclusion by performing a sensitivity analysis.

High-Performance Construction

In the field of engineering, the concept of sustainability refers to designing and managing to fully contribute to the objectives of society, now and in the future, while maintaining the social, ecological, environmental, and economic integrity of the system. Most people would agree that structures such as buildings that have a life span measured in decades to centuries would have an important impact on sustainability, and as a result, these buildings must be viewed as opportunities to build sustainably. When people think about green engineering, what generally comes to mind is solar panels, high-efficiency lighting, green roofs, high-performance windows, rainwater harvesting, and high-efficiency appliances and water fixtures, some of which can be seen in Figure 7.1. This is true, but building green can be so much more.

Figure 7.1 Pine Jog Elementary School LEED® Gold Environmental Education Center (note the solar panels, rainwater cisterns, louvers, recycled aluminum, and other features)

The truth is that the built environment provides countless benefits to society, but it has a considerable impact on the natural environment and human health (USEPA 2010). U.S. buildings are responsible for more carbon dioxide (CO_2) emissions annually than those of any other country except China (USGBC 2011a). In 2004, the total emissions from residential and commercial buildings in the United States were 2,236 million metric tons of CO_2, more than any other sector including the transportation and industrial sectors (USGBC 2011a). Buildings represent 38.9% of U.S. primary energy use, 72% of U.S electricity consumption (and 10% worldwide), 13.6% of all potable water, and 38% of all CO_2 emissions (USGBC 2011b). Most of these emissions come from the combustion of fossil fuels to provide heating, cooling, and lighting and to power appliances and electrical equipment (USGBC 2011b). Buildings have a life span of 50 to 100 years, during which they continually consume energy and produce CO_2 emissions; therefore, if half of the new commercial buildings were built to use only 50% less energy, it would save over 6 million metric tons of CO_2 annually over the life of the buildings. This is the equivalent of taking more than a million cars off the road each year (USGBC 2011b).

The U.S. Green Building Council® (USGBC®) expects that the overall green building market (both nonresidential and residential) will exceed $100 billion in 2015 (McGraw Hill Construction 2008). Despite the economic issues after the 2008 recession, it is expected that green building will support 7.9 million U.S. jobs (USGBC and Booz Allen Hamilton 2009). Local and state governments have taken the lead with respect to green building, although the commercial sector is growing.

Green building, or high-performance building, is the practice of creating structures using processes that are environmentally responsible and resource efficient throughout a building's life cycle, from site assessment, design, construction, operation, maintenance, and renovation to deconstruction (USEPA 2010). High-performance building standards expand and complement the conventional building designs to include factors related to long-term efficiency, economy, utility, durability, sustainability, and comfort. At the same time, green building practices are designed to reduce the overall impact of the built environment on human health and to use natural resources more responsibly by more efficiently using energy, water, and other resources, while also protecting occupant health and improving employee productivity.

7.1 Why Build Green?

High-performance buildings are defined by incorporating attributes such as energy efficiency, durability, life cycle performance, natural lighting, and occupant productivity (USEPA 2010). A high-performance building is constructed from green building materials and reduces the carbon footprint that the building leaves on the environment. The USGBC has developed a checklist program called Leadership in Energy & Environmental Design (LEED®) to certify high-performance buildings. A LEED-certified building uses 32% less electricity and saves around 30% of water use annually (USGBC 2011a). Educated building owners know that there is a return on investment of up to 40% by constructing a green building as a result of savings in energy and water over the life of the project (Northern Arizona University 2012).

The cost per square foot for buildings for which owners seek LEED certification falls into the existing range of costs for conventional buildings for which certification is not sought (Mathiessen and Morris 2007). An up-front investment, called a *premium*, of 2% in green building design, on average, results in life cycle savings of 20% of the total construction costs—more than 10 times the initial investment (Kats et al. 2003), while the sale price for energy-efficient buildings is as much as 10% higher per square foot than for conventional buildings (Miller et al. 2008). At the same time, among the most difficult barriers to green building that must be overcome are real estate and construction professionals who still overestimate the premium costs of building green (World Business Council 2008).

New data indicates that the initial construction cost of LEED-certified buildings some-times can be no more than traditional buildings. A case study by the USGBC showed that the average premium for a LEED-certified Silver building is around 1.9% per square foot more than a conventional building. The premium for Gold is 2.2% and 6.8% for Platinum (the levels of LEED certification are discussed further in Section 7.2.3). These numbers are averaged from all LEED-registered projects, so while the data is limited, it still demonstrates that the additional cost to deliver a LEED-certified project is not prohibitive, and at the same time, certification greatly improves the value of the building and lowers operating costs (Kubba 2010). The authors' experience with the Dania Beach nanofiltration plant indicated the premium was less than 3% to achieve LEED Gold certification compared to standard construction.

One aspect of high-performance building design that is closely monitored is energy consumption. The more power a structure consumes, the more power that must be gen-erated, so major strategies are employed to limit energy consumption, which also reduces water needs because power generation accounts for 39% of all water use in the United States (Lisk et al. 2012). Water and sewer account for 4% of all power use in the United States (Lisk et al. 2012), so water use efficiency saves both water and power. By adopting green building strategies, economic and both built and ecological environmental performance can be maximized.

Beyond energy and water, opportunities exist for reducing waste. For example, when materials are recycled during construction of a building, the amount of waste produced during the construction process is drastically reduced. Building material is a common waste product that accelerates landfill closures. When nearly all materials that are left from con-struction are recycled, the amount of space taken up in landfills is reduced.

According to Kubba (2010), 83% of executives say that they are more likely to look toward high-performance buildings in the future because they have a 52% greater return on investment and a 72% higher building value. However, these savings occur over time and can sometimes take a long time to pay off. The first costs (premium) of high-performance buildings are a barrier to construction by developers who flip the buildings after construction, which may be a reason why more of these kinds of buildings are not currently available (Kubba 2010). The fact is that, in most cases, developers will buy a site and build whatever they can, figuring they will sell it quickly. In order to make the largest profit, they tend to construct buildings as cheaply as possible because developers do not want to risk losing profit on the project.

There are other possible savings in tax rebates and other incentives. Some power entities offer rebates on the energy savings per kilowatt-hour. These agencies will come out and do an inspection for free to help the building owner reduce the power bill. Based on what the building is used for, federal grants can credit up to 30% on solar and wind energy in the United States. Another tax incentive is if the building saves at least 50% on the heating and cooling, the federal government credited $1.80 per square foot in 2011 (Energy Tax Incentives 2011). There are many more tax rebates, policies, and government incentives that depend on the types of technologies that are implemented. It is advantageous to research available incentives during the design period by using a clearinghouse like DSIRE™ (Database of State Incentives for Renewables & Efficiency).

Some of the LEED certification credits focus mainly on indoor environmental quality (USGBC 2010). One case study in Australia found that a company made green renovations to its building and experienced a 39% decrease overall in sick days (Sustainability Victoria and the Kador Group 2011). A similar outcome was observed at Michigan State, and based on the case study, an additional week's worth of work per employee was gained just by improving the workplace environment. Such findings are a significant cost savings for the employer every year depending on the size of the company and payroll.

7.2 Agencies That Evaluate Green Building Performance

An issue that arises with a green building is determining how to evaluate its performance. There are several agencies that have created a means to accomplish this goal. The main organizations are the International Organization for Standardization, which produces ISO 14001; the U.S. Environmental Protection Agency, which administers the ENERGY STAR® and WaterSense® programs; the USGBC, which created the LEED program; and the Green Building Certification Institute, which certifies LEED projects. Of course, there are also many other third-party providers that offer similar certification programs.

7.2.1 International Organization for Standardization

The International Organization for Standardization (ISO), located in Geneva, Switzerland, has established over 19,500 standards to verify that materials, products, processes, and services are fit for their purpose (ISO 2011). Some of these standards are used by communities and organizations in an attempt to reduce their environmental footprint and improve their environmental management systems (EMS). The ISO 14001 standards apply primarily to management processes and not buildings, which is why they are not used for certification of buildings. Both ISO and EMS are more applicable to operation and maintenance rather than design and construction.

The ISO 14000 family of standards addresses environmental management to minimize harmful effects on the environment and to achieve continual improvements in environmental performance. ISO 14001 was first established in 1996 and is implemented worldwide (ISO 2011). The ISO 14001 framework provides the requirements for an EMS for any type of industry's strategic approach to environmental policy, plans, and actions. It is a management tool that enables an organization to identify and control the environmental impact of its activities and products/services, to improve its continual performance, and to implement a systematic approach to setting better objectives and attainable goals. The main requirements of an EMS include a commitment to pollution prevention, continuous improvement, and compliance with any environmental or other applicable standards and regulations. The benefits of implementing an EMS include an overall reduction in pollution and energy costs, which, in turn, increases appeal to the consumers (ISO 2011).

The ISO 14001 standards do not set specific levels of performance or improvements. Instead, ISO 14001 provides an audit process that demonstrates that the organization is compliant as long as the systemic, policy, planning, operational, checking, and review requirements are met. It is the framework that organizations develop as a result of this checklist that allows them to devise their own EMS. This flexibility is what allows ISO 14001 to be implemented in any scenario.

The planning requirement is to identify and set objectives for the organization's environmental impact from an operational standpoint. The legal and operational requirements are designed to ensure ongoing training and environmental awareness throughout the organization. Finally, the managers see the results of the environmental management reviews and are required to address areas that need further improvement (ISO 2011).

ISO 14001 is a tool that can be used to meet internal and external objectives. The internal objectives provide assurance to management and external stakeholders that the organization complies with environmental regulations and has a working internal communication system for its own environmental policies, plans, and actions. The EMS also lets employees know that they are working for an environmentally responsible organization (ISO 2011).

7.2.2 U.S. Environmental Protection Agency

The U.S. Environmental Protection Agency (USEPA) has created two programs that evaluate key aspects of high-performance building: (1) ENERGY STAR and (2) WaterSense. In 1992, the USEPA established the ENERGY STAR program as a voluntary labeling system to help promote energy-efficient products to reduce greenhouse gas emissions. The labeling (see Figure 7.2) started off with computers and monitors, further expanded to office equipment and residential heating and cooling equipment by 1995, and since then has grown to include most household appliances, lighting, and many other electronics devices in residential, commercial, and industrial buildings. ENERGY STAR has partnerships with over 20,000 private and public sector organizations and provides them, along with consumers, the technical information and tools necessary to obtain the most energy-efficient products and provide the best management practices.

ENERGY STAR not only designates products and appliances that are more energy efficient, but it also has established an energy efficiency rating system for commercial buildings (ENERGY STAR 2011). The building rating system is based on a scale from 1 to 100; the energy efficiency of a building is compared to that of a similar facility (USGBC 2010).

Figure 7.2 ENERGY STAR label (USEPA 2012a)

Buildings that earn a rating of 75 or greater qualify for the ENERGY STAR label. Data is entered using the ENERGY STAR Portfolio Manager® and is self-reported. According to ENERGY STAR (2011), Americans saved $20 billion in energy during 2009 using ENERGY STAR products while preventing the greenhouse gas emissions of an equivalent 38 million vehicles.

Whereas ENERGY STAR focuses on energy efficiency, the USEPA WaterSense partnership program, created in 2006, takes aim at water use efficiency. WaterSense promotes the value of water by using a label for products that provides consumers with useful information and by encouraging innovation in manufacturing. WaterSense endorses products that are more water efficient, such as waterless urinals and low-flow faucets and toilets, in an attempt to decrease the amount of water used in everyday activities (USEPA 2012b).

The WaterSense program consists of labels and logos, each of which has its own use. The WaterSense label (Figure 7.3) identifies a water-efficient product that has been independently tested and certified to meet the USEPA WaterSense criteria for efficiency and performance. The WaterSense promotional label (Figure 7.4) advertises the availability of WaterSense-labeled products and encourages consumers to look for them. The WaterSense Partner logo (Figure 7.5) indicates an organization's commitment to the promotion of water conservation and efficiency signified by a signed partnership between the organization and the USEPA.

To earn the WaterSense label, a manufacturer must first sign an agreement with the USEPA based on a draft specification for the product it plans to produce. As part of this agreement, the manufacturer will then have a one-year period in which to obtain certification that meets or exceeds the WaterSense specification from a third-party certifying agency. According to the USEPA (2012b), if WaterSense-labeled faucets or faucet accessories were installed in the bathrooms of just 1 in every 10 homes, 6 billion gallons of water per year could be saved, along with over $50 million in energy costs associated with the treatment, heating, and supply of that water. As of 2012, WaterSense also has helped consumers save 125 billion gallons of water and over $2 billion in water and energy bills, as well as reduce electricity use by 16.7 billion kilowatt-hours and CO_2 emissions by 6 million metric tons (USEPA 2012b).

Figure 7.3 WaterSense label (USEPA 2012b)

Figure 7.4 WaterSense promotional label (USEPA 2012b)

Figure 7.5 WaterSense Partner logo (USEPA 2012b)

7.2.3 U.S. Green Building Council®

The USGBC is a nonprofit organization whose mission is to transform the way buildings and communities are designed, built, and operated to enable an environmentally and socially responsible, healthy, and prosperous environment that improves the quality of life (USGBC 2011b). The USGBC is made up of volunteer members, whose main focus is to influence the design of new buildings and communities through energy efficiency, water efficiency, sustainable materials, indoor air quality, and innovations in design. Verification of high-performance building is accomplished through the use of a checklist developed by USGBC members, known as the LEED certification checklist. Buildings and communities that are designed to be LEED certified not only are more efficient than conventional buildings, but the USGBC suggests that their carbon footprint will be reduced (USGBC 2011b) along with their energy and water consumption (Kubba 2010). LEED-certified buildings are designed to lower operating costs, increase asset value, reduce waste sent to landfills, conserve energy and water, maximize occupant health and comfort, reduce harmful greenhouse gas emissions, and qualify for tax rebates, zoning allowances, and other incentives in hundreds of cities across the United States (USGBC 2011b).

LEED certification provides independent, third-party verification through the Green Building Certification Institute (GBCI®), which was established in 2008 as a separately incorporated entity with the support of the USGBC (USGBC 2010) specifically to confirm that a given building was designed and built using strategies aimed at achieving high performance in key areas of human and environmental health, such as sustainable site development, water savings, energy efficiency, materials selection, and indoor environmental quality (USGBC 2011b). Depending on whether the project is new construction, an existing building, or other category, LEED points are awarded to achieve one of four levels of certification: LEED Certified, Silver, Gold, or Platinum, in order of increasing number of points. To earn LEED certification, projects must be registered with the GBCI, and a minimum number of points must be attained. Projects also must meet all of the prerequisites in the checklist (USGBC 2010). Additional points can be earned for Innovation and Design, which addresses sustainable building expertise as well as design measures, and Regional Priority, which acknowledges the importance of local conditions in determining environmental design and construction practices (USGBC 2010).

Once LEED is determined to be an appropriate guideline for the project and serves as a declaration of intent for certification, the project should be registered with GBCI. This registration allows team members to complete and submit credit templates, track credit interpretation rulings, track key project details, and communicate with customer service and reviewers of LEED credits (Kubba 2010). Along with registering the project, certification fees must be paid. The fees vary depending on whether the project is new construction, existing building renovation, or if the applicant is a USGBC member. Expedited reviews, credit interpretation rulings, and the square footage of the project also help determine the final costs for certification (GBCI 2011).

Once the project is registered and the fees paid, the design team then selects the LEED points that they are aiming to achieve and prepares the documentation in the form of credit templates for submission. These documents should be thoroughly reviewed by the team to ensure that each credit deserved is awarded (GBCI 2011). After the appropriate supporting documentation and credit templates are provided, the points can be awarded in the checklist to estimate the total points. The application is then submitted, and the formal application review is started. The GBCI then makes a decision on certification and the level achieved (i.e., Certified, Silver, Gold, or Platinum).

7.3 LEED® CERTIFICATION

A two-phase application process is used to attain LEED certification. The preliminary submission is for comment and review, and the final submission is to receive credits and certification. In order for a submission to be processed, the following are included with the application: the LEED rating system that is being used; project contact information; type, size, and number of occupants and projected date of completion; a project narrative that describes and highlights the project; a LEED project checklist with anticipated score; the LEED credit templates and supplemental documentation; copies of credit interpretation rulings that were used (all of which are available using LEED online); drawings, photos, and diagrams of the project; and payment of certification fees (Kubba 2010). The USGBC will then issue a final LEED review report that declares the status of all attempted credits and the level of certification. The project team then has 25 days to submit an appeal, with a fee charged per credit in dispute. This is the process used by Dr. Bloetscher to certify Dania Beach's nanofiltration plant as LEED Gold in 2011, with no credit disputes.

7.3.1 Requirements

Minimum program requirements (MPRs) are defined in order to give clear guidance to customers, to protect the integrity of the program, and to reduce the challenges during the certification process. One MPR for new construction states that the project building or space must comply with all federal, state, and local environmental laws. This condition must be met from the earliest date of either the LEED project certification or the schematic design and continue until the date that the building is in use. Another MPR states that the project must be a complete, permanent building or space. The project must be located on existing land and be a building or space that is not designed to move locations. The project must be located on a reasonable boundary and must include all the land that is associated with and supports normal operations. In no way can the project site boundary be altered to include or exclude certain parcels of land in order to more easily comply with the LEED rating system, and all the lands included in the project must be owned by the party sponsoring the project. The LEED project must contain a minimum of 1,000 square feet of gross floor area. A minimum occupancy rate applies, which means that the building must serve one or more full-time-equivalent occupants. A commitment to sharing energy and water usage data with the USGBC and/or GBCI for at least 5 years must be agreed to, even if the building or space changes ownership or lessee. The gross floor area of the subject building must not be less than 2% of the gross land area within the LEED project boundary (USGBC 2010).

While the LEED checklist changes periodically, the basic checklist is broken down into major categories, including sustainable sites, water efficiency, energy and atmosphere, materials and resources, and indoor environmental quality. Each has an established number of potential points that when totaled provide the certification rating. In the following sections, some of the finer details of each category are explored.

7.3.1.1 Sustainable Sites

The prerequisite for this category is to reduce possible construction pollution. It requires that USEPA construction general permit or local development codes (whichever is more stringent) be met. Compliance with the Clean Water Act will be met through the National Pollutant Discharge Elimination System program. Water runoff and control of dust and particulates are of major concern. This prerequisite can be met by using silt fences and covering existing stormwater structures with silt material. Also, seeding, mulching, and sediment traps can be implemented where needed. Points are assigned in this category for alternative modes of transportation, brownfield redevelopment, stormwater retention, community connectivity, heat island effect control, and light pollution reduction (Table 7.1).

7.3.1.2 Water Efficiency

This section addresses water use efficiency. In order to qualify for any of the possible points, the prerequisite is to reduce water consumption by 20% compared to the water use baseline found by calculating typical water usage of conventional water fixtures. This can be achieved through the use of WaterSense-certified fixtures, high-efficiency fixtures, or gray water, reclaimed water, or stormwater in nonpotable applications. Most of the points can be earned by exceeding the prerequisite and reducing water consumption by 40% or better.

7.3.1.3 Energy and Atmosphere

The intent of this section is to address energy efficiency and the use of renewable energy sources. The first prerequisite requires the designation of a commissioning authority (CxA). This individual must have experience as a CxA and cannot be a part of the design team. The CxA will oversee the heating, ventilating, air conditioning, and refrigeration (HVAC&R)

Table 7.1 Example of sustainable sites portion of LEED version 2009 checklist for new construction and major renovations

17 points			Sustainable sites		Possible points: 26
Y	?	N			
Y			Prereq 1	Construction activity pollution prevention	
1			Credit 1	Site selection	1
5			Credit 2	Development density and community connectivity	5
		N	Credit 3	Brownfield redevelopment	1
6			Credit 4.1	Alternative transportation—public transportation access	6
1			Credit 4.2	Alternative transportation—bicycle storage and changing rooms	1
3			Credit 4.3	Alternative transportation—low-emitting and fuel-efficient vehicles	3
	?		Credit 4.4	Alternative transportation—parking capacity	2
1			Credit 5.1	Site development—protect or restore habitat	1
	?		Credit 5.2	Site development—maximize open space	1
	?		Credit 6.1	Stormwater design—quantity control	1
	?		Credit 6.2	Stormwater design—quality control	1
	?		Credit 7.1	Heat island effect—nonroof	1
	?		Credit 7.2	Heat island effect—roof	1
	?		Credit 8	Light pollution reduction	1

systems; lighting controls; hot water systems; and renewable energy systems. Also, the owner must submit documentation of the project requirements for the CxA to review. The CxA is responsible for including commissioning requirements in construction documents, designing a commissioning plan, verification of systems, and completing a summary report (USGBC 2010). Another prerequisite defines the minimum level of energy efficiency for the project. There are three options to satisfy this prerequisite: (1) demonstrate a 10% decrease in energy use for the building compared to a baseline, (2) abide by the measures defined by the appropriate American Society of Heating, Refrigerating, and Air Conditioning Engineers (ASHRAE) *Advanced Energy Design Guides* for the project, or (3) comply with the standards in the Advanced Buildings Core Performance Guide. One more prerequisite emphasizes the reduction of stratospheric ozone depletion and requires that no chlorofluorocarbon refrigerants are used in the HVAC&R systems (USGBC 2010).

Points in this section are available for increasing the minimum energy cost savings from 12 to 48% (nearly a third of the points needed to get to Gold certification). Points can be earned by implementation projects such as installing motion detectors on lighting, using ENERGY STAR appliances, and other relatively simple energy reduction strategies, but other points in this section can only be earned by careful planning, such as installing a holistic energy management system or positioning the building in such a way as to utilize daylight for natural lighting. Points also are assigned for installing an on-site renewable energy system. These points are awarded on a graduated scale based on the amount of renewable energy created per the building's annual energy cost. This category also has points for enhanced commissioning, enhanced refrigerant management, measurement and verification, and green power.

7.3.1.4 Materials and Resources

The intent of this section is to deal with materials used, reused, and recycled. The mandatory prerequisite is the storage and collection of recyclables. This requirement states that there must be an easily accessible dedicated area for recycling. The recyclable materials must include at least paper, corrugated cardboard, glass, plastics, and metals to reduce the amount of waste sent to landfills by diverting it to a recycling facility. This easily can be achieved by simply designating an area to collect and store the materials, as well as instructing occupants on the recycling procedures. After assuring that the prerequisite is met, the credits available in this section generally are determined by percentages of the materials used or recycled. When pursuing points in the category, pay particular attention to how the percentages are calculated when filling out the credit templates. The remainder of the credits in this section deal with reusing portions of an existing building, minimizing construction waste, and using regional materials or materials with recycled content throughout the building process. Other points are assigned for using rapidly renewable materials like bamboo and certified wood obtained from responsibly managed forests as defined by the Forest Stewardship Council. Note that Forest Stewardship Council certification is one area of controversy, as several states have prohibited the use of LEED as a green building certification program for state-funded projects due to lobbying by the lumber industry.

7.3.1.5 Indoor Environmental Quality

The goal of this section is to increase the comfort and well-being of the occupants by controlling and monitoring indoor air quality, thermal comfort, and effective lighting. This section contains prerequisites that address minimum indoor air quality performance and environmental tobacco smoke controls. To achieve minimum indoor air quality for mechanically ventilated spaces, the ventilation system must meet the ventilation rate procedure or local code, whichever is more stringent. To ensure that this prerequisite is met, the ventilation systems should be designed to meet or exceed the ASHRAE standards while balancing cost and energy use at the same time. In terms of the control of environmental tobacco smoke, many state legislatures already have banned smoking indoors, so this requirement may not be a burden. This goal can be achieved by either completely prohibiting smoking on the premises or banning smoking at least within 25 feet of entries, outdoor air intakes, and operable windows. About half of the credits in this section deal with the indoor air quality, through the use of either low-emitting materials (paint, sealants, carpet, adhesives, etc.) or materials with no or low volatile organic compounds, proper ventilation protection, and proper planning during construction. The remaining credits are concentrated on the comfort of the occupants. These include individual control of lighting and temperature systems, as well as the availability of natural light and views of the outside (Kubba 2010).

7.3.1.6 Innovation and Design Process and Regional Priority Credits

The final two sections of the checklist consist of Innovation and Design Process and Regional Priority Credits, which combine to offer "bonus" credits that can be earned in two ways: by achieving significant performance using a strategy that is not addressed in the rating system or by achieving exemplary performance in any of the prerequisites for credits in the checklist. These can be earned by doubling another credit requirement, for instance. Additionally, one credit will be awarded if at least one principal member of the project team is a LEED Accredited Professional (Kubba 2010). The last opportunity for points is associated with the Regional Priority Credits depending on the region in which construction will take place (Kubba 2010). Specifics are based on the zip code of the project site as entered into the GBCI website.

7.4 Triple Bottom Line

A concept for all design professionals to understand is the "triple bottom line," which promotes environmental, economic, and social improvements in the way business is done. It merges these concepts into one theme, with the idea of optimizing all three, while realizing that they compete, to a degree, for the same resources. The thought is that dollars can be saved and productivity improved while enhancing working environments (as opposed to just ecosystems).

Research continues to indicate that while outdoor air quality has improved substantially in the past 30 years, the quality of the indoor environment of many of the buildings in which we live and work is poor; as a result, it affects the health and well-being of the occupants. Indoor levels of pollutants may be 2 to 5 times higher, and occasionally more than 100 times higher, than outdoor levels (USEPA 2009). "Sick buildings" are an example. Many of these buildings were not designed to foster social interaction and community connectivity, but such features are necessary to provide a healthy life for building users and their surrounding communities. The environmental imperative takes into account the way humans and industries are negatively affecting the planet.

The second factor that is taken into account is the economic imperative. The green design and building movement has gained significant popularity among owners, builders, and government agencies. An important driver for this popularity is making technology for high-performance buildings more affordable. Evidence is mounting that sustainable construction practices are here to stay and that they will be considered "best practices" for builders within the foreseeable future. A report by the New Buildings Institute indicated that LEED Certified and LEED Silver buildings achieved energy use savings averaging 30% better than conventional buildings (Montoya 2011). LEED Gold and LEED Platinum buildings averaged about 50% savings in energy consumption (Montoya 2011).

Many of the projects that are built consider only the economic and environmental impacts on the building users and owners, but these projects also affect the overall community and society in which they are constructed. Social responsibility considers and promotes interaction and cultural enrichment for building users and nearby communities. All projects should consider the basic needs of people, such as happiness, health, safety, freedom, sustainability, and other positive attributes. This outreach can also encourage individuals, families, and communities to lead more environmentally sustainable lifestyles (Montoya 2011). Balancing and adopting these three imperatives can have a positive impact on the quality of life.

7.5 References

ENERGY STAR (2011). "Understanding EPA's Energy Star energy performance scale," <http://www.energystar.gov/ia/business/evaluate_performance/The_ENERGY_STAR_Energy_Performance_Scale_3.28.11_abbrev.pdf?caff-0db5> (accessed April 10, 2012).

Energy Tax Incentives (2011). "The Tax Incentives Assistance Project (TIAP)," <http://energytaxincentives.org/business/commercial_buildings.php> (accessed April 20, 2010).

GBCI (2011). "Current certification fees," Green Building Certification Institute, <http://www.gbci.org/main-nav/building-certification/resources/fees/current.aspx> (accessed Jan. 31, 2012).

ISO (2011). "ISO 14000 environmental management," International Organization for Standardization, <http://www.iso.org/iso/iso_catalogue/management_and_leadership_standards/environmental_management.htm> (accessed Jan. 22, 2012).

Kats, G., Alevantis, L., Berman, A., Mills, E., and Perlman, J. (2003). The Costs and Financial Benefits of Green Buildings: A Report to California's Sustainable Building Task Force.

Kubba, S. (2010). *Green Construction Project Management and Cost Oversight*, Elsevier Science.

Lisk, B., Greenberg, E., and Bloetscher, F. (2012). *Implementing Renewable Energy at Water Utilities*, Web Report #4424, Water Research Foundation, Denver, CO.

Mathiessen, L.F., and Morris, P. (2007). *Cost of Green Revisited: Reexamining the Feasibility and Cost Impact of Sustainable Design in the Light of Increased Market Adoption*, Davis Langdon Publishing, <http://www3.cec.org/islandora-gb/en/islandora/object/islandora%3A948/datastream/OBJ-EN/view>.

McGraw Hill Construction (2008). Green Outlook 2009: Trends Driving Change, SmartMarket Report.

Miller, N., Spivey, J., and Florance, A. (2008). "Does green pay off?" *Journal of Real Estate Portfolio Management*, 14(4), 385–400.

Montoya, M. (2011). *Green Building Fundamentals*, Pearson Education, Upper Saddle River, NJ.

Northern Arizona University (2012). "Applied research and development building fact sheet," <http://www.mggen.nau.edu/ARD.pdf> (accessed Jan. 21, 2012).

Sustainability Victoria and the Kador Group (2011). "Employee productivity in a sustainable building, a study commissioned by Sustainability Victoria and the Kador Group," <https://www.propertyoz.com.au/library/Employee%20Productivity%20in%20a%20Sustainable%20Building.pdf> (accessed April 15, 2015).

USEPA (2009). "Estimating 2003 Building-Related Construction and Demolition Materials Amounts."

USEPA (2010). "Green building," U.S. Environmental Protection Agency, <http://www.epa.gov/greenbuilding/pubs/faqs.htm#3> (accessed Jan. 22, 2012).

USEPA (2012a). "History of ENERGY STAR," U.S. Environmental Protection Agency, <http://www.energystar.gov/index.cfm?c=about.ab_history> (accessed Feb. 1, 2012).

USEPA (2012b). "What is WaterSense?" U.S. Environmental Protection Agency, <http://www.epa.gov/WaterSense/about_us/what_is_ws.html> (accessed Feb. 1, 2012).

USGBC (2010). "LEED reference guide for green design & construction," 2009 Edition, <http://www.usgbc.org/Docs/Archive/General/Docs5546.pdf>.

USGBC (2011a). "Buildings and climate change," U.S. Green Building Council, <http://www.documents.dgs.ca.gov/dgs/pio/facts/LA%20workshop/climate.pdf> (accessed Jan. 22, 2012).

USGBC (2011b). "What LEED® is," U.S. Green Building Council, <http://www.usgbc.org/DisplayPage.aspx?CMSPageID=1988> (accessed Jan. 22, 2012).

USGBC and Booz Allen Hamilton (2009). "Green jobs study," 6.

World Business Council (2008). "Energy efficiency in buildings," World Business Council for Sustainable Development.

7.6 Assignments

1. Identify a LEED building in your community. Obtain the appropriate LEED template for this building. Go through the building and identify 15 obvious features that received points.
2. Compare and contrast three green building certification systems. Use LEED as one example. What are the similarities and differences? Why are there differences?
3. Go to the usgbc.org website. Identify five buildings with different certifications nearest to your location. Create a matrix that compares them.
4. Find a LEED-certified building in your community. Identify the contractor and architect. Meet with one or both and discuss the certification process. What were the easy parts? What were the difficult ones? What advice can they give you?
5. Obtain five LEED credit templates under sustainable sites that could be met with your project. Fill them out and submit as a part of your design project.

Environmental Site Assessment

What environmental issues should be of concern when planning to purchase a piece of property for development? Past activities, previous site assessments, impacts of development, long-term sustainability, fiduciary obligations to the public or shareholders, and landowner liability are just some of the issues involved. For instance, past activities (e.g., leaking underground storage tanks with petroleum products) may have contaminated the site and hold long-term liability, which the developer does not wish to inherit. Therefore, the developer should hire a professional, typically an engineer, to compile a history of the property, including past activities that can affect any future development on the site and impact any financial obligations of the new owner. The tool used by the environmental professional is called an environmental site assessment.

8.1 Scope

The procedure for performing an environmental site assessment is outlined by the American Society for Testing and Materials (ASTM) in ASTM E1527. The latest version of this standard should be used as a reference for this type of work. The purpose of conducting a formal environmental site assessment is listed in the standard as follows:

> [T]his practice is intended to permit a user to satisfy one of the requirements to qualify for the innocent landowner, contiguous property owner, or bona fide prospective purchaser limitations on CERCLA [Comprehensive Environmental Response, Compensation, and Liability Act] liability... (42 U.S.C. §9601).

An environmental site assessment is a subjective evaluation conducted on a site where there is reason to believe that an environmental problem may exist. The present condition of the site; signs of possible or past spills, discharges, use and storage of petroleum products or hazardous materials; and other signs are included in this analysis. This information is obtained through a visual site inspection, records review of the history of the property through available documents, and interviews of key site personnel, users, and neighbors. No sampling takes place in a Phase I Environmental Site Assessment. Pieces of information that might prove valuable include historical uses of the property, potential liability or liens related to the property, site history, geology, hydrogeology, zoning, drainage, water and sewer availability, solid waste service, transportation impacts, etc.

The goal of this effort is to identify recognized environmental conditions (RECs), which are defined as the presence or likely presence of any hazardous substances or petroleum products on the property or ground or in groundwater or surface water. This includes existing releases, past releases, and the material threat of a future release. The term "material threat" refers to a physically observable or obvious threat reasonably likely to lead to a release that might impact public health or the environment. An example would be an aboveground diesel storage tank with corroded rivets (see Figure 8.1), such that it may cause structural integrity failure leading to the release of the contents of the tank.

During the course of the investigation, all known or suspected RECs associated with the property are identified. Engineers provide their professional opinion as to the potential impact on the property of these RECs and the rationale for concluding that a condition is or is not currently a REC. Furthermore, engineers must provide recommendation as to any additional investigation required to detect the presence of hazardous substances or petroleum products in regard to identified RECs. If during the process any significant data gaps or failures that affect the ability to identify RECs are encountered, this must be documented.

An environmental condition that in the past would have been considered a REC, but may or may not be one currently, is known as a *historical REC*. A past release that has been addressed

Figure 8.1 Aboveground diesel storage tanks with rusty bolts, indicating potential for leakage

to the satisfaction of the governing regulatory agency and meets the appropriate standards for unrestricted residential use, without the need for any land use controls, is considered a historical REC but not a current one. For example, suppose the release of a hazardous substance (trichloroethylene) from a dry-cleaning establishment in 1991 was fully remediated in 1992 and accepted by the state regulatory agency, which issued a no further action letter on August 1, 1992. This is considered a historical REC and must be included in the findings section of the report. It is up to an environmental professional to give an opinion on the current impact on the property in the opinion section of the report. If deemed that this is a current REC, then it should appear in the list found in the conclusions section of the report (ASTM 2013).

A *controlled REC* refers to a past release that has been addressed to the satisfaction of the governing regulatory agency, but residual contaminants have been allowed to remain, subject to required engineering or institutional controls. In this case, it should be listed in the conclusions section of the report. The term *de minimis* is used when the conditions generally do not represent a threat to human health or the environment and would not be the subject of enforceable action if brought to the attention of the authorities under the law. *De minimis* conditions are not considered RECs.

Examples of RECs include the following:

◆ **Lead in water pipes**—Structures built prior to 1986 must be tested for lead.
◆ **Polychlorinated biphenyls (PCBs)**—A Phase I Environmental Site Assessment must include documentation regarding potential PCB-containing equipment both on-site and off-site.
◆ **Leaking underground storage tanks (LUSTs)**—These storage tanks typically contain petroleum products or hazardous substances and potentially can threaten groundwater in the case of a release.
◆ **Impaired soil or groundwater conditions**—This can be due to historical land use from waste storage/disposal, manufacturing/industrial operations, transportation/railroads, petroleum processing, mining, or agriculture.

Pollutants of concern in a Phase I Environmental Site Assessment include petroleum products and hazardous substances. Petroleum products are defined as those substances included within the meaning of the petroleum exclusion to CERCLA [42 U.S.C. §9601(14)]. It also means any fraction of crude oil that is not specifically listed or designated as a hazardous waste under subparagraphs (A) through (F) of 42 U.S.C. §9601(14).

Examples of petroleum products include the following:

◆ Crude oil
◆ Natural gas
◆ Liquefied natural gas
◆ Synthetic gas usable for fuel
◆ Gasoline
◆ Kerosene
◆ Diesel oil
◆ Jet fuel
◆ Fuel oil
◆ Lubricating oils
◆ Tar
◆ Paraffin wax

A hazardous substance is defined as:

◆ Any substance designated pursuant to 33 U.S.C. §1321(b)(2)(A) or 42 U.S.C. §9602
◆ Any hazardous waste having the characteristics listed pursuant to section 3001 of the Resource Conservation and Recovery Act (42 U.S.C. §6921)
◆ Any toxic pollutant listed under 33 U.S.C. §1317(a)
◆ Any hazardous air pollutant listed under section 112 of the Clean Air Act (42 U.S.C. §7412)
◆ Any imminently hazardous chemical substance or mixture to which the U.S. Environmental Protection Agency administrator has taken action pursuant to 15 U.S.C. §2606

The term hazardous waste does not include any substance specifically designated as a petroleum product. Examples of hazardous substances include the following:

◆ Metals (e.g., mercury, lead, arsenic, chromium, etc.)
◆ Radionuclides (e.g., radon, radium, cobalt-60, strontium-90, technetium-99, etc.)
◆ Volatile organic compounds (e.g., formaldehyde, acetone, ethanol, 2-propanol, etc.)
◆ Solvents (e.g., chloroform, methylene chloride, and many others)
◆ Semi-volatile organic compounds (e.g., phthalates, brominated flame retardants, bisphenol A, triclosan, pentachlorophenol, etc.)
◆ Polychlorinated biphenyls (e.g., Aroclor® and individual chlorinated biphenyl components known as congeners)
◆ Pesticides/herbicides (e.g., DDT, dieldrin, heptachlor, etc.)
◆ Explosives (e.g., trinitrotoluene, C-4, triacetone-triperoxide, etc.)
◆ Dioxin/furans

It is also required to consider the potential for any vapor intrusion from any of these volatile hazardous materials or petroleum products to impact indoor air quality.

8.2 The Environmental Professional

A Phase I Environmental Site Assessment is conducted by an environmental professional (EP) or under the supervision of an EP. At a minimum, an EP must be involved in planning the site reconnaissance and interviews. In addition, the final review and interpretation of the information presented in the report both are performed by an EP. The report must document the qualifications of the EP and of any person(s) who conducts site reconnaissance and interviews. This is accomplished by providing detailed resumes of those persons in the report. According to the ASTM E1527 guidelines, an EP is defined as a person with the following qualifications:

◆ Professional Engineer (P.E.) or Professional Geologist (P.G.) license *and* 3 years of relevant experience, or
◆ Baccalaureate (or higher) degree in engineering or science *and* 5 years of relevant experience, or

◆ A license or certification by a government agency *and* 3 years of relevant experience, or
◆ Ten years of relevant experience

Relevant experience includes previous participation in environmental site assessments, other site inspections or investigations, and remediation engineering activities that involve understanding of surface and subsurface environmental conditions for which professional judgment was used to develop opinions regarding releases or threatened releases of hazardous substances.

8.3 Site Reconnaissance

When performing a Phase I Environmental Site Assessment, one of the most important engineering due diligence items is to visually inspect as much of the site as possible. At least one site visit is required by ASTM 1527, and coordinating interviews with users and occupants at the same time as the site visit is suggested. The goal is to identify areas that warrant further investigation by visually and physically observing the current uses and condition of the property, including the exterior and interior areas of any structures on the site.

8.3.1 Exterior Reconnaissance

A visual survey may be much more enlightening than records. Therefore, it is necessary to walk the site, look at what is on the ground, and describe any other features of note that may warrant additional investigation. When observing the exterior of the building and the grounds, look for the following clues:

◆ Signs of distressed vegetation (Figure 8.2), but be aware that during winter or severe weather conditions, it can be difficult to tell if the vegetation is distressed due to some sort of contamination or if it is caused by natural weather conditions or seasonal changes (Figure 8.3).
◆ Evidence of hazardous substances/petroleum products and storage tanks or drums (Figures 8.4 and 8.5); make sure to record the contents, capacity, and age, if reasonably ascertainable.
◆ Sources of strong, pungent, or noxious odors, as well as pools of liquid (Figure 8.6), PCBs, pits, ponds (Figure 8.7), lagoons, stained soil or pavement (Figure 8.8), solid waste (Figure 8.9), wastewater, piping (Figure 8.10), wells, septic systems, and noise.

The investigator must try to identify specific past uses of the property that could have involved the use, storage, transport, disposal, or generation of hazardous substances or petroleum products and, to the extent observable, the adjoining properties as well. The property should be viewed from all adjacent thoroughfares. If roads or paths with no apparent outlet are found, then an attempt should be made to try to determine if they could have been used to transport or dispose of hazardous substances. The geologic, hydrologic, and topographic features of the site must be determined to better understand how hazardous substances could migrate to the property or within the property boundaries, into the groundwater or soil. All roads on the property, the source of potable water, and the sewage disposal system and its

Figure 8.2 Older house with distressed vegetation in the front yard

Figure 8.3 During winter, it becomes much harder to discern why vegetation is distressed

Figure 8.4 Chemical storage buckets that show evidence of leaking without secondary containment; some are unmarked

Figure 8.5 Aboveground fuel storage tanks with proper secondary containment

Figure 8.6 Unknown liquids draining to storm drain

Figure 8.7 Pipe discharging liquid to a pond should be investigated further

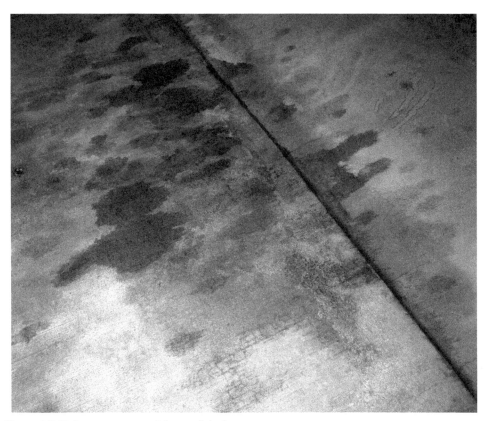

Figure 8.8 Stains on pavement from oil leaks

Figure 8.9 Dumpster on bare ground; liquids could leak out and contaminate the surrounding soil

Figure 8.10 Exposed piping

approximate age should be documented, as this information will be used to develop the property during the design phase after the Phase I Environmental Site Assessment is completed and the property is purchased.

8.3.2 Interior Reconnaissance

If there are any structures on the site, the interior of these buildings should be observed, if possible. All accessible common areas, such as lobbies, hallways, utility rooms, maintenance areas, boiler rooms, and representative occupant spaces, should be checked. It is not necessary to look under floors, above ceilings, or behind walls. Start with a general description of the structures, focusing on the number of buildings, number of stories, and age of the buildings. Then describe the fuel systems for generators or heating, ventilation, and air conditioning (HVAC) systems, such as heating oil, electric, natural gas, etc. Signs of stains or corrosion on floors, walls, and ceilings that were caused by moisture, mildew, and/or mold should be noted. The investigator must locate all floor drains and sumps and describe any odors or physical damage or evidence of hazardous substance disposal.

Some critical questions to answer during the interior site reconnaissance should include, but are not limited, to the following:

◆ Has any environmental survey been done on this building for the presence of asbestos or lead?
◆ What year was the building built?
◆ What is the likelihood of hazardous materials being in the building based on the age of the building?
◆ What kind of interstitial spaces are there in the building, and what materials are in them?
◆ Are plans and specifications available from construction?
◆ What materials were specified for construction?

Examples of corrosion, fuel systems, and material removed from inside a building are shown in Figures 8.11 to 8.14.

Figure 8.11 Corrosion caused by water in pump station dry well

Figure 8.12 Miscellaneous materials stored outside

Figure 8.13 Diesel engine inside building

Figure 8.14 Corrosion damage from steel on concrete

8.4 Records Review

The goal of the records review process is to obtain facility information (e.g., building plans, deeds, permits, prior assessments, compliance records, fire insurance maps, aerial photographs, title records, etc.) to identify prior uses that may have contributed to historical RECs on the property. Photographs must be dated and labeled with a description of the view presented (Figure 8.15). Information regarding contaminant migration pathways (i.e., topographic information, soil and subsurface data, groundwater information) also can be helpful (Figures 8.16 and 8.17).

Sources of historical information include recorded land title records, government records, Internet sites, community organizations, local libraries/historical societies, and any reasonably ascertainable source likely to provide useful information to identify previous uses of the property. *Reasonably ascertainable information* is characterized as being publicly available, obtainable within reasonable cost and time, and practically reviewable. The term *publicly available* means that access is available to anyone upon request. For example, government documents related to environmental regulations would be publicly available. However, corporate records or proprietary secrets would not be. A *reasonable time frame* is typically taken as 20 calendar days from the time of the request. A *reasonable cost* is defined as the typical cost of retrieval and duplication. *Practically reviewable* refers to the form in which the information is available, which should not require any extraordinary analysis to review. For example, if the data is available only by zip code instead of address, then it would take an extraordinary effort to cross-check all of the entries, so this would not be considered practically reviewable (Hejzlar 2007).

Figure 8.15 Aerial photograph properly labeled with date

Figure 8.16 Soil classification map and legend (source: Natural Resources Conservation Service)

Map Unit Legend

Palm Beach County Area, Florida (FL611)			
Map Unit Symbol	Map Unit Name	Acres in AOI	Percent of AOI
9	Beaches	3.1	19.4%
27	Palm Beach-Urban land complex, 0 to 8 percent slopes	10.0	63.4%
100	Waters of the Atlantic Ocean	2.7	17.2%
Totals for Area of Interest		15.8	100.0%

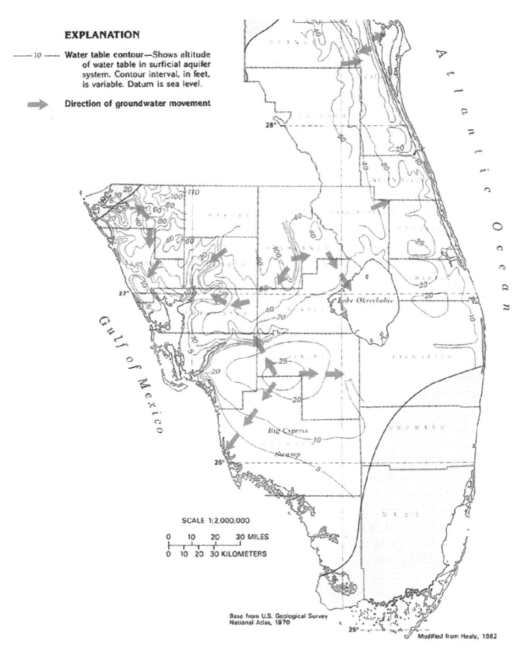

Figure 8.17 Groundwater table contour map (source: South Florida Water Management District as modified from Healy 1982)

If the necessary records are not reasonably ascertainable, then every effort must be made to obtain the information through other methods, such as interviews of owners, occupants, or users. At a minimum, readily available records must be reviewed to help identify likely contaminants and locations where contamination occurred in the past. Start with federal records databases established by the U.S. Environmental Protection Agency (USEPA), such as the National Priorities List; Comprehensive Environmental Response, Compensation, and Liability Information System; aboveground and underground storage tank sites including leaking tanks; Emergency Response Notification System; Resource Conservation and Recovery Act permitted treatment, storage, and disposal facilities; and registered generators of hazardous waste (see Table 8.1). In addition to federal records, state, county, and local environmental regulatory agencies also publish similar databases that must be consulted. Some important databases include netronline.com, Envirofacts, USEPA violations, state environmental divisions, county environmental records, FirstSearch, and Enforcement and Compliance History Online (ECHO online; see Figure 8.18). Be aware that critical information may not be available on websites and may require visiting the local agency office to review hard copies.

The approximate minimum search distances refer to the area for which records must be obtained and reviewed and should be measured from the nearest property boundary. This is not a radius because the property boundary likely is irregularly shaped (see the example in Figure 8.19). The minimum search distances are designed to help assess the likelihood of migration of petroleum products and hazardous substances from areas outside of the property boundary (see Tables 8.2 and 8.3).

Additional environmental record sources that might be useful include the following:

◆ Local brownfield lists
◆ Local lists of landfill/solid waste disposal sites
◆ Local lists of hazardous-waste-contaminated sites
◆ Local lists of registered storage tanks
◆ Local lists of leaking storage tanks (see Table 8.2)
◆ Local lists of dry-cleaning sites (see Table 8.3)
◆ Local land records (for activity and use limitations)
◆ Records of emergency release reports (42 U.S.C. §11004)
◆ Records of contaminated public wells

Some of the best sources to obtain this information include the local health department or environmental regulatory agency, local planning department, local building permit/inspection department, local/regional pollution control agency, local electric utility companies (for records relating to PCBs), U.S. Geological Survey (USGS) office, and county property appraisers offices.

The next set of documents required pertain to sources that provide information about the geologic, hydrogeologic, hydrologic, or topographic characteristics of a property. This is termed the physical setting. The mandatory standard physical setting source is the USGS 7.5-

Table 8.1 Standard Environmental Record Sources for Phase I Environmental Site Assessments (ASTM E1527-13)

Standard environmental record source (where available)	Appropriate minimum search distance (miles)
Federal lists	
Federal NPL site list	1.0
Federal delisted NPL site list	0.5
Federal CERCLIS list	0.5
Federal CERCLIS NFRAP site list	0.5
Federal RCRA CORRACTS facilities list	1.0
Federal RCRA non-CORRACTS TSD facilities list	1.0
Federal RCRA generators list	(property and adjoining properties)
Federal institutional control/engineering control registries	(property and adjoining properties)
Federal ERNS list	(property only)
State and tribal equivalent lists	
NPL	1.0
CERCLIS	0.5
Landfill or solid waste disposal site lists	0.5
Leaking storage tank lists	0.5
Registered storage tank lists	(property and adjoining properties)
Institutional control/engineering control registries	(property and adjoining properties)
Voluntary cleanup sites	0.5
Brownfield sites	0.5

NPL = National Priorities List compiled by the USEPA pursuant to CERCLA 42 U.S.C. §9605(a)(8)(B) of properties with the highest priority for cleanup pursuant to USEPA's Hazard Ranking System.

CERCLIS = Comprehensive Environmental Response, Compensation, and Liability Information System. This is the list of sites compiled by the USEPA that it has investigated or is currently investigating for potential hazardous substance contamination for possible inclusion on the NPL.

NFRAP = No Further Remedial Action Planned.

RCRA = Resource Conservation and Recovery Act.

CORRACTS = Corrective Action Sites. This is a list maintained by the USEPA of hazardous waste treatment, storage, or disposal facilities and other RCRA-regulated facilities (due to past interim status or storage of hazardous waste beyond 90 days) that have been notified by the USEPA to undertake corrective action under RCRA. The CORRACTS list is a subset of the USEPA database that manages RCRA data.

TSD facilities = list kept by the USEPA of those facilities on which treatment, storage, and/or disposal of hazardous wastes takes place, as defined and regulated by RCRA.

Adjoining property = any property that borders the subject property or that would border the subject property but for a street or other public thoroughfare separating them.

Engineering control = physical modification to a site or facility (for example, capping, slurry walls, or point-of-use water treatment) to reduce or eliminate the potential for exposure to hazardous substances or petroleum products in the soil or groundwater on the property. Engineering controls are a type of activity and use limitation.

ERNS = USEPA's Emergency Response Notification System list of reported CERCLA hazardous substance releases or spills in quantities greater than the reportable quantity, as maintained at the National Response Center. Notification requirements for such releases or spills are codified in 40 CFR Parts 302 and 355.

Landfill = a place, location, tract of land, area, or premises used for the disposal of solid wastes as defined by state solid waste regulations. The term is synonymous with the term solid waste disposal site and is also known as a garbage dump, trash dump, or similar term.

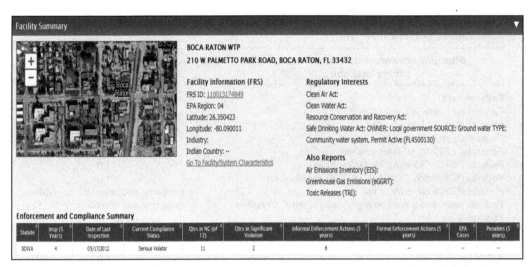

Figure 8.18 Online report of potential REC (source: http://echo.epa.gov/detailed_facility_ report?fid=110013174849)

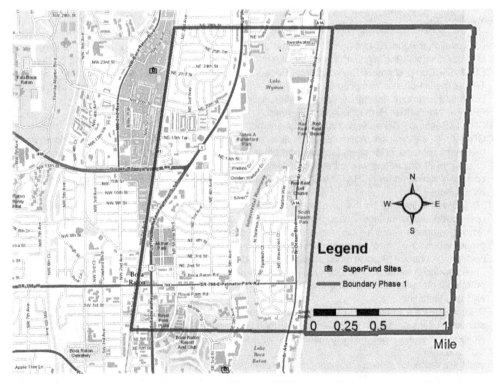

Figure 8.19 Example of boundary for searching for RECs; note that ocean cuts off the eastern half of the search area

Table 8.2 Description of LUSTs found within the search area in Figure 8.19

Facility name	Distance from site (miles)	Contaminant cleanup
The Yacht and Racquet Club	0.60	Petroleum (closed)
The Corniche	0.65	Petroleum (closed)
The Brighton	0.75	Petroleum (reviewed)
Fire Station No. 3	0.80	Petroleum (closed)
The Beresford	0.90	Petroleum (reviewed)
The Aegean	0.90	Petroleum (reviewed)

Table 8.3 Dry cleaners found within the search area in Figure 8.19

Facility name	Distance from site (miles)	Facility status
Suprema Quality Dry Cleaners	0.83	Delisted
Nu Look One Hour Cleaners	0.84	Unassigned

minute topographic map (or equivalent), as shown in Figure 8.20, provided it is reasonably ascertainable. Other nonstandard sources can be obtained from the USGS or state geological survey office, such as groundwater maps, bedrock geology maps, surficial geology maps, and Soil Conservation Service soil maps.

Investigators should try to obtain a legal description of the property and the adjacent properties, which is readily available from the local property appraiser or tax collector's office. It is important to understand the previous uses of the property and the surrounding areas to help identify any past uses that may have led to historical RECs. To this end, historical use information associated with the property should be identified going back to the first developed use or 1940, whichever is earlier. The term "developed use" includes agricultural uses and placement of fill dirt. Even if the property was developed in 1970, for example, it still must be confirmed that it was undeveloped back to 1940. This requires consulting standard historical sources, which include the following:

- ◆ **Aerial photographs**—Visually identify development and potential past uses. These photographs typically are available from government agencies or library collections (see Figure 8.16).
- ◆ **Fire insurance maps**—Indicate uses of properties at specific dates. These maps usually are available from local libraries, historical societies, or the map companies that produced them (see Figure 8.21 for an example).
- ◆ **Property tax files**—Record past ownership, appraisals, maps, photos, and other information specific to a property (see Figure 8.22).
- ◆ **Recorded land title records**—Document historical fee ownership, like leases, land contracts, and activity and use limitations. These are recorded in the local jurisdiction and kept by the municipal or county clerk and are obtained from title companies or directly from the government agency. This source typically only provides names of previous owners, lessees, easement holders, etc. and little to no information about

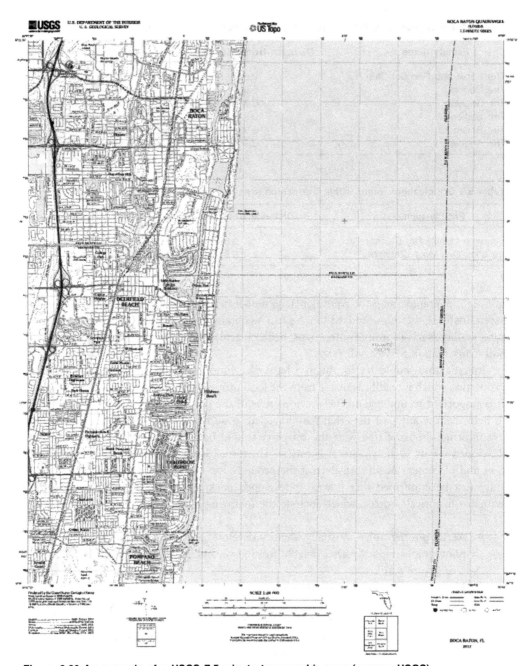

Figure 8.20 An example of a USGS 7.5-minute topographic map (source: USGS)

uses or occupancies, so land title records cannot be the sole source of historical use information consulted. At least one additional standard source must be used.

◆ **Land use and zoning maps**—Used by the local government to indicate permissible uses of the property or adjoining areas or zones. These documents can be maps and/or written records, often available at the local planning department office (see Figures 8.23 and 8.24).

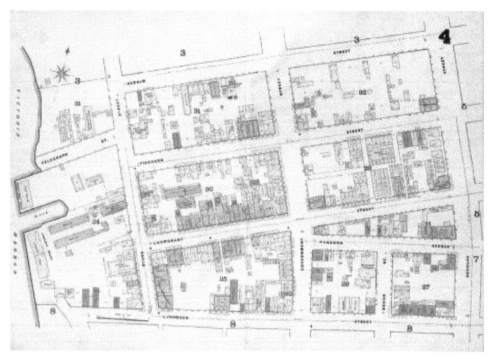

Figure 8.21 Sanborn Fire Insurance Map for Victoria, BC, Canada 1885 (source: http://memory.loc.gov/cgi-bin/map_item.pl?data=/home/www/data/gmd//gmd3m/g3514m/g3514vm/g09794001/sb000040.jp2&style=gmd&itemLink=&title=Victoria,%20BC,%20Canada%201885+-+Image%205)

◆ **Local street directories**—Typically show ownership and use of sites by street address. These documents are available at local libraries, academic institutional libraries, or historical societies.

◆ **Building department records**—Indicate when permission was obtained to renovate, alter, demolish, or construct improvements on the property. These records typically are located in the local building department of the municipality or county.

◆ **Other historical information**—Refers to additional sources not previously listed and includes but is not limited to the following: maps of the property, surveys, newspaper archives, Internet sites, community organizations, historical societies, water billing records, industrial waste permits or inspections, notices of violation, compliance records, occupational or business licenses, census data, current owners or occupants of neighboring properties, and personal knowledge of the property owner and occupants.

It is fortuitous if a prior Phase I Environmental Site Assessment document is available for a property because technically it only needs to be updated. However, it is standard practice to exercise due diligence here and search for more records because there is no guarantee that the previous report was complete. Keep in mind that more data becomes digitally available every year, so a record that was not reasonably ascertainable 10 years ago might be available today. EPs need to find those records and assume prior investigations might have missed them or they were unavailable. Review of standard historical sources at

Site Address	W DANIA BEACH BOULEVARD, DANIA BEACH		ID #	5042 33 00 0591
Property Owner	CITY OF DANIA BEACH		Millage	0413
Mailing Address	100 W DANIA BEACH BLVD DANIA BEACH FL 33004-3643		Use	94

Abbreviated Legal Description	33-50-42 N 50 OF E3/4 OF N1/4 OF SE1/4 OF SE1/4 OF SEC 33-50-42

The just values displayed below were set in compliance with Sec. 193.011, Fla. Stat., and include a reduction for costs of sale and other adjustments required by Sec. 193.011(8).

Property Assessment Values
Click here to see 2013 Exemptions and Taxable Values to be reflected on the Nov. 1, 2013 tax bill.

Year	Land	Building	Just / Market Value	Assessed / SOH Value	Tax
2014	$23,520		$23,520	$23,520	
2013	$23,520		$23,520	$23,520	
2012	$23,520		$23,520	$23,520	

2014 Exemptions and Taxable Values by Taxing Authority				
	County	School Board	Municipal	Independent
Just Value	$23,520	$23,520	$23,520	$23,520
Portability	0	0	0	0
Assessed/SOH	$23,520	$23,520	$23,520	$23,520
Homestead	0	0	0	0
Add. Homestead	0	0	0	0
Wid/Vet/Dis	0	0	0	0
Senior	0	0	0	0
Exempt Type 14	$23,520	$23,520	$23,520	$23,520
Taxable	0	0	0	0

Sales History					Land Calculations		
Date	Type	Price	Book/Page or CIN		Price	Factor	Type
11/21/2003	EAS	$100	36507 / 1906		$21,780	1.08	AC
					Adj. Bldg. S.F.		

Special Assessments								
Fire	Garb	Light	Drain	Impr	Safe	Storm	Clean	Misc
04								
X								
108								

Figure 8.22 Title information from county property appraiser site (source: http://www.bcpa.net/ RecInfo.asp?URL_Folio=504233000591)

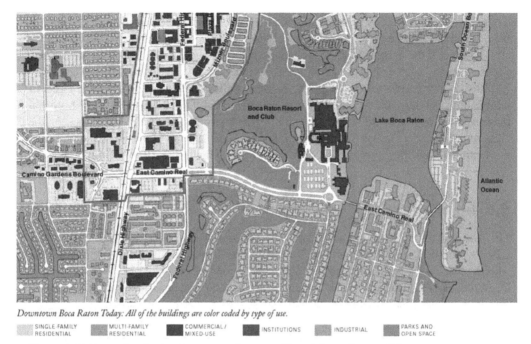

Downtown Boca Raton Today: All of the buildings are color coded by type of use.

| SINGLE-FAMILY RESIDENTIAL | MULTI-FAMILY RESIDENTIAL | COMMERCIAL / MIXED-USE | INSTITUTIONS | INDUSTRIAL | PARKS AND OPEN SPACE |

Figure 8.23 Example of a land use map (source: City of Boca Raton, FL)

Figure 8.24 Example of a zoning map (source: City of Boca Raton, FL)

less than 5-year intervals is not required if uses have not changed. To identify the type of use, it is sufficient to describe obvious uses in general terms, such as office, residential, retail, etc. However, if the usage is industrial or manufacturing, then it is necessary to identify the specific activity (such as paper milling, coal mining, petroleum refining, etc.).

After conducting the search, there may be gaps or difficulty in obtaining specific pieces of information. The search is considered complete when all of the standard sources of historical information have been consulted over the appropriate intervals to determine prior usage on the property or when data failure is encountered. If data failure leads to a data gap, then the report must comment on the impact of this data gap on the ability of the EP to identify RECs.

8.5 Interviews

The goal of speaking with past and present owners, occupants, operators, and users of the property and neighboring properties is to obtain information on possible RECs associated with the site. These interviews can be conducted in person, over the phone, or in writing. Questions should focus on determining the past and current uses and conditions of the property. The key players who should be interviewed include the following:

- ◆ **Site owner and/or key site manager**—Before the interview process, the identity of a key site manager must be determined. This person may be identified by the owner and may be the property manager, chief supervisor, head maintenance person, or in some circumstances the owner himself or herself. Every attempt should be made to interview the key site manager in conjunction with the site visit.
- ◆ **Site occupants**—A reasonable attempt must be made to interview a number of occupants or users. If the number of occupants is less than five, then try to interview all of them, but if the number is more, try to interview those who have been there the longest. Make sure to identify each one and determine their duration of occupancy.
- ◆ **For abandoned properties, neighboring property owners/users**—If there is evidence of potentially unauthorized uses or uncontrolled access, interviews with one or more owners or occupants of neighboring properties should be conducted.
- ◆ **Government officials**—A reasonable attempt must be made to interview at least one staff member of the local fire department, health department, or building or groundwater permitting agency. Questions should focus on the past uses of the property, specific chemicals that are present or once were present, spills or other chemical releases, any environmental cleanups or liens, engineering controls, and land use restrictions or institutional controls that are in place at the site and/or have been filed or recorded in a registry.

There are a series of questions that should be asked. Before conducting the interview with the key site manager and prior to the site visit, ask if any of the following helpful documents exist and are available:

- ◆ Previous environmental site assessment reports
- ◆ Environmental compliance audit reports
- ◆ Environmental permits (National Pollutant Discharge Elimination System, solid waste disposal, underground injection, hazardous waste disposal, etc.)

- ◆ Registrations for storage tanks and/or underground injection systems
- ◆ Material Safety Data Sheets
- ◆ Community right-to-know plans
- ◆ Safety plans (spill prevention plan, evacuation plan, countermeasure and control plans, etc.)
- ◆ Hydrogeologic reports
- ◆ Notices of current violations, liens, or other correspondence with environmental regulatory agencies
- ◆ Hazardous waste generator notices or reports
- ◆ Geotechnical reports
- ◆ Risk assessments
- ◆ Recorded activity and use limitations
- ◆ Environmental impact assessment reports

Prior to the site visit, the owners or key site managers should be asked if they know of any pending, threatened, or past litigation; administrative proceedings; violation notices; or liability relevant to hazardous substances or petroleum products relating to the property. The person conducting the interview has a responsibility to ask the appropriate questions, but the interviewee has no obligation to answer or may not know the answer. If the answers to questions are unknown or only partially known, then this section is considered incomplete. Also, persons interviewed as part of a previous assessment do not need to be questioned again about their responses provided at that time; however, they may be asked about any new information learned since the last assessment.

8.6 Evaluation and Report

The report documents the findings from the records review, site visits, and interviews. Key pieces of information must be supported by documentation. It is easier to follow the report from the reader's perspective if maps and photographs are included in the body of the report, not just in the appendix. The purpose of the report is to identify the presence of RECs and to provide the EP's opinion on the potential impacts of contaminants. The report should identify and list the qualifications of the EP and others who conducted site visits or interviews. The report should indicate if any users provided relevant information or specialized knowledge. It is important to clearly specify the scope of services performed so that the findings can be confirmed. In the findings section, all known or suspected RECs, historical RECs, and *de minimis* conditions should be listed and discussed. The vapor intrusion pathway also should be discussed if appropriate. In the opinion section, the EP must use logical reasoning to evaluate the documents and other information obtained during the investigation to assess if suspect RECs should be listed in the conclusion section of the report as current RECs. Those suspect RECs that do not pose a current threat to the property's environmental integrity are not included in the conclusion. This section should provide the rationale for this opinion. If additional investigation or greater certainty is required or is called for by the EP, then additional appropriate testing to detect hazardous substances or petroleum products can be recommended in a more intrusive Phase II investigation.

If all reasonably ascertainable sources have been reviewed but the objectives of the records review still have not been met, then the Phase I report must give reasons why sources were excluded from the analysis and identify any document data failures which occurred. The report also must include a section on data gaps, particularly if a data gap raises reasonable concerns. A *data gap* occurs if, despite all good faith efforts, there remains an inability to obtain required information. The EP must determine if the data failure represents a significant data gap and if the data gap impacts the ability to identify RECs. For example, if a building on the property is inaccessible during the site visit and in the EP's experience the building involves an activity that typically leads to a REC, then the inability to inspect the interior of the building would warrant a significant data gap comment. In the conclusions/recommendations section, all RECs connected with the property should be summarized, along with a summary statement which indicates that all appropriate inquiries have been conducted and no evidence of RECs was encountered except for those listed in this section, in the opinion of the EP preparing the report.

Any additional services that are beyond the scope of the ASTM E1527 practice but were included in the negotiated scope of services should be summarized in a separate section. This could include more detailed conclusions, liability and risk evaluations, Phase II recommendations, remediation techniques, nonscope items, etc. The standard practice for the Phase I Environmental Site Assessment report includes a recommended format, which should be followed according to the most recent version of ASTM E1527. The most commonly omitted items in the report are the following:

◆ Failing to include the report date on the cover
◆ Omitting the EP's resume, qualifications, and letters of reference
◆ Excluding the property log
◆ Failing to include a section with the EP's opinion, conclusions, and recommendations in the summary
◆ Leaving out the date of the site reconnaissance visit
◆ Forgetting to conduct the appropriate interviews
◆ Failing to provide the names and titles of interviewees
◆ Omitting supporting documentation, interview transcripts, and attempts to interview in the appendices
◆ Not including key photographs with dates and descriptions of the views presented

Other key pieces of information that must not be left out include geological investigations, floodplain maps, wetlands maps, location map indicating groundwater flow, map of well field protection zones, locations of transformers, maps locating all suspect RECs identified in the regulatory review, discussion of each property identified in the environmental regulatory database(s) and rationale(s) for determining whether or not each property presents a current REC to the subject property, and noise assessments.

8.7 Nonscope Considerations

Some items are not included in the standard practice of ASTM E1527 reporting but are relevant to the future development of a property and are helpful to include in any investigation prior to preliminary design. Nonscope considerations refer to asbestos-containing building materials (Figure 8.25), radon, lead-based paint (Figure 8.26), lead in drinking

Figure 8.25 Asbestos-containing materials in the ceiling

Figure 8.26 Lead paint primer on pipe

Figure 8.27 Historic building: Paul Revere House in Boston, MA

water, regulatory compliance, cultural and historic resources (Figure 8.27), industrial hygiene, health and safety, ecological resources, endangered species (Figures 8.28 and 8.29), wetlands (Figure 8.30), indoor air quality, biological agents, and mold (Figure 8.31). It is important to note that nonscope items must not be included in the body of the report. Instead, they should appear in their own section entitled "nonscope considerations."

An asbestos survey must be performed on all buildings scheduled for demolition or renovation. Use of asbestos can be assumed in all buildings constructed before 1980. Representative sampling must be performed, unless a comprehensive asbestos survey was performed on or after January 1, 1986.

A radon test must be performed for renovation or use of an existing structure unless a USEPA-approved test has been performed in the lowest areas within the last 6 months demonstrating levels at or below 4.0 pCi/L. If new construction is planned, then the Phase I report must indicate that the building will be constructed in accordance with USEPA requirements.

For lead issues, it may be assumed that lead-based paint or lead pipes were used if the current or former structure was constructed before 1979. If such structures are not planned for demolition, sampling must be performed on the interior and exterior of the buildings. If such structures are planned for demolition, soil sampling is required, with some exceptions.

For wetlands, state-protected waters, and floodplains, the EP must provide documents to show that no more than 0.1 acre of wetlands will be disturbed, and any potential waters of the state and applicable buffer areas are shown on the site map in the report and remain undisturbed. Furthermore, the Federal Emergency Management Agency flood map for the property must be included, even if the property is not within the 100-year floodplain. If any

Figure 8.28 Threatened burrowing owl

Figure 8.29 Endangered Florida panther

Figure 8.30 Wetlands

Figure 8.31 Mold growing indoors

buildings are within the 100-year floodplain, then evidence must be provided that the lowest existing floor elevation is 6 inches above the floodplain elevation or that the new construction will be reclassified out of the floodplain prior to completion.

For every unit in which the EP is assessing lead-based paint, radon, or asbestos, moisture and mold problems, water leaks, fungi, and microbial growth also should be evaluated. The following areas, if readily accessible, should be visually evaluated: HVAC systems and areas behind walls, under floors, and above ceilings.

For noise issues, determine if the property is within 5 miles of a civil airport, 15 miles of a military airport, 1,000 feet from a major highway or busy road (greater than 10,000 average daily vehicle count), or 3,000 feet from a railroad or rail line. If so, a noise assessment for the property using 10-year projections for traffic, aircraft, and railway noise should be conducted. Where not available for traffic noise, 3% yearly growth should be estimated to determine the projection. Where not available for aircraft or railway noise, currently available data should be used.

8.8 Phase II and III

In a Phase I Environmental Site Assessment, only reviews of records of the property, site interviews, and a site visit are conducted. No sampling is performed. If the existence of current or historical RECs of concern is identified, the EP should recommend that a Phase II site assessment be conducted, which requires soil and groundwater sampling of the property. However, because analysis costs and sampling are expensive, the EP must identify which contaminants should be targeted and where they are likely to be found (i.e., hot spots). For example, if the site is a gas station, testing for benzene, toluene, ethylbenzene, and xylene (BTEX) and total petroleum hydrocarbons near the underground storage tank is appropriate as these are common constituents found in gasoline. Ordering radon tests would not be useful in detecting gasoline migration on the site.

If the Phase II site assessment comes up "hot," which means the constituents searched for are present, a Phase III site assessment is required. A Phase III Environmental Site Assessment delineates the extent of the contamination. In a Phase III site assessment for the gas station example, tests around the site would continue to be performed until the pollutant was no longer found. This might be 10 feet or 1,000 feet from the site boundary. The Phase III Environmental Site Assessment is designed to show how much contamination there is in order to allow engineers to design a feasible remediation system. Site remediation is a whole industry in itself, the costs of which often dissuade potential buyers from purchasing a property.

8.9 References

ASTM (2013). E1527-13, Standard Practice for Environmental Site Assessments: Phase I Environmental Site Assessment Process 1, no. 1.

Healy, H.G. (1982). "Surficial and intermediate aquifers." In Franks, B.J., ed., *Principal Aquifers in Florida*, USGS Water Resource Investigations Open File Report 82-255.

Hejzlar, Z. (2007). ASTM 1527-05, American Society for Testing and Materials, West Conshohocken, PA.

8.10 Assignments

1. Give three examples of due diligence.
2. Compare and contrast a Phase I and Phase II Environmental Site Assessment.
3. Give three examples of a key site manager.
4. Write a list of questions to ask during the interviews for a Phase I Environmental Site Assessment.
5. Name five specific chemicals that would be considered hazardous materials for a Phase I Environmental Site Assessment.
6. What are the main tasks in a Phase I Environmental Site Assessment?
7. Define REC (spell out what the letters stand for and define the term) and also provide three examples.
8. Define historical REC (spell out what the letters stand for and define the term) and also provide three examples.
9. What kinds of documents are needed to demonstrate that a previously leaking underground storage tank is a known or suspect REC?

The Site Plan
Development Process

Planning improvements for a piece of property involves meeting the needs of the client, complying with all of the requisite codes, and creating as much functionality with as little impact as possible. The key pieces of information needed at the beginning of the process must provide a clear understanding of the site, including limitations on use, zoning, setbacks, geotechnical issues, availability of utilities, restrictions due to easements, hydrologic concerns, topography, flooding/drainage, and transportation. Most of this information should be picked up during the environmental site assessment process, at the beginning of the planning stage. Decisions on location and orientation of the project and suitability of the site result from this analysis. The following items must be included for most projects:

- ◆ Site work (changes in grades, fill, cuts, removal of vegetation, etc.)
- ◆ Access (connection to existing roadways, minimizing disruption to surrounding traffic)
- ◆ Parking requirements based on proposed land use
- ◆ Transportation access for customers, deliveries, and emergency services
- ◆ Utilities (water, sewer, power, cable, phone)
- ◆ Flood control/stormwater drainage (retention, detention, or other means to prevent site flooding or damage from standing water)
- ◆ Structural and geotechnical issues related to the building

For most projects, the design process follows this general plan, with the structural and geotechnical issues coming after the other items have been studied. Lack of access can kill a project, as can the need to extend water and sewer lines long distances or the inability to control stormwater. In addition, the building must integrate with the surrounding commu-

Figure 9.1 A design charrette in progress

nity. Often these are urban planning issues that can be approached in collaboration with the local planning department and city engineer, in compliance with regulations and standards.

Urban planners are trained to focus on community interconnectivity, pathways of change over time, diversity of needs, and distributional consequences of use, all areas that engineers should gain some understanding of as well. The planners' focus is the community's means to establish links between current and future land use. Like engineers, planners must serve the public interest and are focused primarily on the long-term effects and consequences of construction projects. Engineers must interface with planners because planners can provide necessary input for the preliminary site plans during the charrette process (see Figure 9.1). A charrette refers to the collaborative design session in which the architects, engineers, planners, and other stakeholders provide input for design solutions and concepts. If an engineer needs to build a bridge, then planners can help determine where the bridge is best built, if that bridge is in the public interest, and if it is an integral part of the future interconnectivity and sustainability of the community.

9.1 Community Plans and Codes

Every community should have a comprehensive plan that outlines its vision for development. However, too many comprehensive plans purposely are made vague to minimize legal liability. One important aspect of a comprehensive plan is that it creates or delineates zoning districts (see Figure 8.24), which define permissible land use (industrial, commercial, residential, mixed use) and define other concepts like setbacks from property lines, building height limitations, development density, parking, landscaping, and even aesthetic issues. Zoning laws are what prevent factories from being located in residential neighborhoods in most communities.

9.2 Site Development

If the project does not conflict with the zoning code, then the next step is to proceed to creating a site plan. The goal of the site planning process is to maximize the use of the available property area and address the relationships of site elements to the building, streets, and neighbors. The engineer's design must be presented in a way that clearly explains what the project is and what it will look like. This can be accomplished by employing drawing sets, models, 3-D visualization, mock-ups, plan documents, and specifications. Software such as Google SketchUp®, Civil 3D®, AutoCAD®, and Revit® helps make the project more real to the client and allows the public to understand how the pieces of the engineer's design fit together.

A site plan cannot be created without some thought or without visiting the site. Rarely will the initial concept be close to the final version. There are a number of issues that can impact the site, and there will be challenges to overcome, especially in terms of land development codes. SWOT analysis, which is a structured planning tool used to evaluate the strengths, weaknesses, opportunities, and threats of a given option, and design charrettes are two ways to get the project team and its constituencies involved in the site planning process. Topics of importance during this process include how design decisions will affect drainage, water, sewer, transportation, parking, access, egress, and high-performance features. The project team needs to discuss site planning issues within the group, come up with justifications for decisions, and then solicit feedback from groups outside of the design team.

Regarding site planning and land use development, the key question is what should be developed, where, and in what order. As noted previously, the site plan is an output of the project vision and the results of the site reconnaissance. The process proceeds as shown in Figure 9.2. The site plan evolves as an iteration of refining the project vision and matching

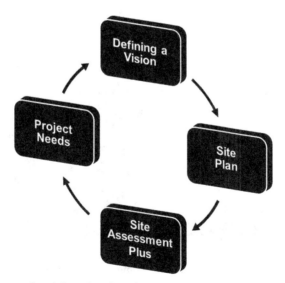

Figure 9.2 The process of arriving at a site plan

the project needs to the characteristics of the site itself in the context of economic, environmental, societal, technical, and regulatory constraints. Without any piece, the site plan will fail to meet the client's goals and fail to maximize the site's potential. The following sections outline site plan development for two different example projects.

Figure 9.3 is a simplified depiction of a conceptual site for a commercial warehouse building. The map is not very revealing, and more information about the site needs to be gathered, such as the surrounding land use, the topography, etc. Figure 9.4 depicts the surrounding property of this conceptual site. A site visit will confirm if the adjacent property is commercial and help to determine why one property has a retention pond (a relatively recent concept) while another does not. Knowing what previous commercial activities may have occurred on and around the site is useful information for future planning and is another reason to review the Phase I Environmental Site Assessment report, if available.

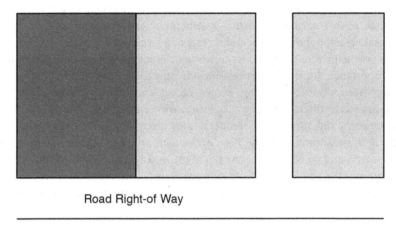

Road Right-of Way

Figure 9.3 Conceptual site for commercial warehouse building

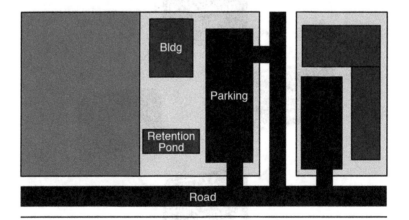

Figure 9.4 The surrounding site conditions for commercial warehouse building

9.3 Easements, Rights-of-Way, and Setbacks

The surrounding conditions are not all that is required. A search needs to be conducted concerning easements, rights-of-way, or other land rights that may limit development on the property. Easements and rights-of-way are reservations of land for purposes such as utilities and roadways. The infrastructure may or may not be in place, but either way, vertical construction over easements and rights-of-way most likely is prohibited. In that case, the presence of an easement will reduce the potential footprint of the proposed building considerably. The site in question has an easement going across the middle (see Figure 9.5).

Local zoning codes or land development codes normally delineate *setbacks*, which are restrictions on building from the edge of the property line to a certain specified distance. Every zoning code is different and is dependent on the type of land use (e.g., commercial, residential, industrial, institutional, etc.) and proximity to differing land uses. Figure 9.6 shows the setbacks on the conceptual site, which are more limiting.

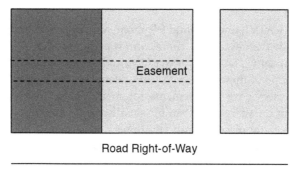

Figure 9.5 Property contains an easement that complicates the location of proposed commercial warehouse building

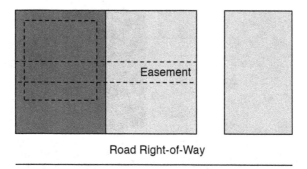

Figure 9.6 Property for commercial warehouse building with setbacks drawn

9.4 Utilities, Parking Requirements, and Roadwork

Figure 9.7 shows the same piece of property with the underground utilities added. The location of utilities is an important piece of the site plan because they will tend to dictate how the building footprint will be oriented so as to reduce the runs of pipe required to connect or to take advantage of topography for gravity pipe flow, frost considerations, etc. Other important items to consider include pipe diameter, pipe material, pipe depth, and pressure in water service lines, as well as pipe diameter, pipe material, and pipe slope for gravity sewers or pressure in sewer force mains. In addition, the proximity of other buried infrastructure is important. The location of the building should be arranged to provide the shortest, most direct link from the utility system to the building, whenever possible. Note that direct links may create challenges in determining the floor plan of the building (to minimize utilities under the slab) and the orientation of the building (to maximize exposure for natural light or renewable energy, to reduce direct sunlight to minimize the heat island effect, or due to client preferences). Furthermore, landscaping should not be placed on top of utility lines because of the affinity of tree roots to seek out leaking or sweating pipes for moisture. Over time, tree roots cause damage to pipes, especially trees that fall over and take the pipes with them.

Cable television (CATV), underground telephone systems (UTS), and power are different at each site, and while the most direct connection is preferred, they do not suffer from the same limitations as water (W), sanitary sewer (SS), and stormwater (SW) systems. However, it is preferable for none of these utilities to run underneath the building slab or pavement, if possible. Many places have requirements for maintaining stormwater on-site or for stormwater treatment on-site, rather than off-site discharge. As a result, there is a limit to the imperviousness of any piece of property. Imperviousness is an indication of how fast water will run off the paved areas of the site. From a property owner's perspective, avoiding

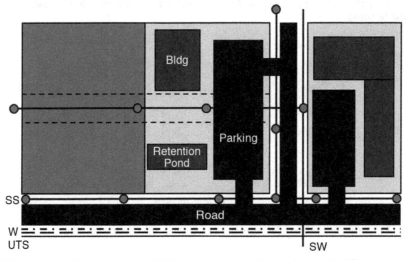

Figure 9.7 Location of underground utilities for commercial warehouse building

flooding is a good thing, but where does the water go? The reality is that it typically is routed to the public storm drainage system. Undeveloped property will have limited runoff, but developed property will have much greater runoff because the area available for percolation is diminished by impervious materials. Because most stormwater systems were built years ago, the capacity is protected by local entities, which may place limits on imperviousness and requirements to hold excess stormwater on-site to limit impacts to downstream stormwater customers. Typically this means that most lots cannot exceed 40% impervious area due to drainage requirements.

The building square footage also specifies the need for parking (in the local zoning codes), which further restricts the allowable percentage of impervious areas. Parking area often is defined by occupancy or square footage. For example, it is common for multifamily housing units to require 1.75 or 2 spaces per dwelling. A typical parking space may be 9 ft × 18 ft, 9 ft × 20 ft, or 10 ft × 20 ft, plus aisle space (24 ft for 90° parking), so parking adds a considerable amount of impervious area to a site. In the case of multifamily construction, parking is often a greater impervious area than the building itself. For commercial buildings, square footage is often used to determine the number of parking spaces, and requirements of one space per 200 to 250 ft^2 are common. The typical parking space including the aisle is about 250 ft^2, so the building floor area will essentially need to correspond with parking square footage. The conflict between imperviousness and parking is a reason why many projects decrease in size and scope from the owner's original expectations. Figure 9.8 shows the buildable area given stormwater drainage restrictions and parking needs. Figure 9.9 shows potential parking and associated stormwater retention, along with some landscaping. Many jurisdictions define required landscaping or other buffering, especially where commercial and residential developments are immediately adjacent to one another. It cannot be assumed that just because there is a stormwater pipe across the property a connection can be made to it.

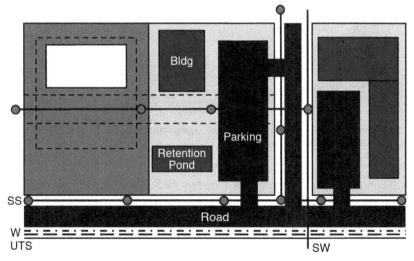

Figure 9.8 Buildable area given stormwater restrictions and parking needs for commercial warehouse building

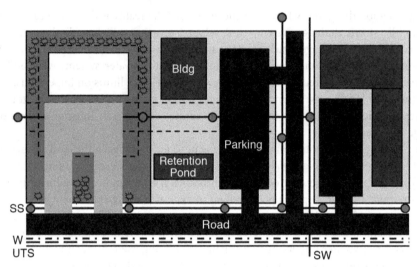

Figure 9.9 Parking and exfiltration trench location on the site plan for commercial warehouse building

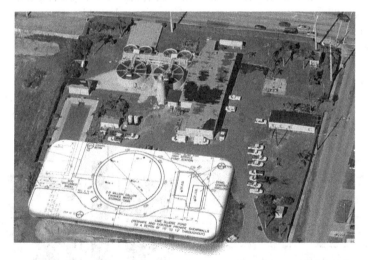

Figure 9.10 Example site plan with a storage tank design superimposed on an aerial photo of the site for commercial warehouse building

Figure 9.10 is an example site plan with a storage tank design superimposed on an aerial photo of the site.

Figure 9.11 shows the service lines for the underground utilities. The floor plan of the building should be designed to permit easy access to the incoming water and sewer to the building. A mechanical room, vertical utility chase (an area in a building that allows direct, unrestricted pipe installation), or other feature normally is used.

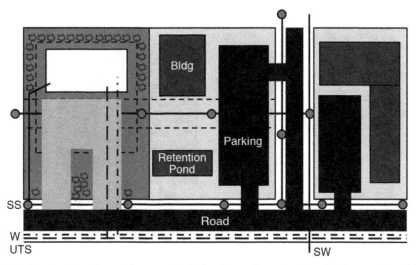

Figure 9.11 Service lines for underground utilities located on the site plan for commercial warehouse building

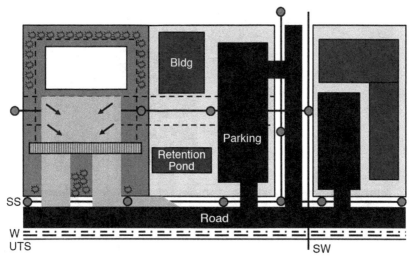

Figure 9.12 Parking and exfiltration trench with turn lane for commercial warehouse building

Many times, commercial activity will require turn lanes, pockets, or other transportation features. Figure 9.12 illustrates parking and an exfiltration trench for drainage with an example turn lane. Figure 9.13 shows an aboveground site plan, and Figure 9.14 shows the same view with access for the garbage dumpster. Note that solid waste haulers, fire trucks, and delivery vehicles may be heavy and require reinforced paving in addition to a larger turning radius compared to normal car traffic. Figure 9.15 shows the final site plan (note how much

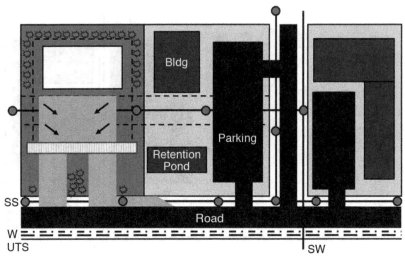

Figure 9.13 Plan view of site plan for commercial warehouse building

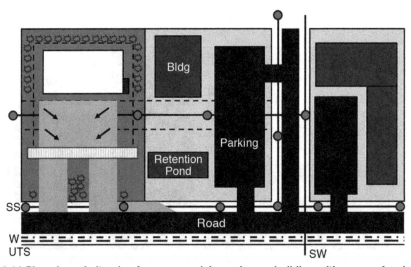

Figure 9.14 Plan view of site plan for commercial warehouse building with access for dumpster

detail is missing). A caveat to this simplistic exercise is that in reality, for any given site, much research must be conducted prior to sketching out the site plan. The actual process is quite a bit more involved than described here.

The engineer needs to address the zoning issues, utility locations, easements, stormwater requirements, topography, and rights-of-way and likely will need to meet with each of the agencies involved to identify challenges and preferences, as permits may be required from each one. Each individual community has different requirements, and in each case they must be carefully studied and accounted for in the preliminary design.

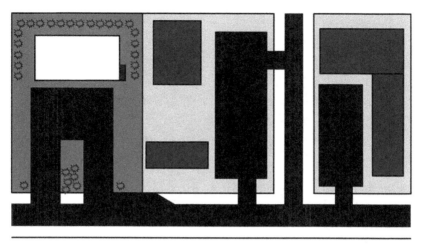

Figure 9.15 Final site plan for commercial warehouse building; note how much detail is missing

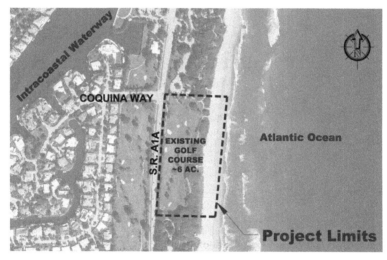

Figure 9.16 Aerial photograph of site boundary for recreational pier

An aerial photograph of the site boundary of a strip of land along the east coast of south Florida, where a proposed recreational pier is to be built, is shown in Figure 9.16. The site has roadway access, and the roads were determined to have low traffic volume most of the day. Because a recreational pier is proposed for the site, there is concern that traffic backing up to turn into the pier site may require turning lanes, an issue to keep in mind as the site progresses and demands are finalized. All aerial maps should be developed into accurate electronic maps, like Figure 9.16, as a part of any design project. This site is bound by setbacks, but no easements. Figure 9.17 is an AutoCAD drawing of the site that illustrates a major limitation of the site: because it is on the beach, crossing the coastal construction control line (CCCL) requires approval by the governor and cabinet. The setbacks and CCCL

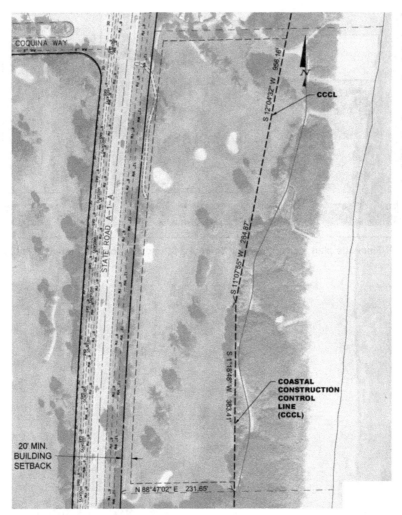

Figure 9.17 AutoCAD map of site for recreational pier

reduce the buildable area, as shown in Figure 9.18. The utilities were included in the initial AutoCAD map and are highlighted in Figure 9.19.

The engineers then needed to determine which dimensions within this property permitted them to develop a site with a pier. The three possible options are shown in Figure 9.20. Option 1 was determined to be the most suitable. The preliminary site plan is shown in Figure 9.21. Keep in mind that the building program (see Chapter 10) for this project will develop concurrently, which may lead to changes in the parking requirements, site access, and utilities. Developing the site and building plans is an iterative process. The site plan should show current and proposed conditions, with changes highlighted.

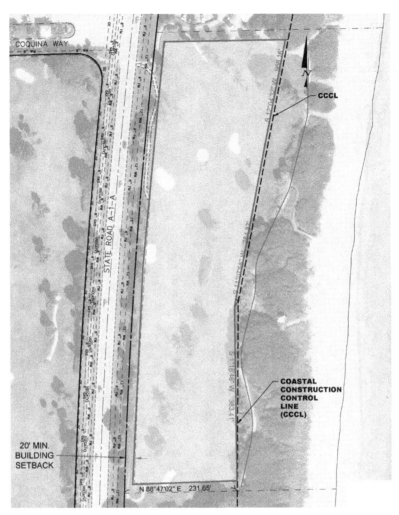

Figure 9.18 Buildable area for recreational pier

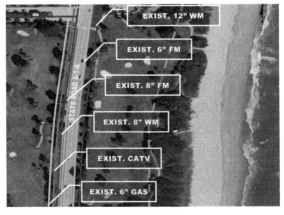

Figure 9.19 Utilities highlighted for recreational pier

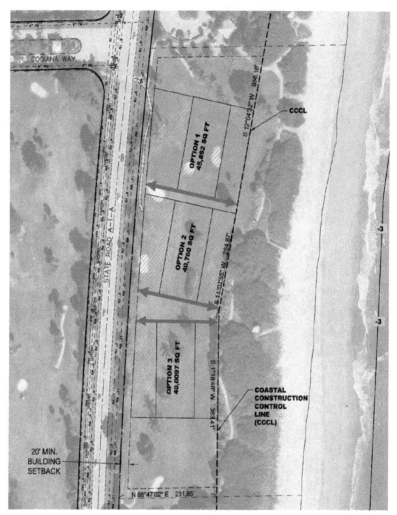

Figure 9.20 Options for building location for recreational pier

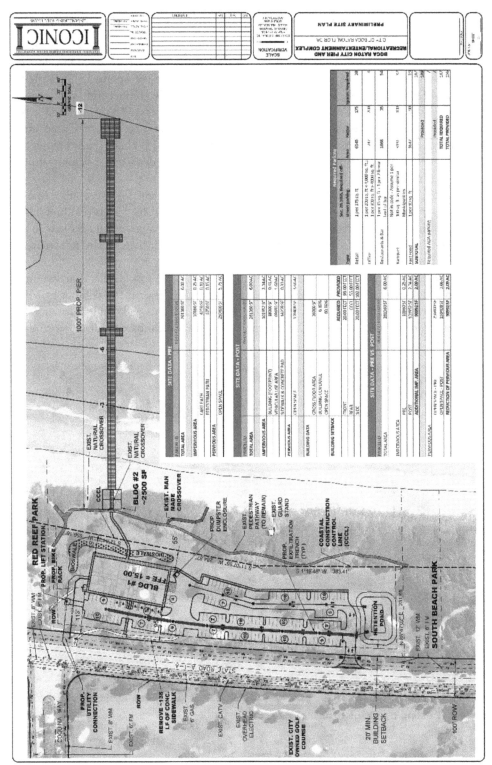

Figure 9.21 Preliminary site plan for recreational pier

9.5 Building Code Requirements and Functionality

Once the preliminary site planning issues have been sorted out, the building codes become the next major consideration. Local building codes will outline materials, methods, design standards, wind loads, foundation requirements, and often types of approvals for certain products and finishes, like glass for windows. Building codes may conflict with local codes, which is an issue that needs to be resolved early on in the process. The Americans with Disabilities Act specifies federal requirements for parking, ramps, number of spaces, and location for disabled access that is part of the building code (and perhaps the zoning code as well). A preliminary planning meeting with the local building department before getting too far along in the design is a great idea.

Plans work well for contractors, but often they are insufficient for the public. Many people are more visual. Figures 9.22 and 9.23 show an example where the current buildings were updated in three dimensions to show what the proposed changes would actually look like. While this has been the purview of architects in the past, there is no reason why engineers cannot incorporate these concepts in their designs.

Finally, site connectivity, access, and circulation need to be considered. If access and circulation create conflict or are just impossible to accomplish, the project will not be successful. Access limitations often are dictated by local officials. This will include distance to intersections, number of access points in a certain road distance, and need for or proximity

Figure 9.22 Existing classroom building

Figure 9.23 Proposed design of building in Figure 9.22

to lights or other traffic control devices. Within the site, access should consider the movement of cars, people, and service personnel. Avoiding conflicts is critical to limit liability of the client. Figure 9.24 shows an access issue for the pier project. Note that most of the parking should be one way. The arrows show how parking should be accomplished, but the design is set up for two-way parking all around. The star shows one conflict point, but there are many more. Sidewalks reduce the conflicts on the east side parking, but the west three lanes of parking will be in direct conflict with vehicles, and people will need to look both ways. Another solution for substantial conflict reduction would be to use angled parking to force the one-way configuration.

A final area of interest is how the site actually works with regard to elevation changes. The topography of the site is a relevant issue when defining parking, access, stormwater runoff, and the like. Figure 9.25 shows a series of cross sections from the site. Selected cross sections (Figures 9.26 to 9.28) show the need for cut-and-fill in order to level out the site to prepare for construction. This need became apparent as the site design was placed on the actual topography. Cut-and-fill can be an important, costly, and sometimes complex part of a project, as in this case. In other cases, the design of the project can work around problems related to topography.

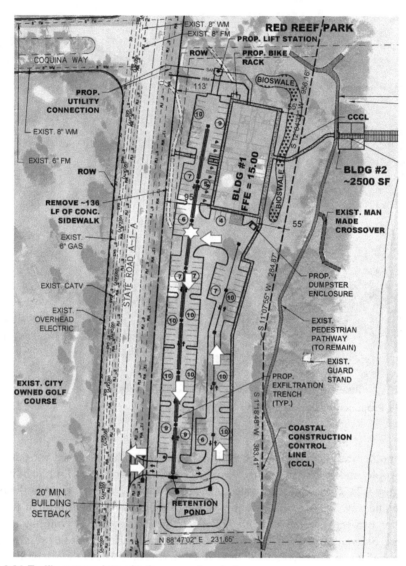

Figure 9.24 Traffic pattern into site for recreational pier

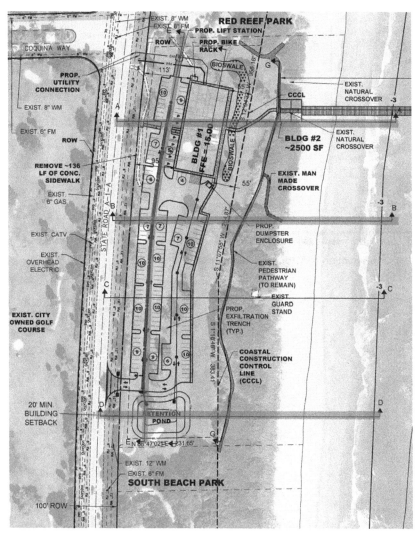

Figure 9.25 Location of cross sections for recreational pier

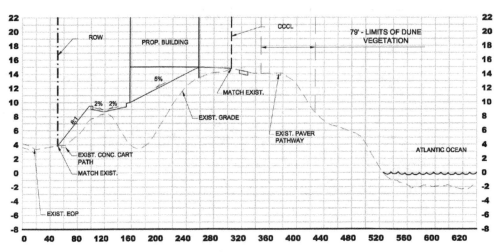

Figure 9.26 Cross-section A-A for recreational pier

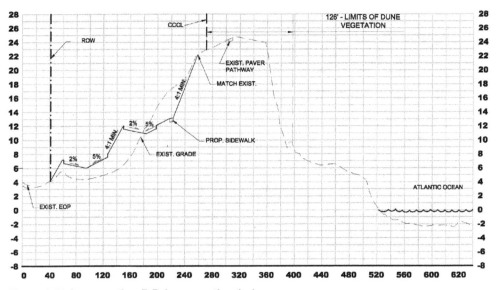

Figure 9.27 Cross-section B-B for recreational pier

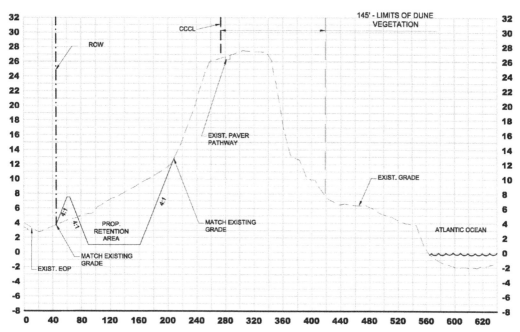

Figure 9.28 Cross-section D-D for recreational pier

9.6 Assignments

1. Identify the key agencies that have jurisdiction for your project.
2. What is the purpose of zoning codes?
3. What is the purpose of a site analysis?
4. Name three things that should be included in the soils report.
5. What are the major differences in design considerations between urban and rural roadways and streets?
6. What factor determines the maximum allowable slope on streets, parking lots, and highways?
7. What are the maximum slopes on sidewalks for disabled access routes?
8. Where should manholes and fire hydrants be placed on your site?
9. What factors influence the quantity of potable water supply and sewage capacity?
10. What is the reason for requiring a minimum velocity in a sewer pipe?
11. Before designing a sewer main pipe, what information is required for the plan and profile view?
12. What is the rainfall intensity for your project's design storm?
13. How is grading used for storm drainage control?
14. Why should water supply and sewer pipes be separated vertically underground?
15. What is the purpose of a water meter?

10

The Floor Plan Development Process

Once the site plan is roughed out, a building footprint will emerge. However, the building footprint cannot be complete without knowing something about what goes on inside the building. The list of functional spaces in the building is termed the *building program*, and it is an essential part of the process that provides specific, detailed information to guide the design of the building and its layout. The client, engineer, and any other decision makers must be in agreement on the building program as it relates to issues like location of parking spaces, impervious area, disabled (accessible) parking requirements, emergency vehicle access, and the like.

10.1 Building Program

Based on the building program, the allocation of spaces, the layout of functions, the circulation between them, and the access and egress points will constitute the building floor plan. However, much thought is required to minimize wasted space and maximize utility. Certain functions need to be located in close proximity to each other. For example, quiet areas should be located away from the front entrance and public rest rooms. Bathroom facilities normally are stacked in buildings with multiple stories to minimize plumbing runs and to be proximal to utility chases. Heating, ventilation, and air conditioning (HVAC) and other mechanical and electrical systems need to be incorporated into the floor plan to maximize efficiency. Staircases are needed to allow people to exit the building from multiple floors according to local or state building codes, but they also must be located in convenient distance for occupants to use. There are many other issues that will arise depending on building use and certain building codes, but much thought needs to be put into whether certain uses are compatible or need to be separated. Keep in mind that every building is different, and often the best way to learn what works and what does not is to visit structures with similar tenants and, more importantly, talk to the users. This also will help with the location of functions, exits, bathrooms, and other important features of the building. Occupants often are the best source of information and usually are very forthcoming about what works, what does not, and how to improve the work flow. Without consulting this important source of information, mistakes are bound to be repeated.

In addition to functionality, the floor plan is an integral part of the structural design. Most engineers use load-bearing walls and columns hidden in walls to support the building. If the floor plan presents issues that cannot be overcome from a structural perspective, the building cannot be constructed efficiently. Structural support columns located in the middle of rooms, hallways, or doorways are not acceptable. Designs should proceed so as to eliminate these user conflicts. One way to look at finding appropriate solutions is to think about the floor plan as a "puzzle," where the engineer must figure out how to incorporate the "puzzle pieces" into the layout and ensure appropriate flow and location for the various functions within the building (Figure 10.1). Keep in mind that hallways are wasted space—they provide no useful accommodation other than circulation. The engineer needs to limit wasted space because the client is paying for all costs on a square foot basis. Thus, the less space assigned to hallways, the less cost to the client. The engineer can help in this regard.

Many clients think floor plans are the purview of architects, but there is no reason why floor plan design should be limited to architects, particularly in cases where integrated design is needed. Architects tend to have more external contact with the client and, as a result, tend to solicit more information about client preferences than engineers. If engineers are more involved with clients, these preferences can be determined, and better decisions about structures can be developed. Floor planning is best accomplished through an integrated design

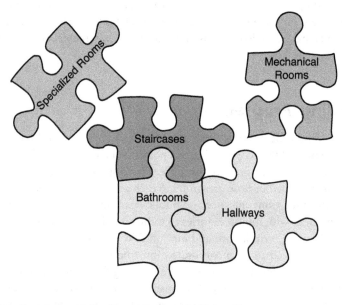

Figure 10.1 How the pieces of the floor plan puzzle fit together

Figure 10.2 Integrated design team meeting

process that involves the engineers, architects, clients, users, and any other important stakeholders (Figure 10.2).

Keep in mind that efficient use of space involves three key aspects: (1) the right amount of space for a given function, (2) the most appropriate configuration (which may be more important than the square footage), and (3) easy access to the functional space. In larger buildings, ensuring the movement of people and equipment between functional spaces is critical to efficient use of the property. Consider the case of a stairway where the landing from the first floor to the second floor is located on the opposite side of the building from the entrance to the stairway going up to the third floor (Figure 10.3). This inconvenience will probably force occupants to use the elevators instead of having to walk around the perimeter of the building to go to the third floor. A simple solution is to find a way to connect both staircases. For office buildings, often it is very useful to design for flexible spaces. Many times engineers and architects try to provide a certain number of individual offices. This strategy may work for the current condition, but future needs may dictate a different layout. Flexible space that can use modular walls and furniture resolves this conflict.

10.2 Floor Planning

There are two critical parts of the floor planning process. The first involves access and egress, which means entrances and exits that comply with fire and building codes. For example, stairways must exit outside of the building and be located within a certain distance from any part of the building. The distance will change based on use (schools are typically much different than offices in most states). Building codes define these egress points. Care should be exercised to ensure the correct minimum distances and that the appropriate widths of hallways and doors are provided. Most building codes have specific minimum egress door widths, opening directions, and hall requirements. Per the Americans with Disabilities Act (ADA), accessible egress points and hallways must be planned for. Engineers must become

Figure 10.3 An inconvenient staircase design

familiar with the building codes in the jurisdictions of their designs to ensure compliance with all codes. In most states, the professional engineer's seal attests to compliance with all codes. There can be penalties for noncompliance. Some states actually require that the code sections be placed on the plan sheets.

The second issue that governs floor planning relates to the plumbing and mechanical systems (Figure 10.4) and code requirements for the number of sinks and water closets, HVAC needs (ducts), and chases. The bathrooms must be laid out to ensure compliance with ADA requirements (noted in Chapter 9) and are most efficient when stacked (Figure 10.5). Keep in mind that the engineer must design the floor plan so that it is easy to count plumbing fixtures, which is the basis of sizing the piping in most design codes (discussed further in Chapter 12). The bathrooms also must be easily accessible for building users.

Clearly, with so many variables, providing a detailed road map for how to create effective floor plans is complicated, and as a result, it is an iterative process. However, there are some basic guidelines for starting the draft floor plan:

Figure 10.4 Plumbing installed in the ceiling for a chilled beam application

◆ Determine the size of each room
◆ Determine which rooms can be adjacent, need to be adjacent, and cannot be adjacent and therefore must be separated
◆ Determine the number of floors (which drives stairway and elevator needs)
◆ Determine a rough shape of the building footprint
◆ Determine the number of bathroom fixture units (from building codes)
◆ Stack bathrooms
◆ Stack elevators and stairs
◆ Locate large rooms first
◆ Locate smaller, associated rooms in the same vicinity
◆ Think about the structural plan and create a grid spacing for structural members (not sized yet)
◆ Squares and rectangles work far better than odd shapes and sizes
◆ Eliminate as much functionless spaces as possible

Once the concepts for a draft floor plan are in place, a charrette can be used to start the iterations for moving spaces around to create a functional layout. A good place to start is by minimizing hallway space, which the owner pays for. A goal should be to keep hallways under approximately 5% of the floor space. Detailed floor plans are required to complete the exterior building elevations. Floor plans should show where the windows are located in each space. It is important to note that walls have thicknesses. However, certain functions do not require windows (e.g., computer labs, movie theaters, photography darkrooms, etc.). Certain spaces such as offices and study areas should take advantage of windows and daylighting (Figure 10.6). It also is critical to keep entry and orientation of the building in perspective for purposes of minimizing direct sunlight to reduce glare or taking advantage of direct sunlight for energy production.

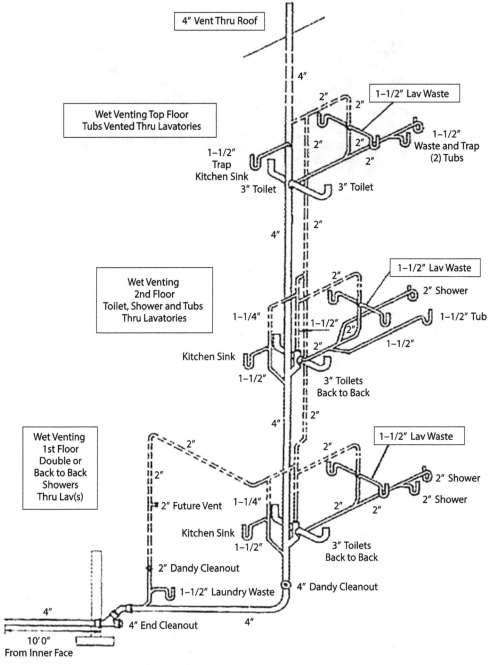

Figure 10.5 Three-dimensional depiction of stacking bathrooms (source: http://www.mass.gov/ocabr/licensee/dpl-boards/pl/regulations/rules-and-regs/248-cmr-1000-contd.html)

Figure 10.6 Aluminum louvers for blocking direct sunlight

There are a few rules for converting floor plans to elevations:

◆ Windows cannot cross walls
◆ Window/door sizes typically are standard unless customization is required (which costs a lot more money)
◆ Natural light is related to building orientation and location of windows
◆ Certain functions within the building are light-limited (e.g., computer rooms)
◆ Columns and shear wall locations may frustrate external aesthetics
◆ Access points must be integrated between the elevations and site plan

It should be obvious in comparing drawings that the details in the elevation view must match the ones in the plan view. The easy way to do this is to use the same scale on all floor plans and elevations, line them up, and compare them (Figure 10.7). Three-dimensional software like Revit® or Civil 3D® can help in this regard.

Floor planning is not a straightforward, obvious exercise. It takes time and experience. Few students have either, which complicates the process. In the rest of this chapter, three different examples of floor plans, developed by capstone design students, are reviewed.

The first project is a middle school. Table 10.1 is the building program. A series of code requirements impacted the floor plan, including exits within 150 ft of all classrooms, the use of "pods" or clusters, and space for up to 1,500 students (but with limits on student:teacher ratios). For example, this building requires extra staircases with building exits to meet the minimum travel distance to means of egress and the minimum number of exits per building story, as dictated by the building code. Plumbing includes toilets, sinks, slop sinks, and

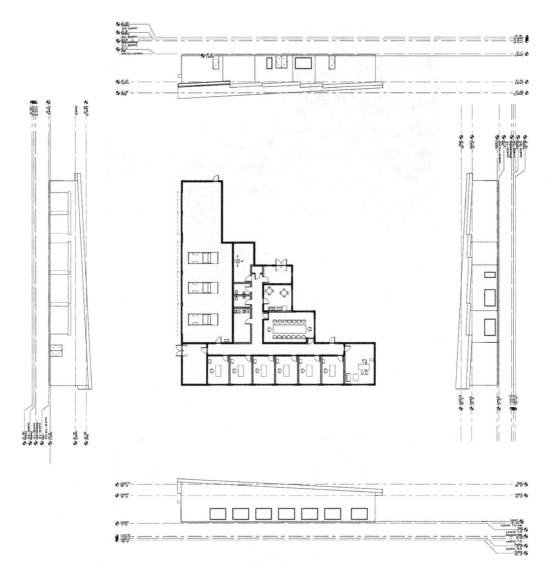

Figure 10.7 Final floor plan with associated elevations

drinking fountains to meet building codes. In addition, for this example, the classroom windows were raised 3 ft to provide useful storage space inside the classroom underneath the windows. Figures 10.8 and 10.9 show the floor plans that were presented to the client. Note that these were not the initial floor plans, but rather floor plans that were developed after several iterations. Figure 10.10 shows the proposed elevation view. The project was converted to a Revit file to address the conflicts associated with windows, doors, and other items. Figures 10.11 to 10.13 are Revit renderings of the middle school building.

The second example involves a hotel project. The existing hotel was comprised of four separate buildings that the owner wanted to convert into one connected structure (see Figure 10.14). Because the project was in a historical protection zone, the façades could not be modified from their original appearance. The biggest issue to solve was that the second floors

Table 10.1 Building program for middle school project

Space type	Net square footage provided	Square footage requested	Notes
ADA rest rooms	1,008	1,600	Adjoining cafetorium
Administration	6,086	7,500	First floor, near entrance
Art studio	1,588	5,125	North-facing windows
Atrium	3,421	As needed	
Cafetorium	15,564	11,000	External wall exposure, first floor with high ceilings
Core classrooms	21,892	17,000	Math, language arts, social studies
Custodial	875	1,800	
Exceptional student education	4,162	4,600	
Family/consumer science lab	2,098	As needed	
Library media center	9,950	8,000	
Mechanical rooms	262	4,000	
Music department	2,804	6,000	
Physical education	13,633	13,500	
Resource rooms	1,734	1,200	
Science demonstration labs	989	As needed	
Science labs	5,048	3,000	
STEM skills labs	3,963	3,000	
Storage	1,489	400	
Teacher planning	2,450	1,200	Second-floor location

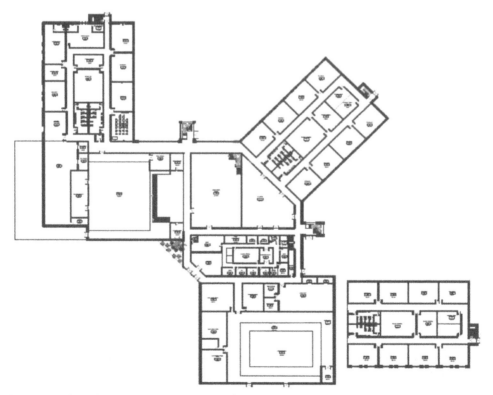

Figure 10.8 First-floor plan for middle school project

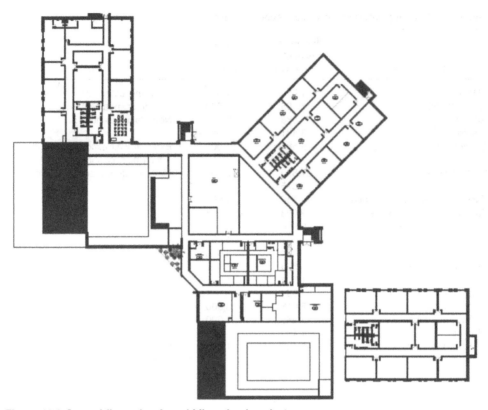

Figure 10.9 Second-floor plan for middle school project

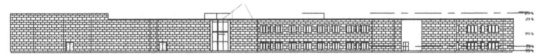

Figure 10.10 Elevation plan for middle school project

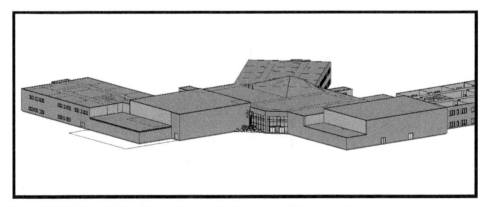

Figure 10.11 3-D rendering view of east elevation for middle school project

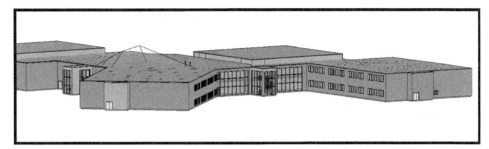

Figure 10.12 3-D rendering view of south elevation for middle school project

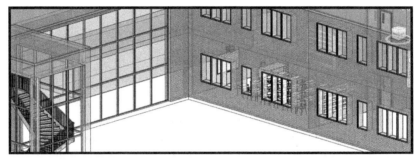

Figure 10.13 Classroom window configuration for middle school project

Figure 10.14 Existing four-building configuration for hotel project

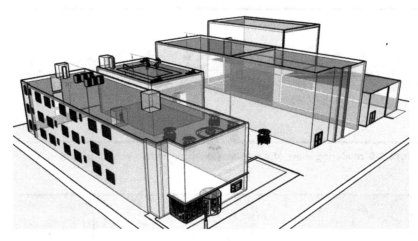

Figure 10.15 Proposed exterior rendering for hotel project

of all four buildings were at different elevations. To make one floor plan, the elevation issue had to be addressed. In addition, the hotel wanted to maximize the number of rooms, even though the façade had to remain the same. This was a severe limitation on the room adjustments that could be made. Figure 10.15 shows the concept of the building elevation after discussions with the hotel owner and a number of employees. In Figure 10.15, the original hotel (structure on the left) is unchanged. The second structure is slightly modified in the rear and connected to the first and third buildings. A floor was added to the third building to match the first building. The external portion of the fourth structure was unchanged except for an addition to match the third building in the rear half of the structure. Note that the historical protection code permitted changes if they were not readily visible from the street, and the additional floor on the fourth building would not be.

Developing floor plans takes time and is not as simple as it sounds. It requires thought, perhaps even more thought than the site planning process. It is unlikely that an engineer will arrive at the best solution on the first few attempts due to the many variables to consider. Designing an effective floor plan requires the design team to think about how people move inside the building and what facilities should go together.

An example of a project that took time to evolve is a medical lab building proposed on a college campus. The building program requirements are shown in Table 10.2. The new building needed to be placed adjacent to an existing medical building. Movement of students and faculty was an important consideration. However, there were a number of issues to be addressed. The staff wanted labs on three floors, which meant a three-story building, as well as specified office sizes (10 ft × 12 ft) and windows for certain faculty and staff. There were two office suites that were undefined and a number of computer labs and examination rooms.

Table 10.2 Building program for medical lab building project

Space type	Net square footage provided	Quantity	Notes
Lobby	min 500	2	First floor at each entrance
Flexible lab space	3,000	1	
Clean rooms	150	4	
Research labs	total 10,000	??	
Researcher offices	min 120	12	
Faculty offices	min 135	12	
Graduate student offices	min 50	24	Can be combined or separate
Department support facilities	min 500	2	Mail room, records/additional storage
Department office	min 420	2	Reception area, additional clerical area, program assistant desk
Chair office	min 200	2	Corner office
Associate chair office	min 200	4	2 per department
Computer laboratories	min 1,800	6	40 workstations each
Conference room	min 600	2	1 per department
Seminar room	min 800	2	50 seats
Medium classrooms	min 1,800	2	60 seats with distance learning capability
Loading dock area	min 750	1	
Kitchenette/lunchroom	min 250	1	
Student simulation center	5,000	1	Area for students to work/study between classes and outside of labs and lobby areas

Figures 10.16 to 10.18 outline the first draft of the building based on the initial site plan. This did not work because the long, thin room spaces were not conducive to large open classrooms and study areas and made the structural design overly complex. Faculty offices were spread all over the building, which disrupts cohesion, and flow from classes (or specific subject disciplines) is unclear from this plan. The capstone team's site plan, which was long and narrow to accommodate stormwater issues, figured into the room shapes.

A site plan change brought the team to Figures 10.19 to 10.21 over a period of months. This squared up the building, making the labs, classrooms, and study areas easier to develop. However, the plan still lacked cohesion between the faculty and departments. The odd sizes of the proposed spaces led to complicated layout schemes and rooms with columns in the middle. Structural aspects of this building also were lacking, as columns (represented by the circles in Figures 10.20 and 10.21) were required in the middle of classrooms and lab spaces. Resolving the column issue created eccentric loads all over the building, which made the design far more complicated than it needed to be.

What was needed was an integration of space and a better flow to reduce long hallways. However, the driving factor for the final site plan (Figures 10.22 to 10.24) turned out to be the layout of columns within the building. The engineers had looked at structures as an

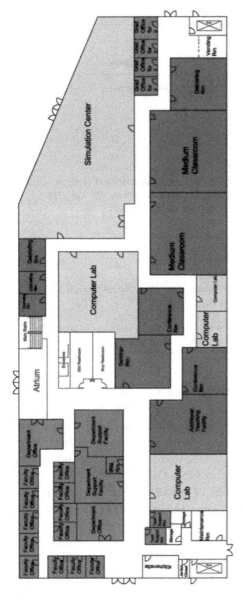

Figure 10.16 First iteration of first-floor plan for medical lab building project

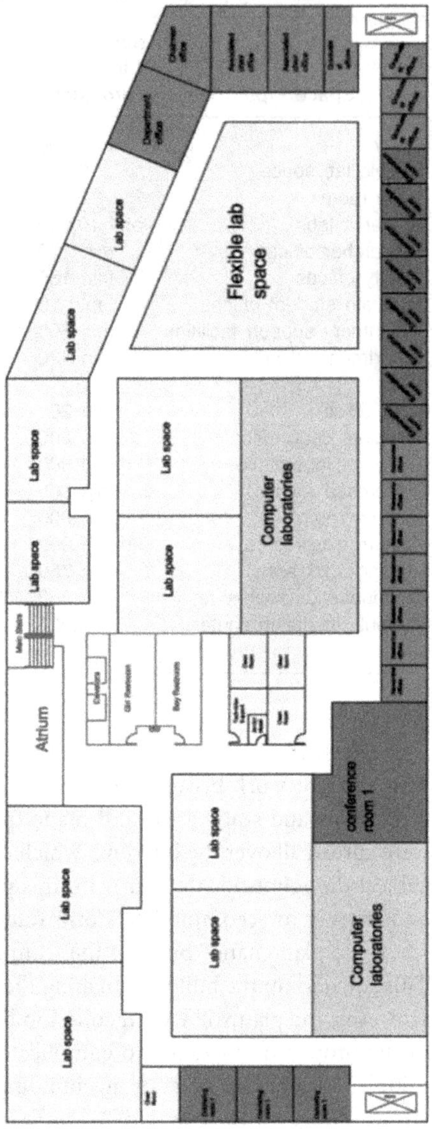

Figure 10.17 First iteration of second-floor plan for medical lab building project

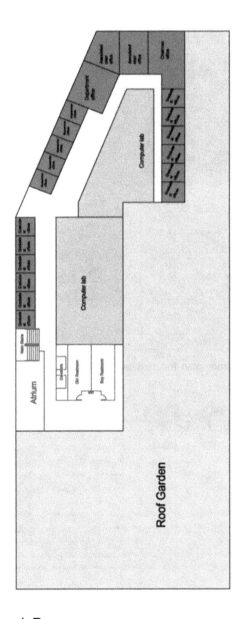

Figure 10.18 First iteration of third-floor plan for medical lab building project

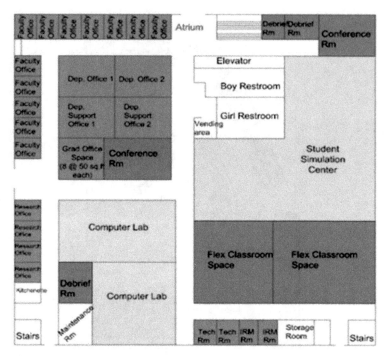

Figure 10.19 Revised first-floor plan for medical lab building project

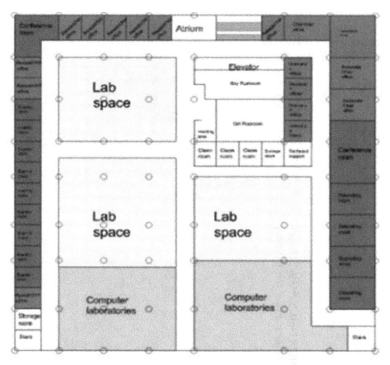

Figure 10.20 Revised second-floor plan with proposed column layouts for medical lab building project

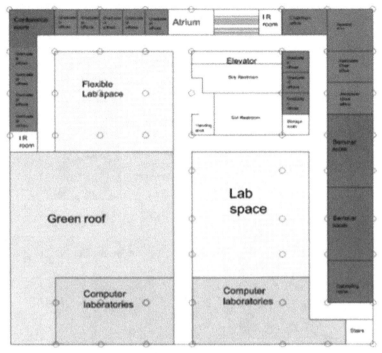

Figure 10.21 Revised third-floor plan with proposed column layouts for medical lab building project

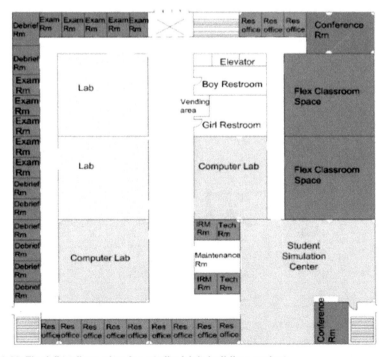

Figure 10.22 Final first-floor plan for medical lab building project

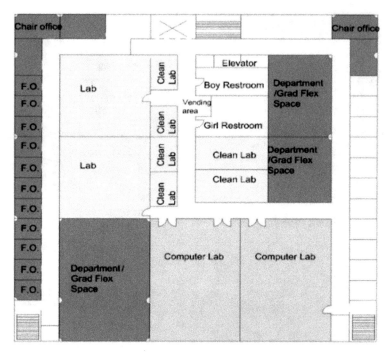

Figure 10.23 Final second-floor plan for medical lab building project

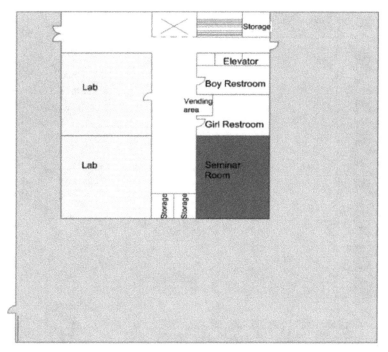

Figure 10.24 Final third-floor plan for medical lab building project with green roof over much of second floor

afterthought, which is not efficient. A column-first approach helped make the floor plan design easier. The project could then move on to elevations only after the floor plans, building shape, and structural columns were defined.

In all of the floor plans for the three different projects discussed, there were substantial challenges, including the fact that none of the capstone students had much experience with floor planning. In the first example, the challenges required the students to talk to middle school teachers and administrators about the needs of the users and incorporate that information into their design. The second example had major issues regarding the different finished floor elevations of the existing buildings that needed to be connected. Extensive efforts were required to resolve the floor plan issue, limited by the maximum allowable ramp slopes and runs, as prescribed by ADA guidelines. The third example had major challenges in putting together a floor plan because of the complexity of the column layout. The capstone students went through multiple iterations of trying to get the structural plan and floor plan to match. In all three examples, an integrated team design process demonstrated how important it was for all members of the group to participate in their respective exercises to create workable floor plans that met the goals of the project.

11

Engineering Economics

Engineering is all about solving problems. Once a problem is defined, additional data and analysis typically are required to identify the alternatives that should be considered in order to make an informed decision (see Chapter 6). Data should include a projection of the future. Forecasts should be made of the appropriate variables to consider in order to meet future needs. Engineers will be asked to evaluate alternatives throughout their careers, and in nearly every decision, the overriding criterion is cost. One of the goals of the engineer, therefore, will be to find cost-effective solutions to present to the client, which does not always mean the option with the least expensive capital outlay. Many clients are more concerned about the most immediate first costs, which include the capital costs and costs associated with design and construction, than they are with operations and maintenance, which is the reason for resistance to the added first costs (premium) of high-performance buildings.

Engineers use several methods to conduct benefit-cost analyses for a project, but the most useful involves calculating the present worth, annual worth, or future worth of an investment. Present worth analyses, also known as present value, calculate the capital costs and the operations and maintenance costs over the life of the project if all lifetime costs were brought to the present (i.e., a lump sum). Future worth, also known as future value, converts the cash flow to a dollar amount a number of years down the road. To understand how to correlate cash flows to a yearly value, the engineer will use annual worth analyses. The current situation, or the "do nothing" alternative, usually is considered the base situation (and likely the least expensive option) for benefit-cost analyses.

Given that an infrastructure item is projected to last 30 to 50 years, the major overall costs during its lifetime may be operations and maintenance. Hence, with the correct economic analysis, the client can be guided to the lowest overall life cycle cost of a project. For example, many of the features associated with high-performance buildings discussed in Chapter 7 are focused on operations cost savings over the life of the item. Sustainable technologies may increase first costs because, as shown before, the LEED® Gold average premium is 2% (USGBC 2011), but major savings in operations can make these investments worthwhile in the long run. The tools used by engineers for such analyses fall under the discipline of *engineering economics*, which is basically what financing agencies, banks, loan officers, and others use for financial projections. Engineers utilize these engineering economics tools to help guide them toward a recommendation as to whether or not to pursue a particular alternative.

Suppose a client must decide between a $100,000 pump with $15,000 in annual operating costs and a $105,000 unit with $12,000 per year in operating costs. Which item is more cost effective in the long term? One way to handle this is to perform a present worth analysis, which is conducted by bringing all anticipated costs of a project to the current time to compute the present value of a cash flow.

On the other hand, if this client is a government agency, which is scrutinized on the basis of annual costs, the client may ask for a projection of the annual costs or the annuities required to be set aside in the annual budget for this item. Engineers can project annual operations costs easily using engineering economics. They also can estimate debt service using the same economic analyses.

Finally, most people, including engineers, are likely to save money in an individual retirement account (IRA) or similar instrument. They probably want to know how much money can be saved over time. The key component in this analysis is time. The future worth (future value) of investments can be calculated by knowing how much money needs to be put away each year for a known number of years to obtain a targeted dollar amount at the end of the investment period.

A major note of caution is that many times the benefits are unknown or not quantifiable at the time the analysis is conducted, so care should be exercised with any financial analysis, and appropriate caveats, assumptions, and missing data should be noted. There may be future benefits, or the alternative may encourage other actions which are difficult to decouple from the original decision. Water/sewer utilities, for example, typically are funded by the public because the benefit-cost analysis of extending pipelines into undeveloped areas will always fail the benefit-cost test. This failure occurs because the timing for future development is unknown, and there are relatively few customers at the time of construction to offset the initial costs. The benefit of extending piping may be to expand or stimulate development. It may also discourage other utilities that might compete in the future. Loss of service or failure to extend the pipe network could also represent a significant disbenefit to the utility in the form of potential lost income. It takes money to make more money, as is true for the expansion of utilities. All of these economic analyses discussed so far can help the engineer answer the following questions:

◆ How much will it cost to borrow money?
◆ What is the payback period on upgrades or engineering alternatives?
◆ What is the rate of return on the investment?

Engineering economic evaluations involve six common components (Blank and Tarquin 2008):

◆ **Present value** (P)—Refers to the cost of the asset or cash flow at the current time ($t = 0$). Capital costs are present value amounts.
◆ **Annuity or uniform payments** (A)—Refers to the periodic costs of the project converted to a uniform flow stream. Annuities can be used as annual amounts, but often are adapted for use with monthly, weekly, or daily payments by adjusting the number

of payment periods (n) in a year. Debt service for capital components is an example of a uniform cash flow stream. Debt service can be determined from P and the interest rate for a given number of payment periods. For instance, some municipal bonds are issued for a period of 30 years. At a given interest rate, the annual debt service can be found from actuarial tables (more on this a little later). Operations and maintenance costs are annual amounts that rarely are uniform, a fact that engineers need to keep in mind.

◆ **Effective interest rate (i)**—Refers to the rate of borrowed money and also the rate of inflation, the rate of increase in operating costs, and the desired rate of return, which are all basically expressed in the form of interest rates. Higher interest rates mean higher annual costs for borrowing. Higher inflation rates will increase operations costs geometrically. Combined interest rates may exist when inflation and growth occur at the same time. Interest rates should be assumed to compound (usually yearly for economic analyses).

◆ **Number of payment periods or number of payments (n)**—Refers to the length of time over which the analysis will be considered or the number of payments that will be made over a given period of time.

◆ **Future value (F)**—Refers to the value of an investment, such as real estate, that is expected to increase in value over time or the eventual value of a series of payments, such as a retirement account. Future worth can be determined from P and i for a given term n. For instance, if a building is owned for 10 years, the future value can be estimated given P, i, and n.

◆ **Gradient value (G)**—Refers to the value of a cash flow that is increasing in constant increments over time. Operations and maintenance costs are annual amounts, but they rarely are uniform. Instead, they tend to inflate by a given amount each year. Gradients may be useful in situations where arithmetic growth in a given number of customers each year (G) is occurring, but not on a percent per year basis (geometric growth). Many rate analysts mistakenly assume that growth is a given percent each year. This actually means growth accelerates with time, which is rarely the case even in high-growth locations such as Las Vegas, NV or Naples, FL. If growth is exponential, this would be treated more like an interest rate.

Every engineering economics problem will require a minimum of four of these six key variables. At least three must be known in order to arrive at a solution. The approach is to lay out the options as follows:

P = Present value
A = Annuity (or uniform payments, but not necessary annually)
i = Effective interest rate
n = Number of payments
F = Future value
G = Gradient (constant increase each year)

The acronym is easy to remember: P-A-i-n-F-G.

The most important part of engineering economic analysis is creating an appropriate cash flow diagram. The diagram has four possible components, which can occur together and in multiples: current amounts (present worth), future amounts (can be sales, profits, or costs), annuities (ongoing payments or receipts), and gradients, which usually follow annuities. These are the P, A, F, and G in P-A-i-n-F-G. Of importance for students is to decide how to best view incoming versus outgoing amounts. Usually present value is indicated by an arrow pointing up because it is the current value. A future amount can be an increase in value for that same present worth (arrow also pointing up) or a sale, which would mean that cash flow is reversed (arrow pointing down); an example is buying a car versus selling one. Likewise, if someone buys a house, for example, the value of the house (the amount paid) is an upward arrow, but the buyer may wish to characterize the monthly payments (expenses or monthly house payment/annuity) as a downward arrow. Students need to develop a comfortable means to understand the drawings. There are conventions, but often these conflict with how students visualize payments coming in versus going out.

Appropriate interest rates and annual increases in operations and maintenance costs are required for any economic analysis to be valid. Likewise, the same life expectancy should be used for present worth alternatives; alternatives that have different life expectancies cannot be compared in a valid economic analysis. Where the lifetime expectations are different, annual worth analyses may be easier to calculate, but this may be a more difficult concept to understand.

11.1 Interest Rates

Before proceeding further, a brief discussion of interest rates is necessary. There are two types of interest rates: simple and compound. *Simple interest* rates rarely are used in commercial businesses, banks, the stock market, or virtually anywhere else. Simple interest is defined as follows:

$$\text{Simple interest} = \text{Principal} \times \text{Periods} \times \text{Interest rate} = P \times n \times i \qquad (11.1)$$

The concept is as follows. If a principal of $P = \$1,000$ is borrowed from a friend for a period of $n = 5$ years and simple interest of $i = 10\%$ is charged each year, the borrower would be responsible for paying only $100 per year in interest. Over the 5-year period, $500 in total interest would be paid.

Businesses and banks typically use *compound interest* rates for borrowing or other financial instruments. These rates are based on interest charged over time. In other words, interest also is paid on the accrued interest for each period (just like home mortgages, credit cards, or student loans). This is the method used most often and certainly by all financial institutions. The concept forms the basis for the actuarial or interest tables in Appendix 11A. Compound interest is defined as follows:

$$\text{Compound interest} = (\text{Principal} + \text{Accrued interest}) \times \text{Interest rate} \qquad (11.2)$$

The difference between the two can be significant after just a few years. For the same $P = \$1,000$ from the previous example borrowed from a friend for $n = 5$ years but this time using compound interest, Table 11.1 compares the amount of simple versus compound interest that will accrue given an interest rate of $i = 10\%$ per year. At the end of 5 years, the compound interest garners $110.51 more, which is equivalent to 22.1% more than the simple interest case over the life of the loan. It is clear why lenders prefer this method.

Table 11.1 Example comparison of compound and simple interest

Year	Value of investment	Simple interest at 10% per year	Value of investment	Compound interest	Difference
0	$1,000.00	—	$1,000.00		
1	$1,100.00	$100.00	$1,100.00	$100.00	—
2	$1,200.00	$100.00	$1,210.00	$110.00	$10.00
3	$1,300.00	$100.00	$1,331.00	$121.00	$21.00
4	$1,400.00	$100.00	$1,464.10	$133.10	$33.10
5	$1,500.00	$100.00	$1,610.51	$146.41	$46.41
Total		$500.00		$610.51	

11.2 Single-Payment Present Worth

By far, the most common economic analysis approach is present worth analysis, in which all costs associated with an engineering alternative over its useful life are brought to a single payment in current-day dollars (present value). Essentially, this calculation determines the lifetime value of the investment if paid all at once in today's money. To organize payments and receipts in a visual representation, create a *cash flow diagram* (Figure 11.1). The horizontal line represents the timeline of the investment. A tick mark represents one payment period (n). Arrows represent payments and receipts. All payments should be represented as up arrows, and all receipts should be represented as down arrows (or vice versa as long as it is consistent). A simple cash flow diagram shows the current (present value) and the future situation, as illustrated in Figure 11.1.

Returning to P-A-i-n-F-G, we know the following information:

P = Present value; this is the variable that must be determined in this analysis (P = ?)
A = Annuity (this value is not used in this problem)
i = Effective interest rate (i = 6%)
n = Number of payments (n = 1)
F = Future value (F = $1,000)
G = Gradient value (this value is not used in this problem)

Note that A and G are not used in this analysis. Depending on whether the goal is to determine the present or future value, the scenario is different. For example, to find the

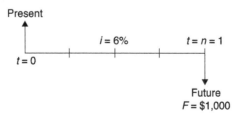

Figure 11.1 Cash flow diagram for single-payment present worth of a future payment of $1,000 1 year from now at 6% interest per year

Table 11.2 Partial factor table for *i* = 6% interest

n	P/F	P/A	P/G	F/P	F/A	A/P	A/F	A/G
1	0.9434	0.9434	0.0000	1.0600	1.0000	1.0600	1.0000	0.0000
2	0.8900	1.8334	0.8900	1.1236	2.0600	0.5454	0.4854	0.4854
3	0.8396	2.6730	2.5692	1.1910	3.1836	0.3741	0.3141	0.9612
4	0.7921	3.4651	4.9455	1.2625	4.3746	0.2886	0.2286	1.4272
5	0.7473	4.2124	7.9345	1.3382	5.6371	0.2374	0.1774	1.8836
6	0.7050	4.9173	11.4594	1.4185	6.9753	0.2034	0.1434	2.3304
7	0.6651	5.5824	15.4497	1.5036	8.3938	0.1791	0.1191	2.7676
8	0.6274	6.2098	19.8416	1.5938	9.8975	0.1610	0.1010	3.1952
9	0.5919	6.8017	24.5768	1.6895	11.4913	0.1470	0.0870	3.6133
10	0.5584	7.3601	29.6023	1.7908	13.1808	0.1359	0.0759	4.0220

present value given a future value, the mathematical formula to solve this question would be as follows:

$$P = F \left[\frac{1}{(1 + i)^n} \right]$$ (11.3)

While this is a relatively simple mathematical formula, the values are more easily determined by using the engineering economics interest tables (actuarial tables) in Appendix 11A, using the notation of engineering economics, as follows:

$$P = F(P/F, i, n)$$ (11.4)

The present value (P) is equal to the future value (F) multiplied by the factor ($P/F, i, n$) from the actuarial tables in Appendix 11A. The factor is the present value given the future value (P given F, or P/F) for a set interest rate (i) and number of payment periods (n). In our example, the factor would be (P/F, 6%, 1) or 0.9434, as shown in Table 11.2, which is a subset of data taken from Appendix 11A. The factor is arrived at by using the 6% interest table and entering at $n = 1$ for the P/F column.

Now the solution looks like:

$$P = F(P/F, 6\%, 1) = \$1,000 \times (0.9434) = \$934.40$$

Excel® also can be used to solve a present worth analysis, as follows:

$$PV(i\%, n, , F)$$ (11.5)

11.3. Future Value or Single-Payment Compound Amount

If instead one wanted to calculate the future value of money invested today, the mathematical formula would be as follows:

$$F = P(1 + i)^n$$ (11.6)

While this is a similarly simple formula, the values are more easily obtained by using the engineering economics interest tables in Appendix 11A, as follows:

$$F = P(F/P, i, n) \tag{11.7}$$

Excel also can be used, as follows:

$$FV(i\%, n, , P) \tag{11.8}$$

For example, assume it is 2012. In 1932, a quarter would have bought two matinee tickets to the movies, which would have included cartoons, newsreels, mini-features, and then the feature film—probably 4 hours in all. For that quarter, the two moviegoers also could have purchased a large tub of popcorn and a drink to share. They would have been entertained for hours. The question is what the purchasing power of that same quarter from 1932 is in the year 2012, given that the rate of inflation over that same time period is estimated to be an average of 3% per year. Would it be possible to take a date to the movies with this kind of money in 2012?

First create the cash flow diagram, as shown in Figure 11.2. Then, returning to the P-A-i-n-F-G model, the following pieces of information are known (note: A and G are unnecessary for this analysis):

P = Present value (P = 0.25)
A = Annuity (this value is not used in this problem)
i = Effective interest rate (i = 3%)
n = Number of periods (n = 2012 − 1932 = 80 years)
F = Future value; this is the value that we want to calculate (F = ?)
G = Gradient value (this value is not used in this problem)

The solution is to determine the future value (F) by multiplying the present value (P) by the factor from the actuarial tables in Appendix 11A. The factor is the future value given the present value (F given P, or F/P) for a set interest rate (i = 3%) and number of payment periods (n = 80). In this example, the factor would be (F/P, 3%, 80) or 10.6409. Now the solution looks like:

$$F = P(F/P, 3\%, 80) = \$0.25(10.6409) = \$2.70$$

This means that the value of $0.25 in 2012 dollars is $2.70, which certainly would not cover the cost of admission or any refreshments at a movie theater today. What this example

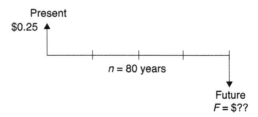

Figure 11.2 Cash flow diagram for future value 80 years from now given a present payment of $0.25

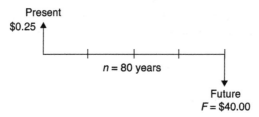

Figure 11.3 Cash flow diagram for single-payment present worth of a future payment of $40.00 80 years from now with a present value of $0.25 to determine the average interest rate over the period

demonstrates is that the cost of a movie theater ticket has increased faster than the rate of inflation over the past 80 years.

Next, let's find out what that actual rate of inflation has been over that 80-year period, assuming that two movie tickets, popcorn, and two drinks cost $40.00 in 2012 (refer to Figure 11.3). Then, returning to the P-A-i-n-F-G model, the following pieces of information are known (note: in this instance, A and G are unnecessary):

P = Present value (P = $0.25)
A = Annuity (this value is not used in this problem)
i = Effective interest rate (i = ?)
n = Number of periods (n = 2012 − 1932 = 80 years)
F = Future value (F = $40.00)
G = Gradient value (this value is not used in this problem)

In this case, the interest rate is unknown. The solution is to determine the interest rate by dividing the future value by the present value to find the value of the factor from the tables in Appendix 11A. The factor is the future value given the present value (F given P, or F/P) for a set interest rate (i) and number of payment periods (n). In our example, the factor would be (F/P, ?%, 80). Now the solution looks like:

$$F = P(F/P, ?\%, 80)$$

$$(F/P, ?\%, 80) = F/P = \$40.00 \div \$0.25 = 120$$

Examining the tables in Appendix 11A, at 80 years in the 6% interest table, the value of (F/P, 6%, 80) is 105.7960, and in the 7% table, the value is 224.2344. Because the value 120 is in between these two table values, the actual inflation rate over those 80 years is between 6 and 7%, actually much closer to 6% because 120 is closer to 105.7960 than it is to 224.2344. A more accurate value is determined by *linear interpolation*:

$$\frac{\text{Rise}}{\text{Sun}} = \frac{224.2344 - 120}{224.2344 - 105.7960} = \frac{7 - x}{7 - 6}$$

where x = 6.12%.

Thus, the effective equivalent inflation rate over this 80-year period is i = 6.12% using this data, which indicates that the rate of increase for a trip to the movies has exceeded 6%

per year, and that value is double the actual reported rate of inflation over that same time period.

11.4 Annual Worth

In an annual worth analysis, all cash flows are converted to periodic payments called *annuities* (not necessarily annual). A cash flow diagram should be constructed to show the current and future situation (Figure 11.4). It is important to note that annuities are paid at the end of the year, so the diagram shows the first payment of A at the end of period $n = 1$.

Think of it as a distributed load (a rectangle) being transferred to another part of a beam. Returning to the concept of P-A-i-n-F-G, the following information is known (ignoring F and G in this case, which are unnecessary for the analysis):

P = Present value
A = Annuity; this is the value that we want to find in this problem (A = ?)
i = Effective interest rate
n = Number of payments
F = Future value (this value is not used in this problem)
G = Gradient value (this value is not used in this problem)

Depending on whether we want to find the present value or the annuity value, the scenario is different. For example, to compute the present value given an annuity value (uniform series present worth), the actual mathematical formula required would be:

$$P = A \left[\frac{(1 + i)^n - 1}{i(1 + i)^n} \right] \tag{11.9}$$

This clearly is not a simple formula and another reason why students should learn to use the actuarial tables as follows:

$$P = A(P/A, i, n) \tag{11.10}$$

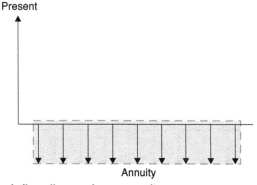

Figure 11.4 Generic cash flow diagram for an annuity

Excel also can be used, as follows:

$$PV(i\%, n, A) \tag{11.11}$$

If instead we want to find the annuity amount given current conditions (capital recovery), the mathematical formula would be:

$$A = P\left[\frac{i(1 + i)^n}{(1 + i)^n - 1}\right] \tag{11.12}$$

This is the formula used for bonds, debts, annual or periodic payments, and the like. Note, however, that this is a far more complicated formula than relating P and F, which means far more potential for mathematical errors. In the tables in Appendix 11A, the quantities already have been calculated, saving the engineer time. The values are easily found in engineering economics interest tables as follows:

$$A = P(A/P, i, n) \tag{11.13}$$

Excel also can be used, as follows (the Excel command for annual payments is PMT):

$$PMT(i\%, n, P) \tag{11.14}$$

A sample annual worth analysis best illustrates the usefulness of determining annuity payments. Kaly is an engineer who is doing very well for herself, and she wants to buy an expensive house. The purchase price is $800,000. The interest rate is 3% per year. If she makes only annual payments for 30 years, what is the payment amount? First, create the cash flow diagram (Figure 11.5). Then list the variables as follows using the P-A-i-n-F-G concept (note: F and G are not used in this analysis):

P = Present value (P = $800,000)
A = Annuity; this is the value that we want to find in this problem (A = ?)
i = Interest rate (i = 3%)
n = Number of payments (n = 30 years)

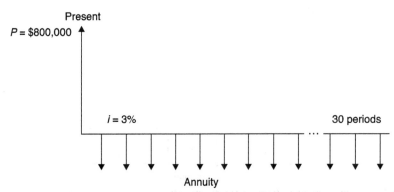

Figure 11.5 Cash flow diagram to determine an annuity from a present cash payment of $800,000 over a period of 30 years at 3% interest per year

F = Future value (this value is not used in this problem)
G = Gradient value (this value is not used in this problem)

The solution is to determine the annuity by multiplying the present value by the factor, which is the value of the annuity given the present value (A given P, or A/P) for the 3% interest rate and 30 payment periods. In this example, the factor would be $(A/P, 3\%, 30)$. Now the solution looks like this:

$$A = P(A/P, 3\%, 30)$$

$$A = \$800{,}000 \times (0.051) = \$40{,}800 \text{ per year}$$

Note: Mortgages are never calculated annually like this. They are computed monthly, but we will come back to that soon.

11.5 Future Worth Given an Annuity

There are times when the future worth (future value) is needed instead of the present value or annuity. One such case is to determine the future value of an IRA or a savings account to accumulate a down payment for a house. The scenario looks like Figure 11.6. Returning to the concept of P-A-i-n-F-G, the following pieces of information are known in this case:

P = Present value (this value is not used in this problem)
A = Annuity; this is the value that is given in this problem
i = Effective interest rate
n = Number of payments
F = Future value; this is the value that we want to find in this problem (F = ?)
G = Gradient value (this value is not used in this problem)

Note that P and G are not necessary for the analysis.

Depending on whether we want to calculate the annuity or the future value, the scenario is different. For example, to find out the annuity given some future value (uniform series sinking fund), the actual mathematical formula required is as follows:

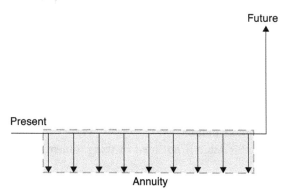

Figure 11.6 Generic cash flow diagram to determine future worth from an annuity

$$A = F\left[\frac{i}{(1+i)^n - 1}\right] \tag{11.15}$$

Although this also is not a simple formula, the values are easily found in engineering economics interest tables as follows:

$$A = F(A/F, i, n) \tag{11.16}$$

Excel also can be used, as follows:

$$\text{PMT}(i\%, n, F) \tag{11.17}$$

If instead we want to calculate the future value of an annuity (uniform series compound amount), the formula would be as follows:

$$F = A\left[\frac{(1+i)^n - 1}{i}\right] \tag{11.18}$$

Note that this complicated formula already is solved in the tables in Appendix 11A. The values are easily found in engineering economics interest tables as follows:

$$F = A(F/A, i, n) \tag{11.19}$$

Excel also can be used:

$$FV(i\%, n, A) \tag{11.20}$$

Again, a sample future worth analysis best illustrates the usefulness of determining annuity payments to reach some future dollar amount for retirement, for instance. Allie will graduate with her master's degree in civil engineering at age 23. She is planning to retire at age 63 and enjoy the fruits of her efforts. She estimates that she will need $3 million to retire comfortably. How much should she put away each year if the average rate of return in the market is 7%?

First, construct the cash flow diagram (Figure 11.7). Returning to the P-A-i-n-F-G concept, the following pieces of information are known (note: P and G are not used in this analysis):

P = Present value (this value is not used in this problem)
A = Annuity; this is the value that we want to find in this problem (A = ?)
i = Effective interest rate (i = 7%)
n = Number of payments (n = 63 − 23 = 40 years)
F = Future value (F = $3 million)
G = Gradient value (this value is not used in this problem)

The solution is to determine the annuity by multiplying the future value by the factor, which is the value of the annuity given the future value (A given F, or A/F) for the 7%

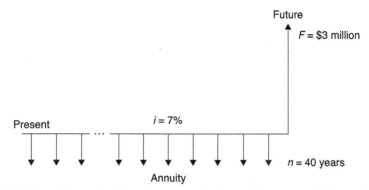

Figure 11.7 Cash flow diagram to determine an annuity from a future expected payout of $3 million at 7% interest over 40 years

interest rate and 40 payment periods. In this example, the factor would be $(A/F, 7\%, 40)$. Now the solution looks as follows:

$$A = F(A/F, 7\%, 40)$$

$$A = \$3,000,000 \times (0.00501) = \$15,030 \text{ per year}$$

Clearly, this is a manageable sum for a young engineer, especially when most of it is tax sheltered in an IRA account, which means that this investment also will lower her tax obligations!

11.6 Gradients

Gradients come in two types: (1) a percent increase for each payment period (g) and (2) a constant increase for each payment period (G). There are limits on how gradients are applied, and there are no Excel functions available at this time to help calculate gradients. There are, however, ways to get to P, F, and A given gradients. Let's start with the present value first. For example, to find the present value of a given constant gradient (G), also known as uniform gradient present worth, the cash flow diagram would look like Figure 11.8.

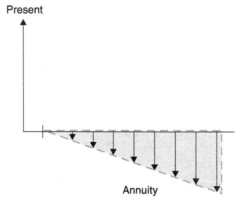

Figure 11.8 Generic cash flow diagram to determine the present worth of a constant gradient of payments

Of critical importance is that the first year of the gradient is *always* zero. This fact is often overlooked and leads to incorrect calculations. The actual formula required is as follows:

$$P = G \left[\frac{(1 + i)^n - 1}{i^2(1 + i)^n} - \frac{n}{i(1 + i)^n} \right] \tag{11.21}$$

This is a complicated formula, but luckily the values are usually found in engineering economics interest tables:

$$P = G(P/G, i, n) \tag{11.22}$$

If we want to find the future value of a given gradient (uniform gradient future worth), the cash flow diagram would look like Figure 11.9. The actual mathematical formula that would be required is as follows:

$$F = G \left[\frac{(1 + i)^n - 1}{i^2} - \frac{n}{i} \right] \tag{11.23}$$

Again, this is a very difficult formula, but the computations can be rather simple using the tables in Appendix 11A with the following factor:

$$F = G(F/G, i, n) \tag{11.24}$$

Note that both F and P are best represented as triangle loads, just like in mechanics, although only present value is normally found in engineering economics interest tables. Therefore, to determine the future value of a gradient, simply obtain the present value and then convert to a future value using the F/P factors.

The conversion of a gradient to an annuity (uniform gradient uniform series) is an uncommon conversion but one that is found in many interest tables. The actual mathematical formula required to perform this operation is as follows:

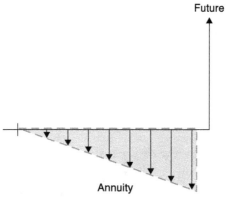

Figure 11.9 Generic cash flow diagram to determine the future worth of a constant gradient of payments

$$A = G\left[\frac{1}{i} - \frac{n}{(1 + i)^n - 1}\right]$$ (11.25)

This complicated formula can sometimes be found in interest tables as follows:

$$A = G(A/G, i, n)$$ (11.26)

One thing to note about gradients is that they usually are applied in conjunction with annuities. An example might help to better illustrate this point. Let's say that Paula wants to purchase a new car, and she wants to know what the present worth of the maintenance costs will be over the next 10 years. The dealer tells her that the first year is free, and after that the projection is that the maintenance cost will rise by $500 per year. What is the present worth of the maintenance costs?

First a cash flow diagram must be created (Figure 11.10). Note that the first year is zero, which helps. The maintenance costs in the cash flow diagram then look like a triangle with a slope of $500 per year). Returning to P-A-i-n-F-G, we know the following (note that A is zero because there are no operations costs and F is not used):

P = Present value; this is the value that we want to find in this problem ($P = ?$)
A = Annuity (this value is not used in this problem)
i = Effective interest rate ($i = 4\%$)
n = Number of payments ($n = 10$ years)
F = Future value (this value is not used in this problem)
G = Gradient value ($G = \$500$)

Remember that the value of the gradient is zero in the first year, so in this case, the number of payment periods is 10 because the first payment is $0, the second payment is $500, the third payment is $1,000, the fourth payment is $1,500, and so on. Now, the solution is to

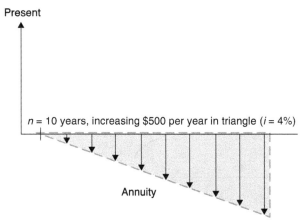

Figure 11.10 Cash flow diagram to determine the present worth of car maintenance payments that increase by $500 per year starting in year 2 and continuing for the next 10 years as a gradient of payments at 4% interest per year

determine the present worth by multiplying the gradient value (G = 500) by the factor (P given G, or P/G) for the 4% annual interest rate and 10 payment periods. In this example, the factor would be (P/G, 4%, 10). Now the solution looks like:

$$P = G(P/G, 4\%, 10)$$

$$P = \$500 \times (33.884) = \$16,942$$

11.7 Shifted Annuities

A common practice, especially with large furniture or car purchases but also with banking for projects that take a long time to construct and as a result have limited cash flow initially, is to delay payments. In this case, the cash flow diagram shows an annuity block that starts after a delay period, as shown in Figure 11.11. The cash flow diagram in Figure 11.11 has the annuity starting at the beginning of term 3 (or the end of term 2). Remember that annuities are paid at the end of the period. This value is P_1, which occurs in year 2, as shown in Figure 11.12. In this case, P_1 does not occur at t = 0. It actually occurs in year 2, so it

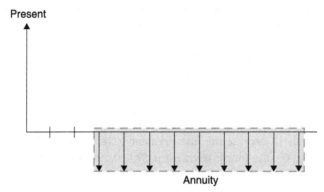

Figure 11.11 Generic cash flow diagram for a shifted annuity of payments delayed to the third period

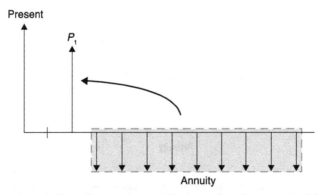

Figure 11.12 Generic cash flow diagram for a shifted annuity of payments delayed to the third period, showing how the present value of the annuity is given as P_1 in year 2, which then must be shifted to the present using P/F

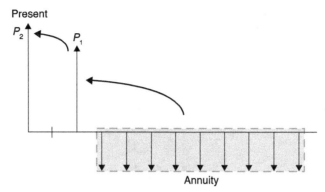

Figure 11.13 Generic cash flow diagram for a shifted annuity of payments delayed to the third period, showing how the present value of the annuity is given as P_1 in year 2 and then, using P/F, the P_1 value is shifted to P_2 for the present value in current dollars

is not the present worth of the annuities. P_1 must be brought back to the present time, which is $t = 0$. To solve this problem, P_1 can be viewed as a future value occurring at $t = 2$, which can then be brought back to the present, which is the actual present value of the annuities (P_2) in this example, as shown in Figure 11.13. Therefore, in this case, the value of the annuities is found to be $P_1 = A \times (P/A, i, 9)$, but this present value is actually in year 2, so it is treated as a future value. Then P is equal to $P_1 \times (P/F, i, 2)$.

Let's try an example to better illustrate this concept. Daniel and his wife purchase a new car. As part of the deal they are promised no maintenance payments until the end of year 3. However, they must keep the car for at least 5 years. For years 3 through 5, maintenance will be $4,000 per year as specified in the purchase agreement. To determine the present worth of this delayed maintenance plan, first create the cash flow diagram as shown in Figure 11.14. The three annuity payments need to be converted into a P_1 in year 2, as shown in Figure 11.15. Assuming the annual interest rate is 5%, the results are as follows:

$$P_1 = A(P/A, i, n)$$

$$P_1 = \$4,000 \ (P/A, 5\%, 3) = \$4,000 \times (2.6347) = \$10,539$$

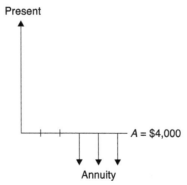

Figure 11.14 Cash flow diagram for the present value of a shifted annuity of delayed payments of $4,000 starting at the end of year 3

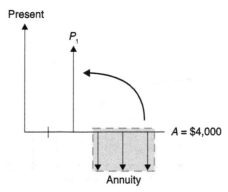

Figure 11.15 Cash flow diagram for the present value of a shifted annuity of delayed payments of $4,000 starting at the end of year 3; the delayed payments have a present value in year 2 of P_1, which must be shifted to the present

But the problem is not done yet; P_1 does not occur in year 0. It is in year 2, so it does not represent the current-day present worth of the cash flow diagram. Thus, the approach will be to first take the annuities and bring them back to year 2 as P_1 and then bring P_1 back to the present as P, as shown in Figure 11.16. The final step, to bring P_1 to $t = 0$, is required. P_1 is essentially a future value, so assuming the annual interest rate is still 5% and $P_1 = F$, the results are as follows:

$$P = F(P/F, i, n)$$

$$P = \$10,539 \ (P/F, 5\%, 2) = \$10,539 \times (0.9070) = \$9,559$$

Hence, the $3 \times \$4,000 = \$12,000$ cost of the annuity in years 3 to 5 is actually worth $9,559 in current-day dollars. This is the time value of money. Money loses value with time as a result of inflation.

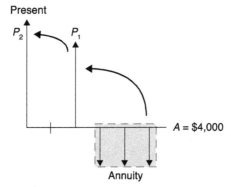

Figure 11.16 Cash flow diagram for the present value of a shifted annuity of delayed payments of $4,000 starting at the end of year 3; the delayed payments have a present value in year 2 of P_1, which must be shifted to the present using P/F to determine the present value (P_2) in current dollars

11.8 More about Interest Rates

Everyone has seen advertisements for loans that show a given annual percent interest rate, followed by some small print with a larger value. The more easily read number is the *nominal interest rate* (*r*). Most people do not know what the number in fine print is, but are suspicious that the lender is not telling them something. That something is that interest rates for loans are rarely calculated on an annual basis by banks; they are calculated on a monthly basis. Most lenders charge interest on a monthly basis because that is how often payments are made. As a result, the interest is compounded monthly, not yearly. That number in the fine print is the equivalent annual percent interest rate or *effective interest rate* (i_a), assuming monthly compounding. And yes, it is always a larger number and is the actual number used to calculate loan payments. Essentially, interest can be compounded in any fashion, and the conversion is relatively easy to do, even without the interest tables in Appendix 11A. The formula for converting to virtually any compounding period is simple and is easily developed:

$$i_a = (1 + r \div m)^m - 1 \tag{11.27}$$

where

i_a = effective interest rate per year in decimal form
r = nominal interest rate per year in decimal form
m = periods per year

You see the two rates expressed next to each other for credit cards and mortgages. For the conversion to, say, a monthly interest rate, the formula would be:

$$i_a = (1 + r \div 12)^{12} - 1 \tag{11.28}$$

As the number of payment periods increases, a limit is reached approaching a natural log:

$$\lim_{m \to \infty} \left(1 + \frac{r}{m}\right)^m - 1 = e \tag{11.29}$$

Therefore:

$$i_a = e^r - 1 \tag{11.30}$$

where *r* is the nominal interest rate.

Let's try an example. Assume a nominal interest rate of 10% per year. What would the effective interest rate be if the compounding period is monthly, weekly, daily, or continuous? Keep in mind that interest rates are actually decimals, so the decimal form (10% = 0.10) is used. The results would look as follows:

Monthly ($m = 12$): $\quad i_a = (1 + r \div m)^m - 1 = (1 + 0.10 \div 12)^{12} - 1$

$$= 0.1047 \text{ (or } 10.47\%)$$

Weekly ($m = 52$): $\quad i_a = (1 + r \div m)^m - 1 = (1 + 0.10 \div 52)^{52} - 1$

$$= 0.1051 \text{ (or } 10.51\%)$$

Daily ($m = 365$): $\quad i_a = (1 + r \div m)^m - 1 = (1 + 0.10 \div 365)^{365} - 1$

$$= 0.1052 \text{ (or } 10.52\%)$$

Continuous ($m = \infty$): $\quad i_a = e^r - 1 = e^{0.10} - 1$

$$= 0.1052 \text{ (or } 10.52\%)$$

From this analysis, it is understandable why monthly compounding is so popular. It generates more money for the lender, for limited extra work. Trying to update all loans and accounts daily, or even weekly, creates only a marginal increase in revenue. This has a purpose when dealing with loans that are related to monthly compounding. For example, what is the annual cost of a mortgage payment if paid only once per year versus monthly? This comparison is easily evaluated using engineering economics principles.

Let's assume the loan amount is $100,000 at 6% annual interest for 10 years. First draw the appropriate cash flow diagram, as shown in Figure 11.17. Returning to P-A-i-n-F-G, we know the following:

P = Present value (P = $100,000)
A = Annuity; this is the value that we want to calculate in this problem (A = ?)
i = Effective interest rate (i = 6%)
n = Number of payments (n = 10 years)
F = Future value (this value is not used in this problem)
G = Gradient value (this value is not used in this problem)

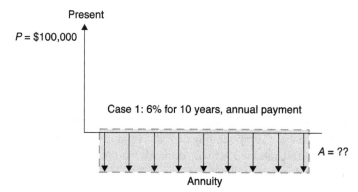

Figure 11.17 Case 1 cash flow diagram for the value of an annuity from a present value of $100,000 at 6% interest for 10 years

The solution is to determine the value of the annuity given the present value, as follows:

$$A = \$100{,}000 \times (A/P, 6\%, 10)$$

$$A = \$100{,}000 \times (0.13587) = \$13{,}587 \text{ per year}$$

If divided up into monthly payments, this value becomes $\$13{,}587 \div 12 = \$1{,}132$ per month.

But mortgages are never paid annually; they are paid monthly, so we use Figure 11.18 to see how this plays out. In this case, things are more complicated. Recalling the effective interest rate formula (Equation 11.27), with $r = 0.06$ and $m = 12$:

$$i_a = (1 + r \div m)^m - 1 = (1 + 0.06 \div 12)^{12} - 1 = 0.0617 \text{ (or } 6.17\%)$$

Unfortunately, there are no interest tables for this number in Appendix 11A. Instead, referring to Equation 11.28, it is clear that the interest rate should be divided by the number of payments in a year ($i = 6\% \div 12 = 0.5\%$), and the number of payments should be multiplied by that same number ($n = 10$ years $\times$ 12 months per year = 120). This results in the following P-A-i-n-F-G answers:

P = Present value ($P = \$100{,}000$)
A = Annuity; this is the value that we want to calculate in this problem ($A = ?$)
i = Effective interest rate ($i = 6\%$ per year $\div$ 12 months per year = 0.5% per month)
n = Number of payments ($n = 10$ years $\times$ 12 months per year = 120)
F = Future value (this value is not used in this problem)
G = Gradient value (this value is not used in this problem)

Now we can use the present value and the factor for A given P for 120 payment periods at an interest rate of 0.5% per month (found in Appendix 11A):

$$A = \$100{,}000 \times (A/P, 0.5\%, 120)$$

$$A = \$100{,}000 \times (0.01110) = \$1{,}110 \text{ per month}$$

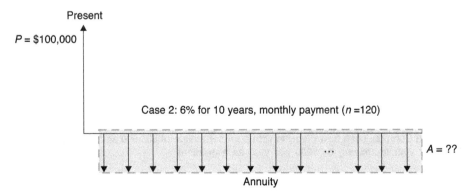

Figure 11.18 Case 2 cash flow diagram for the value of an annuity in monthly payments for 10 years ($n = 10 \times 12 = 120$) from a present value of $100,000 at 6% interest per year

Recall that if paying once per year, the cost would be $13,587. However, the annual cost paying monthly would be $13,320 at $1,110 per month, thus saving $267 per year by paying monthly because the accumulated interest is paid off faster. Therefore, the monthly payments actually save the borrower money as well as get the bank its money faster. It is a win-win situation.

11.9 Dealing with More Complex Cash Flow Diagrams

The cash flow diagrams to this point have been fairly straightforward. In any engineering economics problem, the key to breaking down any given scenario and creating a measurable or comparative analysis is the cash flow diagram.

The next item is to evaluate what happens when multiple A's, G's, or F's are found in the cash flow diagram. Let's try an example. Suppose Paulina wants to buy a new car. The deal is $P = \$35,000$. Maintenance will be $A = \$500$ per year, increasing at $G = \$200$ per year. If she keeps the car for $n = 10$ years, what is the life cycle cost of the car at an interest rate of $i = 4\%$ per year? The first step is to create the cash flow diagram, as shown in Figure 11.19. Using the P-A-i-n-F-G approach:

P = Present value ($P_1 = \$35,000$)
A = Annuity ($A = \$500$)
i = Effective interest rate ($i = 4\%$)
n = Number of payments ($n = 10$ years)
F = Future value (this value is not used in this problem)
G = Gradient value ($G = \$200$)

The approach to use here is to sum up the present values, such that $PW = P_1 + P_2 +$ .... In this example, P_1 is the value of the car. P_2 is the present value of the maintenance costs, and P_3 is the present value of the gradient. The calculations are as follows:

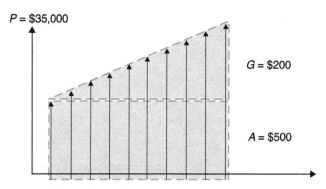

Figure 11.19 Complex cash flow diagram with a present value of $35,000 and a maintenance cost of $500 per year increasing by $200 per year as a gradient

$$PW = P_1 + P_2 + P_3$$

$$P_1 = \$35,000$$

$$P_2 = A \times (P/A, 4\%, 10) = \$500 \times (8.1109) = \$4,055.45$$

$$P_3 = G \times (P/G, 4\%, 10) = \$200 \times (33.8814) = \$6,776.28$$

$$PW = P_1 + P_2 + P_3 = \$35,000 + \$4,055.45 + \$6,776.26 = \underline{\underline{\$45,831.73}}$$

Paulina is offered another deal for the same new car, but this time the sticker price is $40,000. The maintenance is capped at $500 per year, but it does not start to increase until year 5. What is the present worth of this competing offer over a lifetime of 10 years? Again, the first step is to create the cash flow diagram, as shown in Figure 11.20. Using the P-A-i-n-F-G approach:

P = Present value (P_1 = $40,000)
A = Annuity (A = $500)
i = Effective interest rate (i = 4%)
n = Number of payments (n = 10 years)
F = Future value (this value is not used in this problem)
G = Gradient value (G = $200 but does not start until year 5)

Again our approach will be similar, that is, to sum up the present values for each payment:

$$PW = P_1 + P_2 + PG$$

$$P_1 = \$40,000$$

$$P_2 = A \times (P/A, 4\%, 10) = \$500 \times (8.1109) = \$4,055.45$$

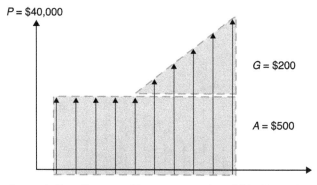

Figure 11.20 Complex cash flow diagram with a present value of $40,000 and a maintenance cost of $500 per year increasing by $200 per year after year 5 as a gradient

The term PG is where things start to change. The gradient does not start until after year 5. Remember that the first year of any gradient is zero, so this means that the gradient goes from year 4 to year 10, which is 6 years. Thus, the calculation looks as follows:

$$PG = G \times (P/G, 4\%, 6) = \$200 \times (12.5062) = \$2,501.24$$

Remember that this PG value does not occur at $t = 0$, but actually at $t = 4$. Hence, for our purposes, this is truly an F value, and it must be brought back to the present before it can be added to the other P values (see Figure 11.21). Thus, PG is written as F, and then the new expression is solved for P_3:

$$P_3 = (PG = F) \times (P/F, 4\%, 4)$$

Note that the years $(6 + 4 = 10)$ equal the total time owned. Now the final present value term (P_3) can be solved for:

$$P_3 = (PG = F) \times (P/F, 4\%, 4) = \$2,501.24 \times (0.8548) = \$2,137.31$$

Next we add up all of the present value terms:

$$PW = P_1 + P_2 + P_3 = \$40,000 + \$4,055.45 + \$2,137.31 = \underline{\mathbf{\$46,192.76}}$$

Recall that in the previous deal, Paulina would pay a present value of $45,831.73. That deal saves $361.03, so she should take the first deal, not just because the sale price is less but because the life cycle costs are lower too.

To take this a step further, let's compare the two options, but include a salvage value price for each. We can assume that the salvage value price at the end of the ownership period is 20% of the initial value. Remember that the salvage value is not a payment; it is money received, so for calculation purposes, it is depicted as a negative value (down arrow in the cash flow diagram). Now which is the better deal?

Let's start with the deal for the first car and create the cash flow diagram (Figure 11.22). Then, using the P-A-i-n-F-G approach:

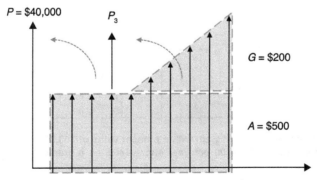

Figure 11.21 Complex cash flow diagram with a present value of $40,000 and a maintenance cost of $500 per year increasing by $200 per year after year 5 as a gradient, showing how the gradient is first brought back to P_3 and then back to the present

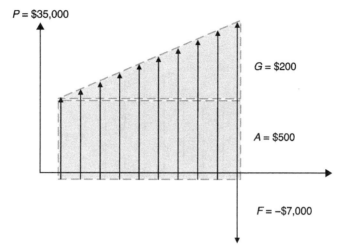

Figure 11.22 Complex cash flow diagram with a present value of $35,000 and a maintenance cost of $500 per year increasing by $200 per year as a gradient but with a salvage value of $7,000

P = Present value (P_1 = $35,000)
A = Annuity (A = $500)
i = Effective interest rate (i = 4%)
n = Number of payments (n = 10 years)
F = Future value (F = −$7,000)
G = Gradient value (G = $200)

Again, the approach will be to add up the present worth values of each payment:

$$P_1 = \$35,000$$

$$P_2 = A \times (P/A, 4\%, 10) = \$500 \times (8.1109) = \$4,055.45$$

$$P_3 = G \times (P/G, 4\%, 10) = \$200 \times (33.8814) = \$6,776.28$$

$$P_4 = F \times (P/F, 4\%, 10) = -\$7,000 \times (0.6756) = -\$4,729.20$$

Now the final present worth for the first deal is computed as follows:

$$PW = P_1 + P_2 + P_3 + P_4 = \$35,000 + \$4,055.45 + \$6,776.28 - \$4,729.20 = \underline{\mathbf{\$41,102.53}}$$

Next, let's take a look at the second deal (Figure 11.23). Using the P-A-i-n-F-G approach:

P = Present value (P_1 = $40,000)
A = Annuity (A = $500)
i = Effective interest rate (i = 4%)
n = Number of payments (n = 10 years)
F = Future value (F = −$8,000)
G = Gradient value (G = $200 but does not start until year 5)

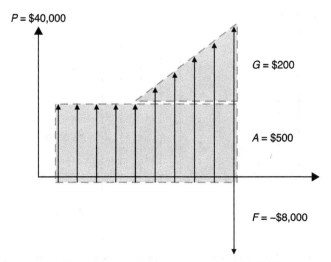

Figure 11.23 Complex cash flow diagram with a present value of $40,000 and a maintenance cost of $500 per year increasing after year 5 as a gradient of $200 per year with a salvage value of $8,000

Again the solution approach is to sum up the P's, so we add up the present worth value of each payment:

$$P_1 = \$40,000$$

$$P_2 = A \times (P/A, 4\%, 10) = \$500 \times (8.1109) = \$4,055.45$$

$$P_3 = G \times (P/G, 4\%, 6) = \$200 \times (12.5062) = \$2,501.24$$

$$P_4 = (F = P_3)(P/F, 4\%, 4) = \$2,501.24 \times (0.8548) = \$2,137.31$$

The final term is the salvage value of $8,000, converted to a present worth:

$$P_5 = F \times (P/F, 4\%, 10) = -\$8,000(0.6756) = -\$5,404.80$$

Now, summing all the terms, the present value for the second deal is computed as follows:

$$PW = P_1 + P_2 + P_4 + P_5 = \$40,000 + \$4,055.45 + \$2,137.31 - \$5,404.80$$

$$= \underline{\$40,787.96}$$

In this case, the first deal ends up costing more ($41,102.53) compared to the second deal ($40,787.96).

11.10 Comparing Options

As demonstrated in the previous car purchase example, present worth analysis is useful for comparing options. Because the costs for capital, operations, and maintenance may be different for each option, the only practical way to analyze them is by bringing all of the payments and revenues to the present value. This is also known as *life cycle cost analysis*. Note that when multiple options with multiple components are compared, signs become important in the cash flow diagram. Just make sure the signs are consistent such that all payments have the same sign and all revenues have the opposite sign. Also, there are several rules that must be adhered to when comparing alternatives in current-year dollars. First, the alternatives must be exclusive. They also must be independent, which means that completing one alternative cannot be a requirement for a second alternative. Keep in mind that the "do nothing" approach is a viable alternative. Both costs and revenues must be included in the analysis.

Let's revisit the first example at the beginning of this chapter. Suppose the client must decide between a $100,000 pump with $15,000 per year in operating costs (option 1) and a $105,000 unit with $12,000 per year in operating costs (option 2). Which item is more cost effective in the long term at 5% interest for 5 years? One way to handle this is to perform a present worth analysis, which is conducted by bringing all anticipated costs of a project to the current time.

In this case, option 1 is calculated as follows:

$$PW_1 = \$100,000 + \$15,000(P/A, 5\%, 5) = \$100,000 + \$15,000(4.3295) = \underline{\mathbf{\$164,843}}$$

and option 2 is calculated as follows:

$$PW_2 = \$105,000 + \$12,000(P/A, 5\%, 5) = \$105,000 + \$12,000(4.3295) = \underline{\mathbf{\$156,954}}$$

Clearly, option 2 costs $7,889 less over the 5-year life of the item.

This approach can best be explained with a set of more realistic and increasingly complex examples. Table 11.3 compares adding more pumps (alternative 1) versus more storage in

Table 11.3 Present worth cost comparison between two viable options

Improvement	Alternative 1 ($)	Alternative 2 ($)
Booster pumps	$875,000	n/a
New tank	$3,255,000	n/a
High-service pumps	$420,000	$1,116,000
New generators for pumps	n/a	$4,742,000
Present worth of total capital	$4,550,000	$5,858,000
Operations and maintenance ($/year)	$54,000	$59,000
Present worth of operations and maintenance (P/A, 3.5%, 20)	$768,600	$839,800
Total present worth	$5,318,600	$6,697,800

Table 11.4 Comparison of infiltration/inflow options

Item	Treating infiltration	Reducing infiltration and inflow
Capital cost	—	$5,000,000
Annual operations cost	$1,500,000	$150,000
Estimated growth rate of flow per year	3.50%	3.50%
Present worth of operations and maintenance (P/A, 3.5%, 20)	$21,350,850	$2,135,085
Present worth	$21,350,850	$7,135,085

the distribution system (alternative 2). The analysis was completed using $i = 3.5\%$ and $n = 20$. The lowest life cycle cost option is alternative 1.

In Table 11.4, the cost of doing infiltration and inflow correction was compared against not removing this excess water ($i = 3.5\%$ interest per year and $n = 20$ years). The "do nothing" alternative clearly costs more money now, but over the 20-year planning horizon this option costs almost $14 million more. Hence, correcting leaky pipes has a significant benefit to the utility by avoiding costs for treating infiltration and inflow.

Keep in mind that when performing a life cycle cost analysis, any values can be brought back to the present worth easily. The procedure uses all of the principles developed so far:

1. Create a cash flow diagram.
2. Convert all payments and receipts to current dollars using the P/G, P/A, or P/F factors from the tables in Appendix 11A.
3. Add up the present worth value of each term ($PW = P_1 + P_2 + P_3 + \dots P_n$).

This analysis method also works for calculating payback periods and bond investments. Also, keep in mind that the focus for engineers tends to be on the design and construction components of a project; however, the operations and maintenance costs may be far larger over the life of the project. Decisions made today in design have a measurable impact on the operations and maintenance costs of the future. This is why the concept of analyzing high-performance buildings on a life cycle basis, as opposed to using construction first costs only, is so important. This means that engineers must consider costs from day one, starting with capital investment and including acquisition (design, assessment, and capital investment) and operations (implementation, construction, maintenance, usage, and phaseout).

Obvious uses of engineering economics in the business world include manufacturing plants where products will create a positive cash flow that will (hopefully) offset their construction and production costs. Another example is to demonstrate that a toll road will eventually pay for itself. It is interesting that in most engineering economics texts, the cost of construction is included in the operations phase, but just the opposite is common practice for public sector infrastructure, where operations and maintenance costs are separated from the initial construction costs!

11.10.1 Break-Even Analysis

In the private sector, analysis often focuses on the break-even point using *net cash flow* (NCF) and neglecting the time value of money. Financing options can be compared with a term called the *payback period* (n_p). This value is the time it takes to recuperate the initial investment if yearly cash flows are equal, using the following approach:

$$P = \text{NCF} \times (P/A, i, n_p) \qquad (11.31)$$

where

n_p = payback period or how long it takes to pay back the investment/loan in full

Let's illustrate this concept with a typical example. A company plans to build a new factory. The annual revenues are expected to be $1 million in current dollars. Expenses are $800,000 per year. The net cash flow is the difference or $200,000. Let's also say the new factory costs $1 million to construct. When will the break-even point occur?

The payback period (n_p) calculates the break-even point in terms of time. Assume inflation is 3% per year, so $i = 3\%$. Therefore:

$$\$1,000,000 = \$200,000 \times (P/A, 3\%, n_p)$$

$$\$1,000,000 \div \$200,000 = 5 = (P/A, 3\%, n_p)$$

Looking at the tables in Appendix 11A:

$$(P/A, 3\%, 5) = 4.5797$$

and

$$(P/A, 3\%, 6) = 5.4172$$

Therefore, by using the factors we can see that the equation equals 5 in between years 5 and 6, so $n \approx 5.5$ years.

11.10.2 Annual Worth Analysis

Annual worth also can also be used instead of present value to compare annual costs by creating an equivalent uniform amount that can be planned for. In this case, the equivalent periodic payments of a cash flow or even the *rate of return* (ROR) on an investment can be determined. Alternatively, the ROR, given some *minimal attractive rate of return* (MARR), can be calculated this way. MARR is the minimum return on investment that would convince an investor to choose this option. This value is effectively just an interest rate. There are five assumptions in this type of analysis:

◆ Plan for one life cycle only.
◆ Services are needed forever.
◆ The same alternatives repeat forever.
◆ Cash flows (revenues) are the same each year (unless there is a reason for a difference).
◆ The annual worth is greater than zero when MARR is met and negative when the interest rate is less than MARR.

ROR is defined as the rate paid on the unpaid balance of borrowed money or the unrecovered balance of an investment. As a loan is paid off, the amount of interest due declines as more and more principal is paid off. This is why the mortgage amount remains constant, but the interest due each month decreases as the principal paid increases over time. ROR is the actual result given a set of circumstances, as opposed to MARR, which is a desired rate of return or a target interest rate. It is important to keep the concepts of ROR and MARR separate as there may be several options that will meet MARR by having an ROR greater than MARR. Note that there is no computer solution for ROR given annual worth. The procedure for annual worth comparisons is as follows:

◆ Rank the alternatives by decreasing initial investment cost.
◆ Develop cash flow diagrams for each alternative, assuming reinvestment.
◆ Verify signs for payments and revenues.
◆ Create a present worth equation for each alternative.
◆ Convert the present worth value of each alternative to annual worth.
◆ Compare annual costs; note that if the present worth analysis is created first, the lifetime is automatically the same for all options.

Thus, basically, a present worth analysis is done and then converted to annual amounts. There can be multiple interest rates. If the interest rate of, say, alternative B is less than MARR, one of the other alternatives is chosen. If the interest rate of, say, alternative C is greater than MARR, the investment is justified. Using annual worth, the intent is to determine which cash flow alternative gives the greatest bang for the buck in terms of periodic revenues.

11.11 Inflation Adjustment

Inflation is a decline in purchasing power. It is an increase in the costs of goods to be purchased compared to those same costs in a prior period. Inflation is sensitive to economic strength, consumer spending, borrowing rates, energy costs, and supply and demand. Also, it can be impacted by political turmoil, war, material shortages, stock market volatility, debt default, fiscal policy, and public perception. What inflation does is devalue money with time! Most people are familiar with annual inflation rates, but the federal government actually tracks inflation quarterly, and the published inflation rates may or may not include certain goods like food or energy, depending on the index used. These are very complicated formulas that go well beyond the scope of this text. There are a variety of inflation indices available. One is for construction (Construction Cost Index published by *Engineering News-Record*). Others exist for various industries. The *Consumer Price Index* (CPI) is the most common and the one most people assume is used. The CPI is used by the U.S. Bureau of Labor Statistics to measure the average change over time in the prices paid by consumers for a common basket of goods and services. The CPI is based on prices of food, clothing, housing, fuel, transportation, medical care, drugs, and other goods and services that people purchase for day-to-day living. The CPI is calculated by taking price changes for each item in the predetermined composition (weighted influence) of these goods and services and averaging them according to their importance. Prices are collected each month in 87 areas across the country from about 6,100 housing units and approximately 24,000 retail establishments (department stores, supermarkets, hospitals, gas stations, and other types of stores and service establishments).

Since the 1920s, the United States has tracked inflation (see Figures 11.24 and 11.25). What is immediately obvious from Figures 11.24 and 11.25 is that the rate of inflation

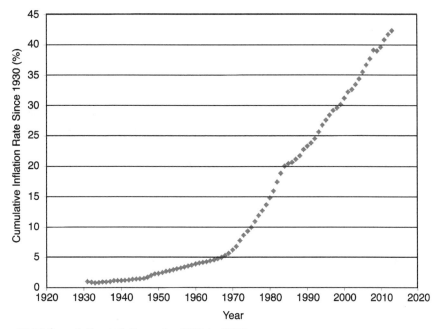

Figure 11.24 Cumulative inflation rate: 1930 to 2013

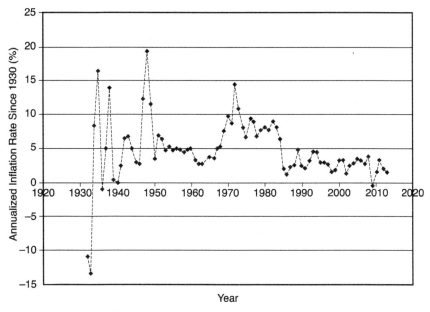

Figure 11.25 Annualized inflation rate: 1930 to 2013

decreased after 1970, when the Federal Reserve Bank made inflation control a monetary policy of the federal government. Also, the Federal Reserve tinkered with how the CPI was calculated.

In the business world, in order for the net value of an investment to grow over time, the growth rate must exceed the rate of inflation. However, expenses increase with time (inflation) as well, and that must be accounted for as a part of the economic analysis by using an F given P approach, as follows:

$$F = P(1 + f)^n \qquad (11.32)$$

where f is the inflation rate.

From this analysis, it is easy to see that performing inflation calculations is identical to F/P calculations, where the goal is to convert constant-value dollars (today's dollars) to future dollars. The caveat is that inflation usually is accompanied by some growth or other interest value (which is why inflation is denoted by f instead of i).

Combining interest rates and inflation is slightly trickier. Let's assume a compounding rate for each. The interest equation is:

$$P = F \div (1 + i)^n \qquad (11.33)$$

With adjustment for inflation, the equation becomes:

$$P = F \div (1 + i)^n \div (1 + f)^n \qquad (11.34)$$

With inflation-adjusted interest rates, the new effective rate (i_a) is given as follows:

$$i_a = i + if + f \qquad (11.35)$$

Inflation and the gradient are added when the items compound one another. An example would be predicting power costs for a utility when power costs are expected to increase by 3% and growth is expected to increase 5%. Those two compound one another. Inflation detracts from an interest rate for income. For example, if you get a 3% raise but inflation is 3%, the relative growth of your purchasing power is zero. As another example, if project costs were expected to increase by 10% per year and inflation was 4%, then:

$$i_a = i + if + f = 0.10 + (0.10 \times 0.04) + 0.04 = 0.144 \text{ (or 14.4\%)}$$

Now, suppose an engineer is asked to project the expenses for a facility's power costs. In this case, power is related to product sales, which are expected to increase at 5% per year for the next several years (note that this is actually the gradient g as opposed to i). If inflation is running at 3%, then the coming year's cost of power can be projected as follows:

$$i_a = i + if + f = 0.05 + (0.05 \times 0.03) + 0.03 = 0.0815 \text{ (or 8.15\%)}$$

The 8.15% is a major difference from either the 3% for inflation or the 5%, which may be the value the finance director wants to use. In either case, the facility ends up exceeding its budget.

Inflation works against investments. As noted previously, investors want their income or profits to increase faster than the rate of inflation. This is so they do not lose purchasing power in the market.

Let's say a buyer is purchasing a building to hold for 10 years. Suppose the buyer expects the building to increase in value at 6% per year. What is the value of the building if it is purchased for $100,000? To answer this question, we construct a cash flow diagram (Figure 11.26) and calculate just like any P/F scenario, but adjust for the two interest rates:

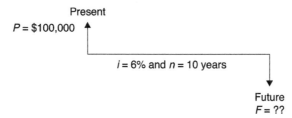

Figure 11.26 Cash flow diagram for future value given a present worth of $100,000 at 6% interest for 10 years

$$F = P \times (F/P, 6\%, 10) = \$100,000 \times (1.7908) = \$179,080$$

While the value of the building is $179,080, inflation has eaten away at this value. In this case, the inflation value would be negative because it is lowering the value of the investment over time:

$$i_a = i + if + f = 0.06 + (0.06 \times -0.03) - 0.03 = 0.0282 \text{ (or 2.82\%)}$$

The new future value is determined as follows:

$$F = P \times (F/P, 2.82\%, 10)$$

$$F = \$100,000 \times (1.3325) = \$133,253$$

This is a big difference ($45,827), but at least still a profit because the value is positive. The bigger problem occurs when inflation exceeds the rate of growth.

Let's take a look at a period of slow growth. Suppose a client is buying a building to hold for 10 years. Let's assume the value of the building will increase 2% per year, with inflation at 3%. What is the value of the building if it is purchased for $100,000? The answer is computed as follows:

$$F = P \times (F/P, 2\%, 10)$$

$$F = \$100,000 \times (1.219) = \$121,900$$

While the value of the building is $121,900, inflation has eaten away at this value. In this case (see Figure 11.27), inflation will make the effective interest rate negative, dragging down the comparative value of the building value (adjusted for inflation), as follows:

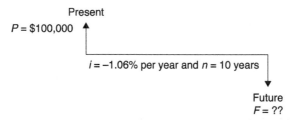

Figure 11.27 Cash flow diagram for future value given a present worth of $100,000 at –1.06% annual interest for 10 years

$$i_a = i + if + f = 0.02 + (0.02 - 0.03) - 0.03 = -0.0106 \text{ (or } -1.06\%)$$

$$F = P \times (F/P, -1.06\%, 10)$$

$$F = \$100,000 \times (0.9071) = \$90,705$$

Adjusted for the heavy inflation rate, the building actually loses value, and the client would be better off finding a different investment because this one ends up costing money even though the value of the building increased.

As a final note, the gradient interest rate (g), as was used for the growth in the utility's customers, is treated exactly the same way as i and f; i or f is just replaced with g as necessary.

11.12 Depreciation

The key to infrastructure management is making and properly maintaining appropriate capital expenditures. In a time when there are limited dollars available from customers, spending large sums of money on capital infrastructure that may be of limited value in the future reflects poorly on management. The long term is evaluated in two ways. First is the useful life of the asset before it needs to be replaced. Second is the annual cost associated with maintaining the asset (Figure 11.28).

Figure 11.28 shows two curves; the lower curve reveals a lower initial cost, but the life expectancy of the asset is significantly less than the asset in the upper curve. The upper curve shows a higher initial cost, but the asset lasts longer. There is an incentive for utilizing capital to improve quality so as to decrease the long-term asset requirements. A present worth analysis of the system shows that a higher quality item (longer lasting with minimal maintenance costs) has a lower life cycle cost than a cheaper item that needs to be repaired more often or may need more frequent replacement. In both cases, however, the asset still depreciates in value.

Keep in mind that depreciation is an accounting tool used to project the loss in value of an asset with time. It does not reflect the actual or market value of the asset. This is an important point. Depreciation is used as a tax deduction for private entities. To improve economic growth after a recession, governments typically will pass legislation allowing for

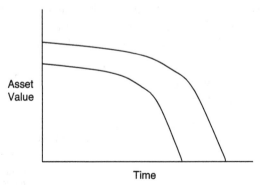

Figure 11.28 The lower curve shows a lower initial cost, but the life expectancy of the asset is significantly less than the asset in the upper curve (Bloetscher 2011)

accelerated depreciation of capital assets, for things like machinery. This means that for tax purposes, depreciation can be deducted faster for big-ticket items or large pieces of equipment. It is an incentive to make big purchases, which are the backbone of the economy. For people and governments, depreciation is not tax deductible, so its real importance is in evaluating the life cycle analysis.

How is depreciation calculated? Actually, it is easy. Because depreciation is defined as the reduction in value of an asset, it generally is based on initial cost, also called *first* or *adjusted cost*. The remaining value is termed *book value*. At the end of the useful life of the asset, the book value is called the *salvage value*. The life of the asset, or the amount of time it will be in service, is termed the *recovery period*. The depreciation rate is the fraction of value removed each year.

There are a variety of depreciation models of varying complexity, including:

◆ Straight line
◆ Declining balance
◆ Modified accelerated cost recovery

The last two are complicated business practices used to accelerate depreciation for tax purposes. As a result, they are more important for tax purposes than for engineering applications, so we will not discuss them further; refer to Blank and Tarquin (2008) for more information. The method of depreciation should account for the diminishing value, usefulness, or life of the asset, but while used for tax purposes, depreciation is *not an expense*, although the depreciation amount may be budgeted as an expense in the operating fund to be transferred to a capital recovery fund that is used to fund new capital!

Figure 11.29 shows a typical cost curve for depreciation and illustrates that while most finance experts depreciate the infrastructure using straight-line depreciation, the reality is that the in-service condition generally stays significantly above that, as illustrated by the curve. Moreover, as a result of the deterioration occurring more slowly than anticipated with straight-line depreciation, infrastructure condition is more likely to be maintained if improvements are made or the system is rehabilitated at the appropriate time (see Figure 11.30). However, rehabilitation can never return the system to its initial asset value unless it is completely replaced.

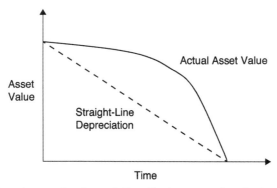

Figure 11.29 Typical cost curve for depreciation: the top curve is actual value, while the dotted line is the commonly used concept of straight-line depreciation, where the asset value decreases the same amount each year

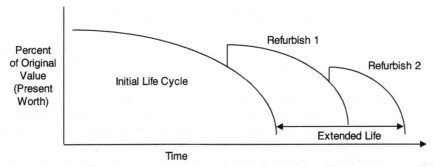

Figure 11.30 Extension of life of asset with refurbishment (Bloetscher 2011); note that refurbishment may occur later in life before full devaluation takes place, but refurbishment never reaches the initial condition, and extension of the life of an asset decreases with each refurbishment

With pipelines and roadways, this may not be as critical a concern, given that the technology does not change significantly with time. In a water treatment plant, for example, upgrades generally replace older, outdated mechanical equipment and controls for new equipment, which allows the systems to operate much more efficiently. Upgrades generally are required to deal with changes in technology; address concerns with age; meet regulatory requirements to improve treatment quality, efficiency, or reliability; and improve the operation and maintenance of the system. This generally is accomplished by lowering costs and simplifying operation to create conditions that are less likely to need repairs.

Straight-line depreciation is the most frequently used and simplest method to calculate depreciation. The concept is shown in Figure 11.31. The formula for straight-line depreciation is as follows:

$$D_j = (C - S_n) \div n \tag{11.36}$$

where

D_j = annual depreciation cost
C = initial cost or initial book value
S_n = salvage value
n = recovery period

Using Excel:

$$SLN(BV, S_n, n)$$

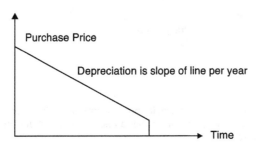

Figure 11.31 Example of a straight-line depreciation curve with a salvage value

Let's define each of these terms. *Book value* (*BV*) reflects the depreciated value at a given point in time. It is unlikely to reflect the true market value of the asset, as that depends more on what a willing buyer and a willing seller can agree upon for a sale or exchange, which is related more to supply and demand or emotional attachment than book value.

Recovery period (*n*) is the depreciable life of the asset in years. Often there are different *n* values for book value and tax depreciation, and both values may be different from the asset's estimated productive life. *Salvage value* (*S_n*) is the estimated trade-in or market value at the end of the asset's useful life. Normally for depreciation purposes, salvage value is estimated at the time of purchase. It can be negative due to dismantling and removal/disposal costs, but typically an asset does not depreciate below its estimated salvage value.

Let's say an asset is purchased for $50,000, and the owner wants to keep it for 5 years before selling it for $10,000 (see Figure 11.32 for the cash flow diagram). What is the total depreciation over the life of this asset, the average depreciation per year, and the book value at the end of each year? The amount this asset depreciates over its life is calculated as follows:

$$C - S_n = \$50,000 - \$10,000 = \$40,000$$

$$D_j = (C - S_n) \div n = \$40,000 \div 5 = \$8,000 \text{ per year}$$

Using the results of the calculation, a table of the book value over the life of the asset can be created, as shown in Table 11.5. Ultimately, the owner will need to gauge the condition of the infrastructure, as depreciation does not tell the full story. Only the operations staff understands the full nature of the assets that it operates, and that understanding may not be very thorough. Nationally, however, there are indications that significant infrastructure problems exist, particularly in older developed areas in the northeast and midwestern United States (USEPA 2002). Newer systems in the southeast and in the mountain states are not as critical yet as they may be newer systems, but their time is coming.

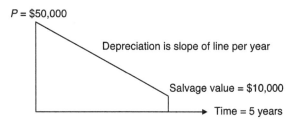

Figure 11.32 Cash flow diagram for an asset purchased for $50,000, kept for 5 years, and sold for a salvage value of $10,000 using straight-line depreciation

Table 11.5 Straight-line depreciation and book value over the lifetime of an example asset

t	*D_j*	*BV_i*
0		$50,000
1	$8,000	$42,000
2	$8,000	$32,000
3	$8,000	$26,000
4	$8,000	$18,000
5	$8,000	$10,000

11.13 A Word of Caution

Despite their usefulness, economic evaluations never should be the only deciding factor, because they tend to ignore the social, environmental, and development impacts of an alternative. These benefits (or disbenefits) are very difficult to quantify, yet they may have the effect of eliminating an alternative due to the potential damage that may occur. For instance, if a community is dependent on salmon fishing, evaluating a series of alternatives and determining that a dam is the most cost-effective alternative for future water supply may ignore the reality that the dam will devastate the salmon industry, thereby eliminating most of the jobs in the community and making the additional water supply unnecessary. Table 11.6 shows that the salmon were clearly the highest priority from an economic perspective, but they are not the priority from a regulatory perspective; agriculture is. Therefore, measuring the positive and negative impacts of a project or program, as compared to the current condition, can be instructive but does not always tell the full story.

Table 11.6 Economic priority based on value

Industry	2000 value ($ million/year)	Potential value ($ million/year)	Current priority water rights	Economic priority based on value
Agriculture	$200	$200	1	4
Tourism	$700	$1,500	n/a	2
Commercial fishery	$70	$4,500	n/a	1
Timber	$250	$250	n/a	3
Tribes	n/a	n/a	2	n/a

11.14 References

Blank, L.T., and Tarquin, A.J. (2008). *Basics of Engineering Economy*, McGraw-Hill Higher Education.

Bloetscher, F. (2009). *Water Basics for Decision Makers: What Local Officials Need to Know about Water and Wastewater Systems*, America Water Works Association, Denver, CO.

Bloetscher, F. (2011). *Utility Management for Water and Wastewater Operators*, American Water Works Association, Denver, CO.

USEPA (2002). The Clean Water and Drinking Water Infrastructure Gaps Analysis, Report EPA-816-R-02-020, U.S. Environmental Protection Agency, Washington, DC.

USGBC (2011). "What LEED® is," U.S. Green Building Council, <http://www.usgbc.org/DisplayPage.aspx?CMSPageID=1988> (accessed Jan. 22, 2012).

11.15 Assignments

1. How many years would it take for an investment of $50,000 in year 1 with increases of 10% per year to reach a present worth of $1 million at an interest rate of 7% per year?

2. An investment of $600,000 increases to $1 million over a 5-year period. What is the rate of return on the investment?

3. After a 7-year period at an interest rate of 15%, how much money would be in a savings account that was started with a deposit of $2,000 in year 1, with each succeeding deposit increasing by $250 per year?

4. What is the present worth of the following cash flow stream at an annual interest rate of 6% per year:

Year 0 = $2,000	Year 2 = $1,600	Year 4 = $1,200
Year 1 = $1,800	Year 3 = $1,400	Year 5 = $1,000

5. For a stated annual interest rate of 10%, find the effective interest rate for quarterly, monthly, weekly, daily, and continuous compounding. Graph the result.

6. You are borrowing money to buy a new $350,000 property. You are offered three options: (1) 30-year mortgage at 6% per year stated interest, (2) 15-year mortgage at 4% per year interest, and (3) 20-year mortgage at 6% per year interest with quarterly payments. Compare each option and discuss the results.

7. A movie ticket worth $15 today would have been worth how much 30 years ago, assuming a 4% per year inflation rate? Has the price increased faster or slower than inflation?

8. If you want $1.5 million in your IRA in 30 years, how much should you contribute each year if your rate of return will be 6% per year? How much less would you need to contribute if you were able to find an investment that made 8% per year?

9. What is the rate of return for an investment that increases by 350% over 15 years?

10. The maintenance cost for a car is $500 per year for the first 3 years. Then the cost will increase $50 each year until year 10, when the car will be sold. The interest rate is 5% compounded annually. What is the present worth maintenance cost over the 10 years?

11. If you make the following series of deposits at an interest rate of 5% compounded annually, what would the total balance be at the end of 10 years:

Year 0 = $800	Years 1 to 9 = $1,500 per year	Year 10 = 0

12. Find the present worth of the following cash flow series at an annual interest rate of 10%:

Year 0 = −$100	Year 2 = $300	Year 4 = $500
Year 1 = $200	Year 3 = $400	

13. If you borrow $50,000 at an interest rate of 12% compounded annually, what is the annual repayment if equal payments are made in years 1, 2, and 4 to 8 but not year 3?

14. Fifteen years ago, $1,000 was deposited in a bank account. Today, there is $2,370 in the account. The bank pays interest quarterly. What is the nominal interest rate paid on this account over the 15 years?

15. A 25-year-old engineer wants to set up a retirement fund. He invests $20,000 now at 6% interest compounded annually and contributes another $5,000 per year. How much will be in the account at retirement at age 65 ($n = 40$ years)?

16. You are borrowing money to buy a new $250,000 property and are offered three options: (1) 20-year mortgage at 6% per year stated interest, (2) 15-year mortgage at 6% interest per year compounded quarterly, and (3) 10-year 5% mortgage with quarterly payments. Which option has the highest cost per year?

17. Find the value of x if the present worth of the following cash flow series at an annual interest rate of 6% per year is $2,000:

Year 0 = −$100	Year 2 = $3x	Year 4 = $700
Year 1 = −$x	Year 3 = $600	

18. A warehouse was purchased for $250,000 10 years ago. The effective annual interest rate in the ensuing 10 years was 8% per year. The cost of operations was $25,000 per year the first year and increased 2.5% per year during the same period. Assuming no depreciation or deterioration, what is the present worth of the building and maintenance?

19. Fifteen years ago, $5,000 was deposited in a bank account. Today, there is $12,700 in the account. The bank pays interest monthly. What is the nominal interest rate paid on this account over the 15 years?

20. For a stated interest rate of 6% per year, find the effective interest rate for quarterly, monthly, weekly, daily, and continuous compounding. Graph the result.

21. A 40-year-old engineer wants to set up a retirement fund to be used starting at age 65. She invests $20,000 now at 6% interest compounded annually. How much will be in the account at retirement? If she contributes another $5,000 a year for the next 10 years, what will the value be? Would she be better off putting $5,000 into the account from age 25 to 35 versus starting now? Show the comparison costs.

22. If you invest $2,000 today in a savings account at an interest rate of 12% compounded annually, what will the value of the investment be in 7 years? If inflation is 4% per year over the same period, what is the actual purchasing value of your money?

23. The annual income from a toll highway is $200,000. If invested at an effective annual interest rate of 6% per year, how much money would be in the account at the end of 10 years:
    a. $2.2 million
    b. $2.6 million
    c. $2.7 million
    d. $2.11 million

24. A warehouse was purchased for $500,000 10 years ago. The effective annual interest rate in the ensuing 10 years was 8% per year. Assuming no depreciation or deterioration, how much is the building worth today:
    a. $427,000
    b. $540,000
    c. $678,000
    d. $691,000

25. For the warehouse in the previous problem, if inflation was 2.5% per year during the same period, what is the actual value of the building in 2000 dollars (assuming it is 2010):
    a. $427,000
    b. $540,000
    c. $678,000
    d. $691,000

26. A warehouse is sold for $1 million. The purchase price 10 years ago was $500,000. What is the rate of return on this investment?

27. A robotics system will be purchased for a factory. The lease-purchase cost is $100,000 per year. Warranties also will be purchased. The warranties start at $20,000 in year 5 and increase by $20,000 per year through the end of the lease. What is the present worth of the lease-purchase project at an interest rate of 4% compounded annually?

28. A stream of payments over a 5-year period with an interest rate of 5% per year compounded annually has a present worth of $100,000. Payments in years 1, 4, and 5 are $15,000, $30,000, and $35,000, respectively. The value in years 2 and 3 must be determined. Year 3 is twice year 2. What is the value of year 2 and year 3 payments?

29. A backhoe is purchased for $60,000. It is owned for 5 years and then sold for $10,000. What is the depreciation each year? What is the book value and the amount of accumulated depreciation in year 3?

30. If the value of a piece of property increases by 350% in 7 years, what is the estimated rate of return?

31. If the interest rate is 8% per year, approximately how long will it take for an investment to double in value?

32. The annual income from a toll highway is $400,000. If invested at an effective annual interest rate of 6% per year, how much money would be in the account at the end of 15 years?

33. Fifteen years ago, a warehouse was purchased for $500,000. The effective annual interest rate in the ensuing 15 years was 8% per year.

    a. Assuming no depreciation or deterioration, how much is the building worth today?

    b. If the building depreciated at 5% per year, then what is the value of the building?

Appendix 11A Interest tables

0.5%

n	P/F	P/A	P/G	F/P	F/A	A/P	A/F	A/G
1	0.9950	0.9950		1.0050	1.0000	1.00500	1.00000	
2	0.9901	1.9851	0.9901	1.0100	2.0050	0.50375	0.49875	0.4988
3	0.9851	2.9702	2.9604	1.0151	3.0150	0.33667	0.33167	0.9967
4	0.9802	3.9505	5.9011	1.0202	4.0301	0.25313	0.24813	1.4938
5	0.9754	4.9259	9.8026	1.0253	5.0503	0.20301	0.19801	1.9900
6	0.9705	5.8964	14.6552	1.0304	6.0755	0.16960	0.16460	2.4855
7	0.9657	6.8621	20.4493	1.0355	7.1059	0.14573	0.14073	2.9801
8	0.9609	7.8230	27.1755	1.0407	8.1414	0.12783	0.12283	3.4738
9	0.9561	8.7791	34.8244	1.0459	9.1821	0.11391	0.10891	3.9668
10	0.9513	9.7304	43.3865	1.0511	10.2280	0.10277	0.09777	4.4589
11	0.9466	10.6770	52.8526	1.0564	11.2792	0.09366	0.08866	4.9501
12	0.9419	11.6189	63.2136	1.0617	12.3356	0.08607	0.08107	5.4406
13	0.9372	12.5562	74.4602	1.0670	13.3972	0.07964	0.07464	5.9302
14	0.9326	13.4887	86.5835	1.0723	14.4642	0.07414	0.06914	6.4190
15	0.9279	14.4166	99.5743	1.0777	15.5365	0.06936	0.06436	6.9069
16	0.9233	15.3399	113.4238	1.0831	16.6142	0.06519	0.06019	7.3940
17	0.9187	16.2586	128.1231	1.0885	17.6973	0.06151	0.05651	7.8803
18	0.9141	17.1728	143.6634	1.0939	18.7858	0.05823	0.05323	8.3658
19	0.9096	18.0824	160.0360	1.0994	19.8797	0.05530	0.05030	8.8504
20	0.9051	18.9874	177.2322	1.1049	20.9791	0.05267	0.04767	9.3342
21	0.9006	19.8880	195.2434	1.1104	22.0840	0.05028	0.04528	9.8172
22	0.8961	20.7841	214.0611	1.1160	23.1944	0.04811	0.04311	10.2993
23	0.8916	21.6757	233.6768	1.1216	24.3104	0.04613	0.04113	10.7806
24	0.8872	22.5629	254.0820	1.1272	25.4320	0.04432	0.03932	11.2611
25	0.8828	23.4456	275.2686	1.1328	26.5591	0.04265	0.03765	11.7407
30	0.8610	27.7941	392.6324	1.1614	32.2800	0.03598	0.03098	14.1265
40	0.8191	36.1722	681.3347	1.2208	44.1588	0.02765	0.02265	18.8359
50	0.7793	44.1428	1035.6966	1.2832	56.6452	0.02265	0.01765	23.4624
60	0.7414	51.7256	1448.6458	1.3489	69.7700	0.01933	0.01433	28.0064
100	0.6073	78.5426	3562.7934	1.6467	129.3337	0.01273	0.00773	45.3613

1%

n	P/F	P/A	P/G	F/P	F/A	A/P	A/F	A/G
1	0.9901	0.9901		1.0100	1.0000	1.01000	1.00000	
2	0.9803	1.9704	0.9803	1.0201	2.0100	0.50751	0.49751	0.4975
3	0.9706	2.9410	2.9215	1.0303	3.0301	0.34002	0.33002	0.9934
4	0.9610	3.9020	5.8044	1.0406	4.0604	0.25628	0.24628	1.4876
5	0.9515	4.8534	9.6103	1.0510	5.1010	0.20604	0.19604	1.9801
6	0.9420	5.7955	14.3205	1.0615	6.1520	0.17255	0.16255	2.4710
7	0.9327	6.7282	19.9168	1.0721	7.2135	0.14863	0.13863	2.9602
8	0.9235	7.6517	26.3812	1.0829	8.2857	0.13069	0.12069	3.4478
9	0.9143	8.5660	33.6959	1.0937	9.3685	0.11674	0.10674	3.9337
10	0.9053	9.4713	41.8435	1.1046	10.4622	0.10558	0.09558	4.4179
11	0.8963	10.3676	50.8067	1.1157	11.5668	0.09645	0.08645	4.9005
12	0.8874	11.2551	60.5687	1.1268	12.6825	0.08885	0.07885	5.3815
13	0.8787	12.1337	71.1126	1.1381	13.8093	0.08241	0.07241	5.8607
14	0.8700	13.0037	82.4221	1.1495	14.9474	0.07690	0.06690	6.3384
15	0.8613	13.8651	94.4810	1.1610	16.0969	0.07212	0.06212	6.8143
16	0.8528	14.7179	107.2734	1.1726	17.2579	0.06794	0.05794	7.2886
17	0.8444	15.5623	120.7834	1.1843	18.4304	0.06426	0.05426	7.7613
18	0.8360	16.3983	134.9957	1.1961	19.6147	0.06098	0.05098	8.2323
19	0.8277	17.2260	149.8950	1.2081	20.8109	0.05805	0.04805	8.7017
20	0.8195	18.0456	165.4664	1.2202	22.0190	0.05542	0.04542	9.1694
21	0.8114	18.8570	181.6950	1.2324	23.2392	0.05303	0.04303	9.6354
22	0.8034	19.6604	198.5663	1.2447	24.4716	0.05086	0.04086	10.0998
23	0.7954	20.4558	216.0660	1.2572	25.7163	0.04889	0.03889	10.5626
24	0.7876	21.2434	234.1800	1.2697	26.9735	0.04707	0.03707	11.0237
25	0.7798	22.0232	252.8945	1.2824	28.2432	0.04541	0.03541	11.4831
30	0.7419	25.8077	355.0021	1.3478	34.7849	0.03875	0.02875	13.7557
40	0.6717	32.8347	596.8561	1.4889	48.8864	0.03046	0.02046	18.1776
50	0.6080	39.1961	879.4176	1.6446	64.4632	0.02551	0.01551	22.4363
60	0.5504	44.9550	1192.8061	1.8167	81.6697	0.02224	0.01224	26.5333
100	0.3697	63.0289	2605.7758	2.7048	170.4814	0.01587	0.00587	41.3426

2%

n	P/F	P/A	P/G	F/P	F/A	A/P	A/F	A/G
1	0.9804	0.9804	0.0000	1.0200	1.0000	1.0200	1.0000	0.0000
2	0.9612	1.9416	0.9612	1.0404	2.0200	0.5150	0.4950	0.4950
3	0.9423	2.8839	2.8458	1.0612	3.0604	0.3468	0.3268	0.9868
4	0.9238	3.8077	5.6173	1.0824	4.1216	0.2626	0.2426	1.4752
5	0.9057	4.7135	9.2403	1.1041	5.2040	0.2122	0.1922	1.9604
6	0.8880	5.6014	13.6801	1.1262	6.3081	0.1785	0.1585	2.4423
7	0.8706	6.4720	18.9035	1.1487	7.4343	0.1545	0.1345	2.9208
8	0.8535	7.3255	24.8779	1.1717	8.5830	0.1365	0.1165	3.3961
9	0.8368	8.1622	31.5720	1.1951	9.7546	0.1225	0.1025	3.8661
10	0.8203	8.9826	38.9551	1.2190	10.9497	0.1113	0.0913	4.3367
11	0.8043	9.7868	46.9977	1.2434	12.1687	0.1022	0.0822	4.8021
12	0.7885	10.5753	55.6712	1.2682	13.4121	0.0946	0.0746	5.2642
13	0.7730	11.3484	64.9475	1.2936	14.6803	0.0881	0.0681	5.7231
14	0.7579	12.1062	74.7999	1.3195	15.9739	0.0826	0.0626	6.1786
15	0.7430	12.8493	85.2021	1.3459	17.2934	0.0778	0.0578	6.6309
16	0.7284	13.5777	96.1288	1.3728	18.6393	0.0737	0.0537	7.0799
17	0.7142	14.2919	107.5554	1.4002	20.0121	0.0700	0.0500	7.5256
18	0.7002	14.9920	119.4581	1.4282	21.4123	0.0667	0.0467	7.9681
19	0.6864	15.6785	131.8139	1.4568	22.8406	0.0638	0.0438	8.4073
20	0.6730	16.3514	144.6003	1.4859	24.2974	0.0612	0.0412	8.8433
21	0.6598	17.0112	157.7959	1.5157	25.7833	0.0588	0.0388	9.2760
22	0.6468	17.6580	171.3795	1.5460	27.2990	0.0566	0.0366	9.7055
23	0.6342	18.2922	185.3309	1.5769	28.8450	0.0547	0.0347	10.1317
24	0.6217	18.9139	199.6305	1.6084	30.4219	0.0529	0.0329	10.5547
25	0.6095	19.5235	214.2592	1.6406	32.0303	0.0512	0.0312	10.9745
30	0.5521	22.3965	291.7164	1.8114	40.5681	0.0446	0.0246	13.0251
40	0.4529	27.3555	461.9931	2.2080	60.4020	0.0366	0.0166	16.8885
50	0.3715	31.4236	642.3606	2.6916	84.5794	0.0318	0.0118	20.4420
60	0.3048	34.7609	823.6975	3.2810	114.0515	0.0288	0.0088	23.6961
100	0.1380	43.0984	1464.7527	7.2446	312.2323	0.0232	0.0032	33.9863

3%

n	P/F	P/A	P/G	F/P	F/A	A/P	A/F	A/G
1	0.9709	0.9709		1.0300	1.0000	1.03000	1.00000	
2	0.9426	1.9135	0.9426	1.0609	2.0300	0.52261	0.49261	0.4926
3	0.9151	2.8286	2.7729	1.0927	3.0909	0.35353	0.32353	0.9803
4	0.8885	3.7171	5.4383	1.1255	4.1836	0.26903	0.23903	1.4631
5	0.8626	4.5797	8.8888	1.1593	5.3091	0.21835	0.18835	1.9409
6	0.8375	5.4172	13.0762	1.1941	6.4684	0.18460	0.15460	2.4138
7	0.8131	6.2303	17.9547	1.2299	7.6625	0.16051	0.13051	2.8819
8	0.7894	7.0197	23.4806	1.2668	8.8923	0.14246	0.11246	3.3450
9	0.7664	7.7861	29.6119	1.3048	10.1591	0.12843	0.09843	3.8032
10	0.7441	8.5302	36.3088	1.3439	11.4639	0.11723	0.08723	4.2565
11	0.7224	9.2526	43.5330	1.3842	12.8078	0.10808	0.07808	4.7049
12	0.7014	9.9540	51.2482	1.4258	14.1920	0.10046	0.07046	5.1485
13	0.6810	10.6350	59.4196	1.4685	15.6178	0.09403	0.06403	5.5872
14	0.6611	11.2961	68.0141	1.5126	17.0863	0.08853	0.05853	6.0210
15	0.6419	11.9379	77.0002	1.5580	18.5989	0.08377	0.05377	6.4500
16	0.6232	12.5611	86.3477	1.6047	20.1569	0.07961	0.04961	6.8742
17	0.6050	13.1661	96.0280	1.6528	21.7616	0.07595	0.04595	7.2936
18	0.5874	13.7535	106.0137	1.7024	23.4144	0.07271	0.04271	7.7081
19	0.5703	14.3238	116.2788	1.7535	25.1169	0.06981	0.03981	8.1179
20	0.5537	14.8775	126.7987	1.8061	26.8704	0.06722	0.03722	8.5229
21	0.5375	15.4150	137.5496	1.8603	28.6765	0.06487	0.03487	8.9231
22	0.5219	15.9369	148.5094	1.9161	30.5368	0.06275	0.03275	9.3186
23	0.5067	16.4436	159.6566	1.9736	32.4529	0.06081	0.03081	9.7093
24	0.4919	16.9355	170.9711	2.0328	34.4265	0.05905	0.02905	10.0954
25	0.4776	17.4131	182.4336	2.0938	36.4593	0.05743	0.02743	10.4768
30	0.4120	19.6004	241.3613	2.4273	47.5754	0.05102	0.02102	12.3141
40	0.3066	23.1148	361.7499	3.2620	75.4013	0.04326	0.01326	15.6502
50	0.2281	25.7298	477.4803	4.3839	112.7969	0.03887	0.00887	18.5575
60	0.1697	27.6756	583.0526	5.8916	163.0534	0.03613	0.00613	21.0674
100	0.0520	31.5989	879.8540	19.2186	607.2877	0.03165	0.00165	27.8444

Appendix 11A Interest tables (continued)

4%

n	P/F	P/A	P/G	F/P	F/A	A/P	A/F	A/G
1	0.9615	0.9615		1.0400	1.0000	1.04000	1.00000	
2	0.9246	1.8861	0.9246	1.0816	2.0400	0.53020	0.49020	0.4902
3	0.8890	2.7751	2.7025	1.1249	3.1216	0.36035	0.32035	0.9739
4	0.8548	3.6299	5.2670	1.1699	4.2465	0.27549	0.23549	1.4510
5	0.8219	4.4518	8.5547	1.2167	5.4163	0.22463	0.18463	1.9216
6	0.7903	5.2421	12.5062	1.2653	6.6330	0.19076	0.15076	2.3857
7	0.7599	6.0021	17.0657	1.3159	7.8983	0.16661	0.12661	2.8433
8	0.7307	6.7327	22.1806	1.3686	9.2142	0.14853	0.10853	3.2944
9	0.7026	7.4353	27.8013	1.4233	10.5828	0.13449	0.09449	3.7391
10	0.6756	8.1109	33.8814	1.4802	12.0061	0.12329	0.08329	4.1773
11	0.6496	8.7605	40.3772	1.5395	13.4864	0.11415	0.07415	4.6090
12	0.6246	9.3851	47.2477	1.6010	15.0258	0.10655	0.06655	5.0343
13	0.6006	9.9856	54.4546	1.6651	16.6268	0.10014	0.06014	5.4533
14	0.5775	10.5631	61.9618	1.7317	18.2919	0.09467	0.05467	5.8659
15	0.5553	11.1184	69.7355	1.8009	20.0236	0.08994	0.04994	6.2721
16	0.5339	11.6523	77.7441	1.8730	21.8245	0.08582	0.04582	6.6720
17	0.5134	12.1657	85.9581	1.9479	23.6975	0.08220	0.04220	7.0656
18	0.4936	12.6593	94.3498	2.0258	25.6454	0.07899	0.03899	7.4530
19	0.4746	13.1339	102.8933	2.1068	27.6712	0.07614	0.03614	7.8342
20	0.4564	13.5903	111.5647	2.1911	29.7781	0.07358	0.03358	8.2091
21	0.4388	14.0292	120.3414	2.2788	31.9692	0.07128	0.03128	8.5779
22	0.4220	14.4511	129.2024	2.3699	34.2480	0.06920	0.02920	8.9407
23	0.4057	14.8568	138.1284	2.4647	36.6179	0.06731	0.02731	9.2973
24	0.3901	15.2470	147.1012	2.5633	39.0826	0.06559	0.02559	9.6479
25	0.3751	15.6221	156.1040	2.6658	41.6459	0.06401	0.02401	9.9925
30	0.3083	17.2920	201.0618	3.2434	56.0849	0.05783	0.01783	11.6274
40	0.2083	19.7928	286.5303	4.8010	95.0255	0.05052	0.01052	14.4765
50	0.1407	21.4822	361.1638	7.1067	152.6671	0.04655	0.00655	16.8122
60	0.0951	22.6235	422.9966	10.5196	237.9907	0.04420	0.00420	18.6972
100	0.0198	24.5050	563.1249	50.5049	1237.6237	0.04081	0.00081	22.9800

5%

n	P/F	P/A	P/G	F/P	F/A	A/P	A/F	A/G
1	0.9524	0.9524		1.0500	1.0000	1.05000	1.00000	
2	0.9070	1.8594	0.9070	1.1025	2.0500	0.53780	0.48780	0.4878
3	0.8638	2.7232	2.6347	1.1576	3.1525	0.36721	0.31721	0.9675
4	0.8227	3.5460	5.1028	1.2155	4.3101	0.28201	0.23201	1.4391
5	0.7835	4.3295	8.2369	1.2763	5.5256	0.23097	0.18097	1.9025
6	0.7462	5.0757	11.9680	1.3401	6.8019	0.19702	0.14702	2.3579
7	0.7107	5.7864	16.2321	1.4071	8.1420	0.17282	0.12282	2.8052
8	0.6768	6.4632	20.9700	1.4775	9.5491	0.15472	0.10472	3.2445
9	0.6446	7.1078	26.1268	1.5513	11.0266	0.14069	0.09069	3.6758
10	0.6139	7.7217	31.6520	1.6289	12.5779	0.12950	0.07950	4.0991
11	0.5847	8.3064	37.4988	1.7103	14.2068	0.12039	0.07039	4.5144
12	0.5568	8.8633	43.6241	1.7959	15.9171	0.11283	0.06283	4.9219
13	0.5303	9.3936	49.9879	1.8856	17.7130	0.10646	0.05646	5.3215
14	0.5051	9.8986	56.5538	1.9799	19.5986	0.10102	0.05102	5.7133
15	0.4810	10.3797	63.2880	2.0789	21.5786	0.09634	0.04634	6.0973
16	0.4581	10.8378	70.1597	2.1829	23.6575	0.09227	0.04227	6.4736
17	0.4363	11.2741	77.1405	2.2920	25.8404	0.08870	0.03870	6.8423
18	0.4155	11.6896	84.2043	2.4066	28.1324	0.08555	0.03555	7.2034
19	0.3957	12.0853	91.3275	2.5270	30.5390	0.08275	0.03275	7.5569
20	0.3769	12.4622	98.4884	2.6533	33.0660	0.08024	0.03024	7.9030
21	0.3589	12.8212	105.6673	2.7860	35.7193	0.07800	0.02800	8.2416
22	0.3418	13.1630	112.8461	2.9253	38.5052	0.07597	0.02597	8.5730
23	0.3256	13.4886	120.0087	3.0715	41.4305	0.07414	0.02414	8.8971
24	0.3101	13.7986	127.1402	3.2251	44.5020	0.07247	0.02247	9.2140
25	0.2953	14.0939	134.2275	3.3864	47.7271	0.07095	0.02095	9.5238
30	0.2314	15.3725	168.6226	4.3219	66.4388	0.06505	0.01505	10.9691
40	0.1420	17.1591	229.5452	7.0400	120.7998	0.05828	0.00828	13.3775
50	0.0872	18.2559	277.9148	11.4674	209.3480	0.05478	0.00478	15.2233
60	0.0535	18.9293	314.3432	18.6792	353.5837	0.05283	0.00283	16.6062
100	0.0076	19.8479	381.7492	131.5013	2610.0252	0.05038	0.00038	19.2337

8%

n	P/F	P/A	P/G	F/P	F/A	A/P	A/F	A/G
1	0.9259	0.9259		1.0800	1.0000	1.08000	1.00000	
2	0.8573	1.7833	0.8573	1.1664	2.0800	0.56077	0.48077	0.4808
3	0.7938	2.5771	2.4450	1.2597	3.2464	0.38803	0.30803	0.9487
4	0.7350	3.3121	4.6501	1.3605	4.5061	0.30192	0.22192	1.4040
5	0.6806	3.9927	7.3724	1.4693	5.8666	0.25046	0.17046	1.8465
6	0.6302	4.6229	10.5233	1.5869	7.3359	0.21632	0.13632	2.2763
7	0.5835	5.2064	14.0242	1.7138	8.9228	0.19207	0.11207	2.6937
8	0.5403	5.7466	17.8061	1.8509	10.6366	0.17401	0.09401	3.0985
9	0.5002	6.2469	21.8081	1.9990	12.4876	0.16008	0.08008	3.4910
10	0.4632	6.7101	25.9768	2.1589	14.4866	0.14903	0.06903	3.8713
11	0.4289	7.1390	30.2657	2.3316	16.6455	0.14008	0.06008	4.2395
12	0.3971	7.5361	34.6339	2.5182	18.9771	0.13270	0.05270	4.5957
13	0.3677	7.9038	39.0463	2.7196	21.4953	0.12652	0.04652	4.9402
14	0.3405	8.2442	43.4723	2.9372	24.2149	0.12130	0.04130	5.2731
15	0.3152	8.5595	47.8857	3.1722	27.1521	0.11683	0.03683	5.5945
16	0.2919	8.8514	52.2640	3.4259	30.3243	0.11298	0.03298	5.9046
17	0.2703	9.1216	56.5883	3.7000	33.7502	0.10963	0.02963	6.2037
18	0.2502	9.3719	60.8426	3.9960	37.4502	0.10670	0.02670	6.4920
19	0.2317	9.6036	65.0134	4.3157	41.4463	0.10413	0.02413	6.7697
20	0.2145	9.8181	69.0898	4.6610	45.7620	0.10185	0.02185	7.0369
21	0.1987	10.0168	73.0629	5.0338	50.4229	0.09983	0.01983	7.2940
22	0.1839	10.2007	76.9257	5.4365	55.4568	0.09803	0.01803	7.5412
23	0.1703	10.3711	80.6726	5.8715	60.8933	0.09642	0.01642	7.7786
24	0.1577	10.5288	84.2997	6.3412	66.7648	0.09498	0.01498	8.0066
25	0.1460	10.6748	87.8041	6.8485	73.1059	0.09368	0.01368	8.2254
30	0.0994	11.2578	103.4558	10.0627	113.2832	0.08883	0.00883	9.1897
40	0.0460	11.9246	126.0422	21.7245	259.0565	0.08386	0.00386	10.5699
50	0.0213	12.2335	139.5928	46.9016	573.7702	0.08174	0.00174	11.4107
60	0.0099	12.3766	147.3000	101.2571	1253.2133	0.08080	0.00080	11.9015
100	0.0005	12.4943	155.6107	2199.7613	27484.5157	0.08004	0.00004	12.4545

6%

n	P/F	P/A	P/G	F/P	F/A	A/P	A/F	A/G
1	0.9434	0.9434	0.0000	1.0600	1.0000	1.0600	1.0000	0.0000
2	0.8900	1.8334	0.8900	1.1236	2.0600	0.5454	0.4854	0.4854
3	0.8396	2.6730	2.5692	1.1910	3.1836	0.3741	0.3141	0.9612
4	0.7921	3.4651	4.9455	1.2625	4.3746	0.2886	0.2286	1.4272
5	0.7473	4.2124	7.9345	1.3382	5.6371	0.2374	0.1774	1.8836
6	0.7050	4.9173	11.4594	1.4185	6.9753	0.2034	0.1434	2.3304
7	0.6651	5.5824	15.4497	1.5036	8.3938	0.1791	0.1191	2.7676
8	0.6274	6.2098	19.8416	1.5938	9.8975	0.1610	0.1010	3.1952
9	0.5919	6.8017	24.5768	1.6895	11.4913	0.1470	0.0870	3.6133
10	0.5584	7.3601	29.6023	1.7908	13.1808	0.1359	0.0759	4.0220
11	0.5268	7.8869	34.8702	1.8983	14.9716	0.1268	0.0668	4.4213
12	0.4970	8.3838	40.3369	2.0122	16.8699	0.1193	0.0593	4.8113
13	0.4688	8.8527	45.9629	2.1329	18.8821	0.1130	0.0530	5.1920
14	0.4423	9.2950	51.7128	2.2609	21.0151	0.1076	0.0476	5.5635
15	0.4173	9.7122	57.5546	2.3966	23.2760	0.1030	0.0430	5.9260
16	0.3936	10.1059	63.4592	2.5404	25.6725	0.0990	0.0390	6.2794
17	0.3714	10.4773	69.4011	2.6928	28.2129	0.0954	0.0354	6.6240
18	0.3505	10.8276	75.3569	2.8543	30.9057	0.0924	0.0324	6.9597
19	0.3305	11.1581	81.3062	3.0256	33.7600	0.0896	0.0296	7.2867
20	0.3118	11.4699	87.2304	3.2071	36.7856	0.0872	0.0272	7.6051
21	0.2942	11.7641	93.1136	3.3996	39.9927	0.0850	0.0250	7.9151
22	0.2775	12.0416	98.9412	3.6035	43.3923	0.0830	0.0230	8.2166
23	0.2618	12.3034	104.7007	3.8197	46.9958	0.0813	0.0213	8.5099
24	0.2470	12.5504	110.3812	4.0489	50.8156	0.0797	0.0197	8.7951
25	0.2330	12.7834	115.9732	4.2919	54.8645	0.0782	0.0182	9.0722
30	0.1741	13.7648	142.3588	5.7435	79.0582	0.0726	0.0126	10.3422
40	0.0972	15.0463	185.9568	10.2857	154.7620	0.0665	0.0065	12.3590
50	0.0543	15.7619	217.4574	18.4202	290.3359	0.0634	0.0034	13.7964
60	0.0303	16.1614	239.0428	32.9877	533.1282	0.0619	0.0019	14.7909
100	0.0029	16.6175	272.0471	339.3021	5638.3681	0.0602	0.0002	16.3711

Appendix 11A Interest tables (continued)

10%

n	P/F	P/A	P/G	F/P	F/A	A/P	A/F	A/G
1	0.9091	0.9091		1.1000	1.0000	1.10000	1.00000	
2	0.8264	1.7355	0.8264	1.2100	2.1000	0.57619	0.47619	0.4762
3	0.7513	2.4869	2.3291	1.3310	3.3100	0.40211	0.30211	0.9366
4	0.6830	3.1699	4.3781	1.4641	4.6410	0.31547	0.21547	1.3812
5	0.6209	3.7908	6.8618	1.6105	6.1051	0.26380	0.16380	1.8101
6	0.5645	4.3553	9.6842	1.7716	7.7156	0.22961	0.12961	2.2236
7	0.5132	4.8684	12.7631	1.9487	9.4872	0.20541	0.10541	2.6216
8	0.4665	5.3349	16.0287	2.1436	11.4359	0.18744	0.08744	3.0045
9	0.4241	5.7590	19.4215	2.3579	13.5795	0.17364	0.07364	3.3724
10	0.3855	6.1446	22.8913	2.5937	15.9374	0.16275	0.06275	3.7255
11	0.3505	6.4951	26.3963	2.8531	18.5312	0.15396	0.05396	4.0641
12	0.3186	6.8137	29.9012	3.1384	21.3843	0.14676	0.04676	4.3884
13	0.2897	7.1034	33.3772	3.4523	24.5227	0.14078	0.04078	4.6988
14	0.2633	7.3667	36.8005	3.7975	27.9750	0.13575	0.03575	4.9955
15	0.2394	7.6061	40.1520	4.1772	31.7725	0.13147	0.03147	5.2789
16	0.2176	7.8237	43.4164	4.5950	35.9497	0.12782	0.02782	5.5493
17	0.1978	8.0216	46.5819	5.0545	40.5447	0.12466	0.02466	5.8071
18	0.1799	8.2014	49.6395	5.5599	45.5992	0.12193	0.02193	6.0526
19	0.1635	8.3649	52.5827	6.1159	51.1591	0.11955	0.01955	6.2861
20	0.1486	8.5136	55.4069	6.7275	57.2750	0.11746	0.01746	6.5081
21	0.1351	8.6487	58.1095	7.4002	64.0025	0.11562	0.01562	6.7189
22	0.1228	8.7715	60.6893	8.1403	71.4027	0.11401	0.01401	6.9189
23	0.1117	8.8832	63.1462	8.9543	79.5430	0.11257	0.01257	7.1085
24	0.1015	8.9847	65.4813	9.8497	88.4973	0.11130	0.01130	7.2881
25	0.0923	9.0770	67.6964	10.8347	98.3471	0.11017	0.01017	7.4580
30	0.0573	9.4269	77.0766	17.4494	164.4940	0.10608	0.00608	8.1762
40	0.0221	9.7791	88.9525	45.2593	442.5926	0.10226	0.00226	9.0962
50	0.0085	9.9148	94.8889	117.3909	1163.9085	0.10086	0.00086	9.5704
60	0.0033	9.9672	97.7010	304.4816	3034.8164	0.10033	0.00033	9.8023

12%

n	P/F	P/A	P/G	F/P	F/A	A/P	A/F	A/G
1	0.8929	0.8929		1.1200	1.0000	1.12000	1.00000	
2	0.7972	1.6901	0.7972	1.2544	2.1200	0.59170	0.47170	0.4717
3	0.7118	2.4018	2.2208	1.4049	3.3744	0.41635	0.29635	0.9246
4	0.6355	3.0373	4.1273	1.5735	4.7793	0.32923	0.20923	1.3589
5	0.5674	3.6048	6.3970	1.7623	6.3528	0.27741	0.15741	1.7746
6	0.5066	4.1114	8.9302	1.9738	8.1152	0.24323	0.12323	2.1720
7	0.4523	4.5638	11.6443	2.2107	10.0890	0.21912	0.09912	2.5515
8	0.4039	4.9676	14.4714	2.4760	12.2997	0.20130	0.08130	2.9131
9	0.3606	5.3282	17.3563	2.7731	14.7757	0.18768	0.06768	3.2574
10	0.3220	5.6502	20.2541	3.1058	17.5487	0.17698	0.05698	3.5847
11	0.2875	5.9377	23.1288	3.4785	20.6546	0.16842	0.04842	3.8953
12	0.2567	6.1944	25.9523	3.8960	24.1331	0.16144	0.04144	4.1897
13	0.2292	6.4235	28.7024	4.3635	28.0291	0.15568	0.03568	4.4683
14	0.2046	6.6282	31.3624	4.8871	32.3926	0.15087	0.03087	4.7317
15	0.1827	6.8109	33.9202	5.4736	37.2797	0.14682	0.02682	4.9803
16	0.1631	6.9740	36.3670	6.1304	42.7533	0.14339	0.02339	5.2147
17	0.1456	7.1196	38.6973	6.8660	48.8837	0.14046	0.02046	5.4353
18	0.1300	7.2497	40.9080	7.6900	55.7497	0.13794	0.01794	5.6427
19	0.1161	7.3658	42.9979	8.6128	63.4397	0.13576	0.01576	5.8375
20	0.1037	7.4694	44.9676	9.6463	72.0524	0.13388	0.01388	6.0202
21	0.0926	7.5620	46.8188	10.8038	81.6987	0.13224	0.01224	6.1913
22	0.0826	7.6446	48.5543	12.1003	92.5026	0.13081	0.01081	6.3514
23	0.0738	7.7184	50.1776	13.5523	104.6029	0.12956	0.00956	6.5010
24	0.0659	7.7843	51.6929	15.1786	118.1552	0.12846	0.00846	6.6406
25	0.0588	7.8431	53.1046	17.0001	133.3339	0.12750	0.00750	6.7708
30	0.0334	8.0552	58.7821	29.9599	241.3327	0.12414	0.00414	7.2974
40	0.0107	8.2438	65.1159	93.0510	767.0914	0.12130	0.00130	7.8988
50	0.0035	8.3045	67.7624	289.0022	2400.0182	0.12042	0.00042	8.1597
60	0.0011	8.3240	68.8100	897.5969	7471.6411	0.12013	0.00013	8.2664

18%

n	P/F	P/A	P/G	F/P	F/A	A/P	A/F	A/G
1	0.8475	0.8475		1.1800	1.0000	1.18000	1.00000	
2	0.7182	1.5656	0.7182	1.3924	2.1800	0.63872	0.45872	0.4587
3	0.6086	2.1743	1.9354	1.6430	3.5724	0.45992	0.27992	0.8902
4	0.5158	2.6901	3.4828	1.9388	5.2154	0.37174	0.19174	1.2947
5	0.4371	3.1272	5.2312	2.2878	7.1542	0.31978	0.13978	1.6728
6	0.3704	3.4976	7.0834	2.6996	9.4420	0.28591	0.10591	2.0252
7	0.3139	3.8115	8.9670	3.1855	12.1415	0.26236	0.08236	2.3526
8	0.2660	4.0776	10.8292	3.7589	15.3270	0.24524	0.06524	2.6558
9	0.2255	4.3030	12.6329	4.4355	19.0859	0.23239	0.05239	2.9358
10	0.1911	4.4941	14.3525	5.2338	23.5213	0.22251	0.04251	3.1936
11	0.1619	4.6560	15.9716	6.1759	28.7551	0.21478	0.03478	3.4303
12	0.1372	4.7932	17.4811	7.2876	34.9311	0.20863	0.02863	3.6470
13	0.1163	4.9095	18.8765	8.5994	42.2187	0.20369	0.02369	3.8449
14	0.0985	5.0081	20.1576	10.1472	50.8180	0.19968	0.01968	4.0250
15	0.0835	5.0916	21.3269	11.9737	60.9653	0.19640	0.01640	4.1887
16	0.0708	5.1624	22.3885	14.1290	72.9390	0.19371	0.01371	4.3369
17	0.0600	5.2223	23.3482	16.6722	87.0680	0.19149	0.01149	4.4708
18	0.0508	5.2732	24.2123	19.6733	103.7403	0.18964	0.00964	4.5916
19	0.0431	5.3162	24.9877	23.2144	123.4135	0.18810	0.00810	4.7003
20	0.0365	5.3527	25.6813	27.3930	146.6280	0.18682	0.00682	4.7978
21	0.0309	5.3837	26.3000	32.3238	174.0210	0.18575	0.00575	4.8851
22	0.0262	5.4099	26.8506	38.1421	206.3448	0.18485	0.00485	4.9632
23	0.0222	5.4321	27.3394	45.0076	244.4868	0.18409	0.00409	5.0329
24	0.0188	5.4509	27.7725	53.1090	289.4945	0.18345	0.00345	5.0950
25	0.0160	5.4669	28.1555	62.6686	342.6035	0.18292	0.00292	5.1502
30	0.0070	5.5168	29.4864	143.3706	790.9480	0.18126	0.00126	5.3448
40	0.0013	5.5482	30.5269	750.3783	4163.2130	0.18024	0.00024	5.5022
50	0.0003	5.5541	30.7856	3927.3569	21813.0937	0.18005	0.00005	5.5428

24%

n	P/F	P/A	P/G	F/P	F/A	A/P	A/F	A/G
1	0.8065	0.8065		1.2400	1.0000	1.24000	1.00000	
2	0.6504	1.4568	0.6504	1.5376	2.2400	0.68643	0.44643	0.4464
3	0.5245	1.9813	1.6993	1.9066	3.7776	0.50472	0.26472	0.8577
4	0.4230	2.4043	2.9683	2.3642	5.6842	0.41593	0.17593	1.2346
5	0.3411	2.7454	4.3327	2.9316	8.0484	0.36425	0.12425	1.5782
6	0.2751	3.0205	5.7081	3.6352	10.9801	0.33107	0.09107	1.8898
7	0.2218	3.2423	7.0392	4.5077	14.6153	0.30842	0.06842	2.1710
8	0.1789	3.4212	8.2915	5.5895	19.1229	0.29229	0.05229	2.4236
9	0.1443	3.5655	9.4458	6.9310	24.7125	0.28047	0.04047	2.6492
10	0.1164	3.6819	10.4930	8.5944	31.6434	0.27160	0.03160	2.8499
11	0.0938	3.7757	11.4313	10.6571	40.2379	0.26485	0.02485	3.0276
12	0.0757	3.8514	12.2637	13.2148	50.8950	0.25965	0.01965	3.1843
13	0.0610	3.9124	12.9960	16.3863	64.1097	0.25560	0.01560	3.3218
14	0.0492	3.9616	13.6358	20.3191	80.4961	0.25242	0.01242	3.4420
15	0.0397	4.0013	14.1915	25.1956	100.8151	0.24992	0.00992	3.5467
16	0.0320	4.0333	14.6716	31.2426	126.0108	0.24794	0.00794	3.6376
17	0.0258	4.0591	15.0846	38.7408	157.2534	0.24636	0.00636	3.7162
18	0.0208	4.0799	15.4385	48.0386	195.9942	0.24510	0.00510	3.7840
19	0.0168	4.0967	15.7406	59.5679	244.0328	0.24410	0.00410	3.8423
20	0.0135	4.1103	15.9979	73.8641	303.6006	0.24329	0.00329	3.8922
21	0.0109	4.1212	16.2162	91.5915	377.4648	0.24265	0.00265	3.9349
22	0.0088	4.1300	16.4011	113.5735	469.0563	0.24213	0.00213	3.9712
23	0.0071	4.1371	16.5574	140.8312	582.6298	0.24172	0.00172	4.0022
24	0.0057	4.1428	16.6891	174.6306	723.4610	0.24138	0.00138	4.0284
25	0.0046	4.1474	16.7999	216.5420	898.0916	0.24111	0.00111	4.0507
30	0.0016	4.1601	17.1369	634.8199	2640.9164	0.24038	0.00038	4.1193
40	0.0002	4.1659	17.3274	5455.9126	22728.8026	0.24004	0.00004	4.1593
50	0.0000	4.1666	17.3563	46890.435	195372.644	0.24001	0.00001	4.1656

Preliminary Site Design and Nonstructural Concepts

The purpose of this chapter is to help students in conceptualizing their designs for portions of the project that are on the site or inside the building, but not part of the structural components (see Chapter 13 for structural and geotechnical design concepts). Note that the intent is not to provide specifics for the design process, nor to provide the details of any specific professional design, as local, state, and other guidelines may dictate differences between jurisdictions. Students and professionals always should consult the appropriate engineering codes that apply in their jurisdictions; however, some general concepts are similar for most designs.

In this chapter, the process for the design of the nonstructural components of a project is outlined. Examples are provided to help students understand the thought process and address some of the important issues that must be considered. All of the examples are based on student design projects and illustrate the methods used to solve specific conditions. They may or may not be applicable to other projects or meet codes outside the jurisdictions where they were done, but the concepts should correspond to other projects and jurisdictions. Therefore, it is important to view these examples in terms of the design process and not necessarily the finer details of the projects.

The size of the building or project may not be indicative of the complexity of design, as site limitations, design needs, or other factors may make the project more or less complicated. Part of the engineer's job is to find ways to limit the challenges imposed by the design—in other words, make it simpler if possible. Ease in design often translates into ease in construction, which means less cost and less time to complete, a double benefit for the client. Ease in construction improves integration of the various components of the project. This means that when designing a building, the floor plans and site plan must be complete, although some interaction between floor plans and the structure is needed to ensure the project is buildable. The site plan and floor plans must integrate with water, sewer, and stormwater utilities, which in turn must integrate with heating, ventilation, and air conditioning (HVAC) and other components internal and external to the building. The following are key concepts to keep in mind (most of which are obvious but tend to get lost during the design process):

◆ Water on roofs should flow to drains as quickly as possible to prevent leaks and moisture intrusion.
◆ Roof drains cannot conflict with interior walls and structures.
◆ Roof drains tie into the stormwater drainage system.
◆ Bathrooms must have both water pipes and sewer pipes.
◆ It is important to think about locations of utilities inside the building and connections to the outside when locating plumbing, electrical, data, telecommunications, mechanical rooms, and other mechanical systems.
◆ Access should be convenient.

Engineers must specify the equipment and materials to be used during construction for lift stations, water lines, sewer lines, stormwater drainage, roadway details, windows, doors, roof systems, structures, HVAC, and a host of other important components of the design. Many of these items will be discussed and illustrated briefly, but design manuals, building codes, textbooks, and product literature should be consulted as a part of the design process.

12.1 Roof Systems

The number one *purpose* of a roof is to keep water out of a building. Water from outside elements can move in an insidious fashion along the roof of a building and will penetrate virtually every opening in the roof and adjacent walls. Moisture inside a building will promote the growth of mold, creating "sick building syndrome," which can lead to various human health effects. Water in combination with air and salts also can promote corrosion, which can both damage the structure and lead to even greater moisture intrusion over time. If water is allowed to stand on a roof for any period of time, the roof will leak regardless of the roofing material, and all leaks get worse with time. Therefore, the design goal for a roof is to get water off the roof as quickly as possible.

Many different materials are used for roofs, including steel, aluminum, concrete, wood, and asphalt, among others. Pressboard or particleboard never should be used for roofs because of immediate degradation when in contact with water. All roofs are sloped, including "flat roofs," which commonly have a minimum slope of 2%. For flat concrete roofs, Styrofoam™ or other building materials are used to build them up. Membranes are commonly used as a seal for commercial buildings, much like shingles are used for residential roofs (Figure 12.1) or metal for commercial roofs (Figure 12.2). Roof drains and scuppers

Figure 12.1 Shingled roof with no gutters

Figure 12.2 Metal roof with gutters

may be required as a backup system for commercial roofs (see Figure 12.3). Design engineers will need to consult all of the local codes for roof system design rules and specifications for the project location. A rule of thumb is to get the water off in less than 5 min.

The demands for pitched roofs will vary by locale, and generally the major requirement is to size gutters to meet the storm intensity of the area. When the design rainfall intensity

Figure 12.3 Roof leaders exiting to rainwater-harvesting system and overflow scuppers along the parapet wall

is exceeded, the gutters simply overflow. Pitched roofs offer a benefit in smaller buildings, but the minimum slope creates a challenge in building height for larger roofs. Consequently, larger building roofs typically are "flat" roofs. The following is the basic procedure for a flat roof stormwater design:

1. Divide the roof into sections of between 2,000 and 10,000 ft^2 each.
2. Identify the desired low points for each roof section.
3. Place the low points adjacent to overflows. Both drains and overflow scuppers are required in case the drains get plugged (see Figure 12.3). Most codes require both on

flat roofs but not on pitched roofs. As a result, it is useful to co-locate them as opposed to allowing water to pool on the roof.

4. Build up the roof slope in sections to get water to flow to the drains.
5. Size the vertical drain pipes based on the local building codes.
6. Size the horizontal pipes connecting the vertical leaders according to the local building codes.
7. Size the overflows based on building codes.
8. Size scuppers and vertical leaders. Note that these pipes cannot be combined.
9. Connect the piping system to the site stormwater system.

An example of designing a roof drainage system is useful at this point, keeping in mind that local codes may change the requirements. The following stormwater design was for an assisted living facility on the Boca Raton campus of Florida Atlantic University (FAU) on several acres of property. Applicable codes used were:

◆ 2010 Florida Building Code
◆ 2010 Florida Building Code: Plumbing
◆ South Florida Water Management District design standards
◆ Florida Department of Transportation design standards

The 2010 Florida Building Code: Plumbing (FBCP) provides the 100-year, 1-hr rainfall rates for the state of Florida (see Figure 12.4). According to FBCP 1106.1, the roof for the structure should provide enough drainage for 5 in. of rainfall in 1 hr. To achieve this goal, the total area of the roof is determined first, and then the roof is divided into subareas to determine the number and size of the vertical drainage leaders and the size of the horizontal drainage pipes. The north and south buildings of the assisted living facility were divided into sections (see Figure 12.5), with the biggest of these subdivided roof drainage areas located on the west side of the north building.

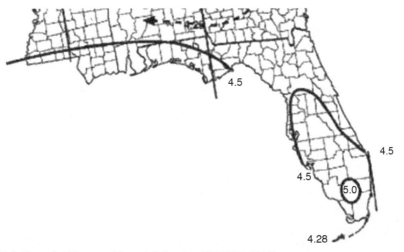

Figure 12.4 Map of 100-year, 1-hr rainfall rates (SFWMD 2014)

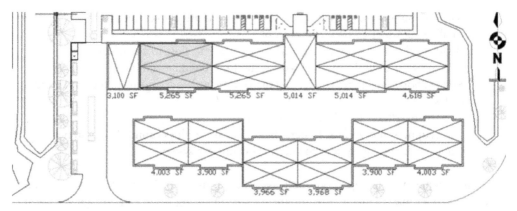

Figure 12.5 Roof area sectioned off for roof drains

The tables in the FBCP were used to determine the number of drains and their appropriate sizes. Using the largest roof drainage section of 5,265 ft^2 and a rainfall rate of 5 in./hr, Table 12.1 (vertical leaders) and Table 12.2 (horizontal piping) indicate that 5-in.-diameter vertical pipes and 5-in.-diameter horizontal pipes are required, but 5-in. diameter is an odd size, so 6-in. pipes are used.

Table 12.3 outlines the sizing for the wall scuppers, and Table 12.4 outlines the sizing of horizontal piping and gutter diameters, based on the same roof area. The proposed piping layout is shown in Figure 12.6.

All but two roof sections have two drains. The drain pipe runs inside the building and into the ground, where horizontal pipes drain to the site's two retention areas along the east and west perimeters.

A second sample project is shown in Figure 12.7. This is a small 6,000-ft^2 building. The roof slopes to the north as a monoslope and is subdivided into three roughly equal sections of 2,000 ft^2 each. Assuming the same rainfall intensity as the previous example and using the same set of FBCP tables, the vertical drain leaders should be 4 in. (Table 12.5) and the horizontal ones (located at ground level) 4 in. (Table 12.6). The scupper drains (Table 12.7), assuming a 2-in. head, dictate the length of the opening to be 16 in. for each of the three roof sections.

Table 12.1 Size of vertical leader pipes by horizontally projected roof area in square feet (FBCP 2007)

Diameter of leader (in.)	Rainfall rate (in./hr)											
	1	2	3	4	5	6	7	8	9	10	11	12
2	2,880	1,440	960	720	575	480	410	360	320	290	260	240
3	8,800	4,400	2,930	2,200	1,760	1,470	1,260	1,100	980	880	800	730
4	18,400	9,200	6,130	4,600	3,680	3,070	2,630	2,300	2,045	1,840	1,675	1,530
5	34,600	17,300	11,530	8,650	6,920	5,765	4,945	4,325	3,845	3,460	3,145	2,880
6	54,000	27,000	17,995	13,500	10,800	9,000	7,715	6,750	6,000	5,400	4,910	4,500
8	116,000	58,000	38,660	29,000	23,200	19,315	16,570	14,500	12,890	11,600	10,545	9,600

Table 12.2 Size of horizontal roof drain pipes by horizontally projected roof area in square feet (FBCP 2007)

Size of horizontal piping (in.)	Rainfall rate (in./hr)					
	1	2	3	4	5	6
1/8 unit vertical in 12 units horizontal (1% slope)						
3	3,288	1,644	1,096	822	657	548
4	7,520	3,760	2,506	1,800	1,504	1,253
5	13,360	6,680	4,453	3,340	2,672	2,227
6	21,400	10,700	7,133	5,350	4,280	3,566
8	46,000	23,000	15,330	11,500	9,200	7,600
10	82,800	41,400	27,600	20,700	16,580	13,800
12	133,200	66,600	44,400	33,300	26,650	22,200
15	218,000	109,000	72,800	59,500	47,600	39,650
1/4 unit vertical in 12 units horizontal (2% slope)						
3	4,640	2,320	1,546	1,160	928	773
4	10,600	5,300	3,533	2,650	2,120	1,766
5	18,880	9,440	6,293	4,720	3,776	3,146
6	30,200	15,100	10,066	7,550	6,040	5,033
8	65,200	32,600	21,733	16,300	13,040	10,866
10	116,800	58,400	38,950	29,200	23,350	19,450
12	188,000	94,000	62,600	47,000	37,600	31,350
15	336,000	168,000	112,000	84,000	67,250	56,000
1/2 unit vertical in 12 units horizontal (4% slope)						
3	6,576	3,288	2,295	1,644	1,310	1,096
4	15,040	7,520	5,010	3,760	3,010	2,500
5	26,720	13,360	8,900	6,680	5,320	4,450
6	42,800	21,400	13,700	10,700	8,580	7,140
8	92,000	46,000	30,650	23,000	18,400	15,320
10	171,600	85,800	55,200	41,400	33,150	27,600
12	266,400	133,200	88,800	66,600	53,200	44,400
15	476,000	238,000	158,800	119,000	95,300	79,250

Table 12.3 Sizing scuppers for a 5-in./hr rate of rainfall by horizontally projected roof area in square feet (FBCP 2007, Table 1106.7)

Head (in.)	Length of weir (in.)						
	4	6	8	12	16	20	24
1	230	346	461	692	923	1,153	1,384
2	641	961	1,282	1,923	2,564	3,205	3,846
3	1,153	1,730	2,307	3,461	4,615	5,769	6,923
4	1,794	2,692	3,589	5,384	7,179	8,974	10,769

Table 12.4 Size of horizontal piping and gutter diameters (FBCP 2007, Table 1106.3)

Piping	Size of horizontal piping (in.) ($\frac{1}{4}$ unit vertical in 12 units horizontal, 2% slope)	Size of scuppers (in.) for a 5-in./hr rate of rainfall (Table 1106.7) assuming 3-in. head
Vertical	6	20
Horizontal	6	20

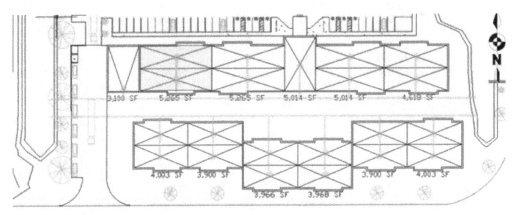

Figure 12.6 Proposed piping layout for stormwater underground

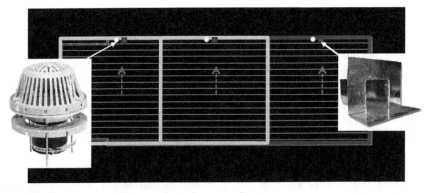

Figure 12.7 Monoslope roof broken into three sections

Table 12.5 Size of vertical leader pipes by horizontally projected roof area in square feet (FBCP 2007)

Diameter of leader (in.)	Rainfall rate (in./hr)											
	1	2	3	4	5	6	7	8	9	10	11	12
2	2,880	1,440	960	720	575	480	410	360	320	290	260	240
3	8,800	4,400	2,930	2,200	1,760	1,470	1,260	1,100	980	880	800	730
4	18,400	9,200	6,130	4,600	**3,680**	3,070	2,630	2,300	2,045	1,840	1,675	1,530
5	34,600	17,300	11,530	8,650	6,920	5,765	4,945	4,325	3,845	3,460	3,145	2,880
6	54,000	27,000	17,995	13,500	10,800	9,000	7,715	6,750	6,000	5,400	4,910	4,500
8	116,000	58,000	38,660	29,000	23,200	19,315	16,570	14,500	12,890	11,600	10,545	9,600

Table 12.6 Size of horizontal roof drain pipes by horizontally projected roof area in square feet (FBCP 2007)

Size of horizontal piping (in.)	Rainfall rate (in./hr)					
	1	2	3	4	5	6
1/8 unit vertical in 12 units horizontal (1% slope)						
3	3,288	1,644	1,096	822	657	548
4	7,520	3,760	2,506	1,800	1,504	1,253
5	13,360	6,680	4,453	3,340	2,672	2,227
6	21,400	10,700	7,133	5,350	4,280	3,566
8	46,000	23,000	15,330	11,500	9,200	7,600
10	82,800	41,400	27,600	20,700	16,580	13,800
12	133,200	66,600	44,400	33,300	26,650	22,200
15	218,000	109,000	72,800	59,500	47,600	39,650
1/4 unit vertical in 12 units horizontal (2% slope)						
3	4,640	2,320	1,546	1,160	928	773
4	10,600	5,300	3,533	2,650	2,120	1,766
5	18,880	9,440	6,293	4,720	3,776	3,146
6	30,200	15,100	10,066	7,550	6,040	5,033
8	65,200	32,600	21,733	16,300	13,040	10,866
10	116,800	58,400	38,950	29,200	23,350	19,450
12	188,000	94,000	62,600	47,000	37,600	31,350
15	336,000	168,000	112,000	84,000	67,250	56,000
1/2 unit vertical in 12 units horizontal (4% slope)						
3	6,576	3,288	2,295	1,644	1,310	1,096
4	15,040	7,520	5,010	3,760	3,010	2,500
5	26,720	13,360	8,900	6,680	5,320	4,450
6	42,800	21,400	13,700	10,700	8,580	7,140
8	92,000	46,000	30,650	23,000	18,400	15,320
10	171,600	85,800	55,200	41,400	33,150	27,600
12	266,400	133,200	88,800	66,600	53,200	44,400
15	476,000	238,000	158,800	119,000	95,300	79,250

Table 12.7 Opening for scupper weir by horizontally projected roof area in square feet (FBCP 2007)

Head (in.)	Length of weir (in.)						
	4	6	8	12	16	20	24
1	230	346	461	692	923	1,153	1,384
2	641	961	1,282	1,923	2,564	3,205	3,846
3	1,153	1,730	2,307	3,461	4,615	5,769	6,923
4	1,794	2,692	3,589	5,384	7,179	8,974	10,769

12.2 On-site Stormwater Drainage

Stormwater collection and transport systems represent one of the most critical parts of the infrastructure of any urbanized area or municipality. Stormwater routing is an important component of any site planning pertaining to a specific project. Stormwater systems are designed to keep the land surface dry and to limit the impact of standing water or flooding on people and vehicular traffic. Depending on the locale and applicable codes, calculations may need to be performed to be able to properly store all of the stormwater that falls on the site, retain a portion of the stormwater on-site, or only detain the water for a period of time and release it during lower flow periods.

Current stormwater management programs typically have significant financial constraints placed on them by local officials who see competing needs for funding (police, fire, parks, etc.). In addition, compliance with federal stormwater permits (MS4) must be maintained, which will require additional record keeping, policy development, inspections, and maintenance compared to what is currently being performed. The MS4 program is a federally mandated program that enacts a portion of the Clean Water Act. The intent is to reduce pollution to waterways that impact aquatic life. As a result, restrictions on off-site discharges have increased with time, resulting in on-site stormwater rules being locally driven. The question is how much stormwater must be retained on-site. Some basic assumptions are used in most stormwater systems:

◆ Storm sewer system capacity should be designed for the conveyance of peak inflow from the applicable catchment area, based on the appropriate 3-day rainfall event, such that the hydraulic gradient of the water does not exceed the grate or cover elevation at any inlet or manhole.

◆ Inlet times assumed for determining required drainage system capacity should not exceed 5 min.

◆ Storm sewer pipe should have a nominal diameter of not less than 15 in. or equivalent oval pipe size.

◆ Storm sewers should be designed to ensure flow velocities of not less than 2.5 ft/sec in all pipe runs serving two or more inlets, nor greater than 10 ft/sec in any pipe run.

◆ A suitable access structure such as a manhole, junction box, or inlet must be installed at each junction or change in pipe size, slope, or direction.

◆ The maximum pipe run between access structures should be (a) 300 ft for 15- and 18-in. pipe, (b) 400 ft for 24- to 36-in. pipe, or (c) 500 ft for 42-in. and larger pipe.

◆ All pipe used in the storm sewer system should conform to current American Society for Testing and Materials, American Association of State Highway and Transportation Officials, or American National Standards Institute standard specifications for materials and fabrication of barrel and joints and should meet current applicable local and state standard specifications.

◆ All concrete pipes should have gasket joints.

◆ When metal pipe is used beneath pavement within a street, it must be designed to provide a joint-free installation or, where joint-free installations are not feasible, be jointed with a 12-in.-wide band with a mastic or neoprene gasket to provide a watertight joint.

◆ Where a drainage pipe ends, such as at the entrance to a retention basin, the edge of a driveway or road, or similar installation, the pipe should be fitted with headwalls, endwalls, inlets, and other appropriate terminating and intermediate structures (see Figures 12.8 to 12.10). Structure design must meet or exceed local or state standards.

Figure 12.8 Stormwater grate in brick paver driveway

Figure 12.9 Stormwater grate in street

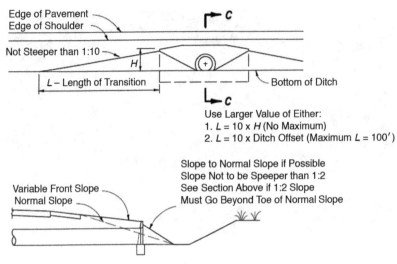

NOTE: Filling or excavation of variable slopes to be done during normal grading operations

SECTION CC

Figure 12.10 Headwall schematic (FDOT 2014)

The basic design process for stormwater conveyance systems is as follows:

- ◆ Determine the appropriate design storm, typically something like a 25-year, 3-day event or 100-year storm event, depending on the local drainage code requirements.
- ◆ Determine the curve number (CN) for the site prior to development. The CN relates to the degree of perviousness of a site. A building, parking lot, or driveway is impervious. Water runs off impervious surface faster than grass or other pervious land. The CN captures this difference.
- ◆ Determine the amount of runoff from the site prior to development.
- ◆ Determine the CN value for the developed site.
- ◆ Determine the amount of runoff from the site after development.
- ◆ Determine the amount of runoff generated with the change in development.
- ◆ Determine the amount of runoff required to be retained on-site.
- ◆ Determine appropriate methods to retain water (ponds, exfiltration, French drains, etc.).

The concept is best illustrated with an example. The assisted living facility building discussed earlier had a requirement that the stormwater design for the site comply with the regulations of the South Florida Water Management District (SFWMD), which specify that the site must store the first 1 in. of runoff from the entire site or the runoff from a 3-day, 25-year storm event, whichever is greater. For a conservative design, the 3-day, 25-year storm event was chosen (see Figure 12.11).

To achieve these criteria, three retention areas were designed to collect the runoff from the impervious areas. Figure 12.12 shows the location of the three retention areas along the north, west, and southeast boundaries of the site, which primarily collect runoff from the

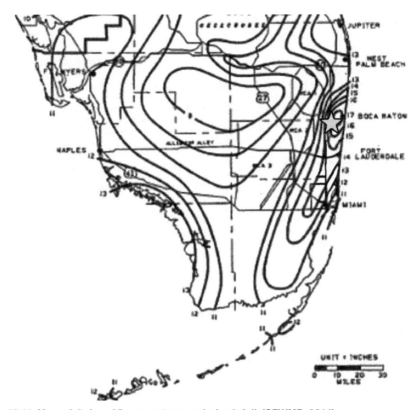

Figure 12.11 Map of 3-day, 25-year return period rainfall (SFWMD 2014)

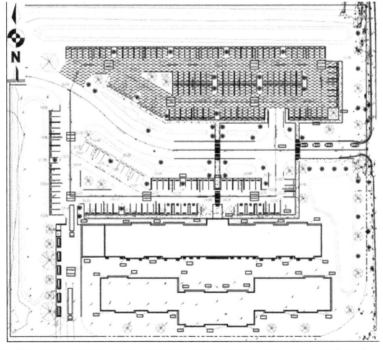

Figure 12.12 Site plan showing three retention areas (north, west, and southeast)

Table 12.8 Runoff CN for urban areas (USDA 1986)

Cover description (cover type and hydrologic condition)	CN for hydrologic soil group			
	A	**B**	**C**	**D**
Fully developed urban areas (vegetation established)				
Open space (lawns, parks, golf courses, cemeteries, etc.)				
Poor condition (grass cover <50%)	68	79	86	89
Fair condition (grass cover 50 to 75%)	49	69	79	84
Good condition (grass cover >75%)	39	61	74	80
Impervious areas				
Paved parking lots, roofs, driveways, etc. (excluding right-of-way)	98	98	98	98
Streets and roads				
Paved: curbs and storm sewers (excluding right-of-way)	98	98	98	98
Paved: open ditches (including right-of-way)	83	89	92	93
Gravel (including right-of-way)	76	85	89	91
Dirt (including right-of-way)	72	82	87	89

roadway running through the site and the parking lots. The CN method was used to calculate the runoff. The CN value is assigned to predict runoff from rainfall and is based on the following factors: land use, soil type, and percentage impervious. Table 12.8 shows CN values for various types of ground cover. Soil borings near this site demonstrated that the subsurface contains mostly sand and virtually no clay, which means the local drainage soil belongs to hydrologic soil group A for design purposes. Next, the predevelopment runoff must be calculated. The total site area was determined to be 372,401 ft^2, and the total impervious area was 295,482 ft^2 (or 79% of the total area), with a CN of 98. The remaining 76,919 ft^2 (or 21% of the total area) is pervious open areas consisting of mostly grass in good condition with a CN of 39.

The composite CN (computed as a weighted average) was used to determine runoff and resulted in a value of 86:

$$\text{CN composite} = \frac{\sum_{i}^{N} A_i CN_i}{\sum_{i}^{N} A_i} = \frac{(285,482 \times 98) + (76,919 \times 39)}{372,401} = 86 \qquad (12.1)$$

where

    A_i = area
  CN_i = curve number

Next, the soil retention capacity was calculated with the following formula:

$$S = \frac{1,000}{CN} - 10 = \frac{1,000}{86} - 10 = 1.63 \text{ in.} \qquad (12.2)$$

where

S = soil retention capacity
CN = composite curve number

The result was found to be 1.63 in. Now the runoff from the site can be determined using the following:

$$R = \frac{(P - 0.2S)^2}{P + 0.8S} = \frac{(16 - 0.2 \times 1.63)^2}{16 + 0.8 \times 1.63} = 14.2 \text{ in.} \tag{12.3}$$

where

R = runoff depth
P = precipitation

The value for the precipitation was determined to be 16 in. using the SFWMD Environmental Resource Permit Information Manual, Volume IV, Figure 2 for a 3-day, 25-year storm event. The predevelopment runoff was determined to be 14.2 in. for the site.

To calculate the postdevelopment runoff, the total impervious area was calculated to be 101,499 ft^2 with a CN of 98. This total is made up of roadways, parking lots, and building roof areas. SFWMD requires the retention areas to be counted as impervious areas in this method, which total 46,305 ft^2 with a CN of 98. In the north parking lot (58,540 ft^2), the engineer elected to use pervious pavers with a CN of 70. The remaining area is pervious space, consisting of mostly grass, with a CN of 39. The composite CN method was used (Equation 12.1) to determine runoff and resulted in a composite CN of 68:

$$\text{CN composite} = \frac{\displaystyle\sum_{i}^{N} A_i CN_i}{\displaystyle\sum_{i}^{N} A_i}$$

$$= \frac{(46,305 \times 98) + (458,540 \times 70) + (101,499 \times 98) + (166,057 \times 39)}{372,401} = 68$$

Next, the soil retention capacity (S) was then calculated using Equation 12.2:

$$S = \frac{1,000}{CN} - 10 = \frac{1,000}{68} - 10 = 4.71 \text{ in.}$$

The soil retention capacity was found to be 4.71 in.

Next, the runoff from the site can be determined using Equation 12.3:

$$R = \frac{(P - 0.2S)^2}{P + 0.8S} = \frac{(16 - 0.2 \times 4.71)^2}{16 + 0.8 \times 4.71} = 11.47 \text{ in.}$$

The postdevelopment runoff was determined to be 11.47 in., which is a smaller amount than the predevelopment runoff (14.2 in.) because the site was an existing airplane runway and some of the existing concrete was removed during construction to create more pervious area and green space. Even though the runoff was decreased, the design engineers decided to maximize usage of the retention areas to reduce discharge into existing off-site infrastructure. The overall surface area of the three retention zones was designed to be 46,305 ft², with an average depth of 4 ft, which gives a retention volume of 185,220 ft³ and accounts for all of the runoff for the site.

The next step was to lay out the drain locations and underground pipe layout. The existing stormwater pipe system consisted of 30- and 24-in. corrugated aluminum pipes connected to curb drainage structures along the roads on the east and south sides of the site, including connecting the overflow structure to the curb structure nearest the east entrance, as shown in Figure 12.13 (marked by an oval). The inlets in the parking lots were designed to slope toward the nearest retention pond at a 1% slope, as shown in Figure 12.14.

The following shows an example of this type of calculation using a spreadsheet as presented in Table 12.8. The time of concentration is determined by dividing the length by the velocity:

$$t_c = \frac{L_o}{V} \tag{12.4}$$

The time of concentration (t_c) was determined by taking the diagonal length of the watershed (L_o) and dividing it by the velocity (V) of 2 ft/sec assuming paved ground:

$$t_c = \frac{L_o}{V} = \frac{106 \text{ ft}}{2 \text{ ft/sec}} = 53 \text{ sec}$$

The piping needed to handle the stormwater drainage is 15-in.-diameter concrete pipe. This time of concentration was used to determine the intensity from intensity-duration-frequency curves (SFWMD 2014) using the following formula:

$$Q - kCiA \tag{12.5}$$

where

Q = flow
k = 1.008 (conversion factor)
C = runoff coefficient
i = intensity (in./hr)
A = area (acres)

The result is:

$$Q - kCiA = 1.008 \times 0.95 \times 6.5 \text{ in./hr} \times 0.46 \text{ acre} = 1.54 \text{ ft}^3/\text{sec}$$

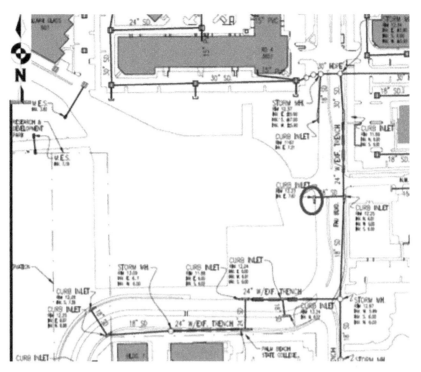

Figure 12.13 Existing stormwater system adjacent to project site (FAU 2014)

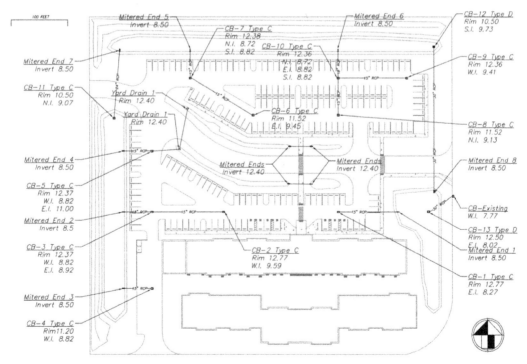

Figure 12.14 Stormwater piping layout

Next, the diameter of the pipe (D) was calculated using a variation of Manning's equation:

$$D = \left(\frac{2.16Qn}{\sqrt{S}} \right)^{0.375}$$

(12.6)

where

n = roughness coefficient (0.013 for concrete pipe)
S = the slope (1%)

The result yields:

$$D = \left(\frac{2.16 \times 1.54 \times 0.013n}{\sqrt{0.01}} \right)^{0.375} = 0.73 \text{ ft (or 8.8 in.)}$$

However, most jurisdictions have a minimum pipe size. In this case, the local code specified a minimum allowed pipe diameter of 15 in. The rest of the results are given in Table 12.9, which shows calculated pipe diameters for the entire stormwater piping system.

All catch basins in parking lots were specified to be Florida Department of Transportation (FDOT) type C basins with 15-in.-diameter reinforced concrete pipe (RCP). Piping connecting yard drains from the treatment bioswales was specified to be 8-in.-diameter high-density polyethylene. Retention ponds were designed with type C basins and 24-in.-diameter RCP. The overflow structure for the east retention pond was specified to be an FDOT type D basin (see Figure 12.15) with a 30-in.-diameter RCP connected to an existing curb structure at FAU

Table 12.9 Calculated peak flows and pipe diameters

Area	Pipe #	k	C	i (in./hr)	Total flow area (acres)	Q (ft³/sec)	D (ft)	Calculated diameter (in.)	Pipe diameter (in.)	Pipe velocity (ft/sec)
South parking	1	1.008	0.95	8.20	0.30	2.32	0.85	10.21	15	1.89
South parking	2	1.008	0.95	8.20	0.32	2.51	0.88	10.52	15	2.04
South parking	3	1.008	0.95	8.20	0.66	5.15	1.15	13.78	15	4.20
South parking	6	1.008	0.95	8.20	0.40	3.14	0.95	11.45	15	2.56
Bioswales	5A	1.008	0.15	8.20	0.80	0.99	0.62	7.43	8	2.84
South parking	5B	1.008	0.37	8.20	1.11	3.44	0.99	11.84	15	2.80
North parking	6	1.008	0.75	8.20	0.26	1.60	0.74	8.88	15	1.30
North parking	7	1.008	0.75	8.20	0.62	3.82	1.03	12.32	15	3.11
North parking	8	1.008	0.75	8.20	0.30	1.86	0.78	9.40	15	1.51
North parking	9	1.008	0.75	8.20	0.25	1.56	0.73	8.80	15	1.27
North parking	10	1.008	0.75	8.20	0.77	4.80	1.12	13.42	15	3.91
Retention connecting pipe	11	N/A	N/A	N/A	N/A	11.72	1.56	18.76	24	3.73
Retention connecting pipe	12	N/A	N/A	N/A	N/A	20.34	1.92	23.06	24	6.48
Outfall pipe	13	N/A	N/A	N/A	N/A	22.66	2.00	24.02	30	4.62

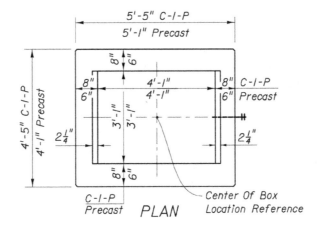

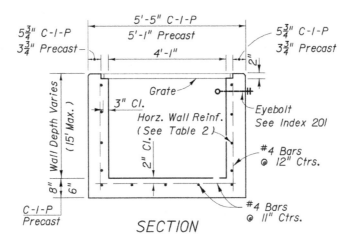

HORIZONTAL WALL REINFORCING SCHEDULES (TABLE 2)

WALL DEPTH	SCHEDULE	AREA (in.²/ft.)	MAX. SPACING	
			BARS	WWF
0'-6'	A12	0.20	12"	8"
6'-10'	A6	0.20	6"	5"
10'-13'	A4	0.20	4"	3"
10'-15'	B5.5	0.24	5½"	5"

TYPE D

Recommended Maximum Pipe Size:
3'-1" Wall - 24" Pipe
4'-1" Wall - 36" Pipe

Figure 12.15 FDOT type D basin (FDOT 2014)

Boulevard. The north parking lot was proposed to be pervious pavers (see Figure 12.16 for paver detail).

The second example project was located on an industrial site (see Figure 12.17). Because the property is compact and crowded with numerous structures, the site needed to be broken into sections. The engineers reviewed the site before and after the new construction (see Figures 12.18 and 12.19). The site also was divided into subbasins (see Figure 12.20). In this case, the runoff volume increases from 36,187 ft^3 to 44,915 ft^3 postredevelopment. That means an additional 8,730 ft^3 of stormwater needs to be retained on-site. If a new pond is constructed as dry retention (water not in pond except after rainfall), the maximum depth is 4 ft based on the local groundwater depth (5 ft from the surface). In this case, the pond needs to have a surface area of 2,200 ft^2. Note that there will be local sediment control requirements to limit runoff of particulates to the pond. Also, the pond will have maximum slopes (e.g., 1:4 vertical:horizontal) to limit the potential hazard of falling into the pond.

The next example is a three-story mixed-use building with a roof area of 28,800 ft^2 (see Figure 12.21). The engineers divided the roof into six main areas, including a smaller area on the east end. The roof zones can be divided equally into areas of approximately 4,666 ft^2. Based on Table 12.5, this requires vertical leaders of 4 in. However, the design engineer intended for the drains to exit on the south side of the building, so the horizontal leaders will actually carry 9,333 ft^2 to the vertical leaders. Reviewing Tables 12.5 and 12.6, the vertical leaders are required to be 6 in., but the horizontal leaders must be 8 in. Design engineers never downsize pipe, so all leaders to the ground are 8 in. where access to the stormwater system could be accomplished (Figure 12.22). The project also included a scupper system as an overflow. Roof drains, like any drain, can become clogged with debris. The overflow scuppers ensure that the roof will not hold water because water is also a heavy load. The scupper locations are shown in Figure 12.23. Note that these are not the same locations as the drains because the scuppers must be located on the exterior of the building, whereas the drains can be located conveniently. As a result, for this example, the number of drains and scuppers is different. The designers needed to avoid windows, doors, points of access, etc., which required the use of more scuppers in this case.

Designers should be aware that in most building codes, the roof drains and scuppers often are prohibited from being connected to the same vertical pipe, so the local codes should be checked. Ultimately, both piping systems are tied to the storm sewer system (Figure 12.24), which was designed as shown in the previous example.

This project incorporated the use of a cistern (see Figure 12.25) underneath the parking lot to store water for irrigation purposes (and to earn LEED® points). Because cisterns will overflow, a backup system is needed. In some cases, this can be a gravity well where there is saltwater intrusion of an aquifer that is not under pressure (see Figure 12.26). Other solutions are retention ponds, detention basins, French drains, and off-site discharges, where permitted.

Local or state building codes, along with rainfall intensity, define the pipe sizes, number of drains/scuppers, cistern sizes, and ancillary requirements like baffles, pumps, stormwater pipe material, minimum pipe sizes, and concepts like inflow wells and French drains. Design engineers must be familiar with these codes.

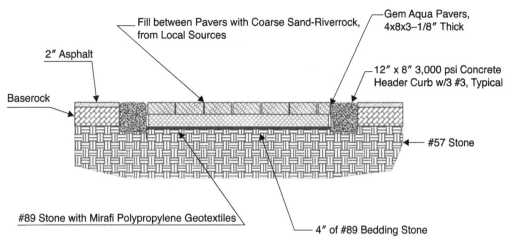

Figure 12.16 Cut sheet for pervious pavers

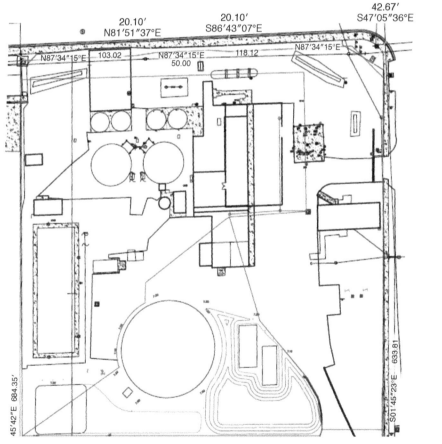

Figure 12.17 Current site plan for Dania Beach Water Plant as proposed, divided into drainage subbasins

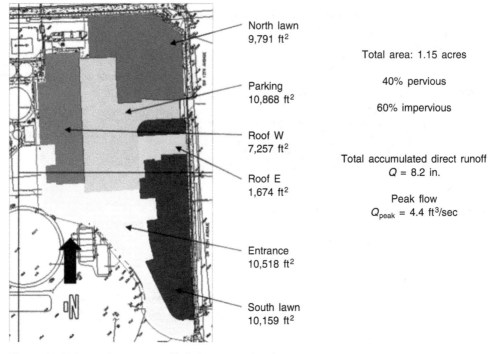

North lawn
9,791 ft²

Total area: 1.15 acres

40% pervious

60% impervious

Parking
10,868 ft²

Roof W
7,257 ft²

Total accumulated direct runoff
Q = 8.2 in.

Roof E
1,674 ft²

Peak flow
Q_{peak} = 4.4 ft³/sec

Entrance
10,518 ft²

South lawn
10,159 ft²

Figure 12.18 Impervious area calculations preredevelopment

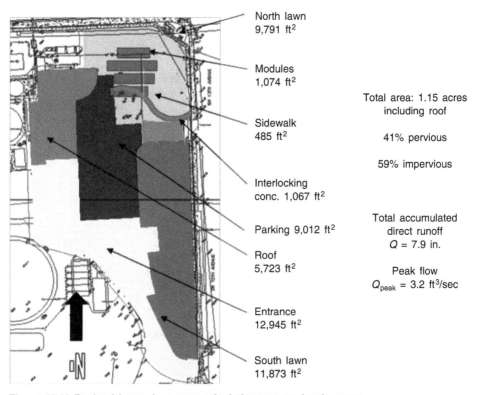

North lawn
9,791 ft²

Modules
1,074 ft²

Total area: 1.15 acres
including roof

41% pervious

59% impervious

Sidewalk
485 ft²

Interlocking
conc. 1,067 ft²

Total accumulated
direct runoff
Q = 7.9 in.

Parking 9,012 ft²

Roof
5,723 ft²

Peak flow
Q_{peak} = 3.2 ft³/sec

Entrance
12,945 ft²

South lawn
11,873 ft²

Figure 12.19 Revised impervious area calculations postredevelopment

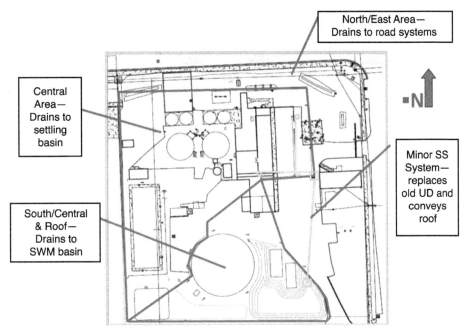

North/East Area—
Drains to road systems

Central
Area—
Drains to
settling
basin

Minor SS
System—
replaces
old UD and
conveys
roof

South/Central
& Roof—
Drains to
SWM basin

Figure 12.20 Basins for the site (SWM = stormwater master plan, SS = site stormwater, UD = under drain)

Figure 12.21 Proposed mixed-use development; note solar panels on roof

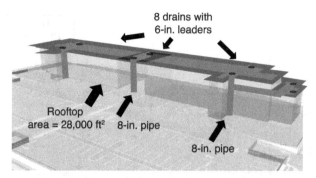

Figure 12.22 Rooftop area and drains based on calculation of roof area

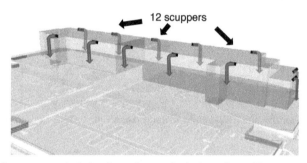

Figure 12.23 Rooftop area and drains based on calculations from Figure 12.15

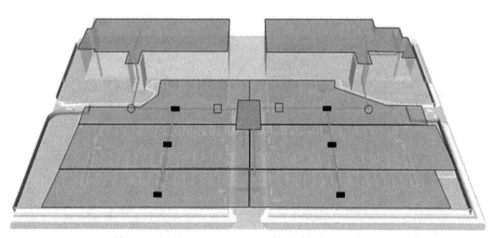

Figure 12.24 Parking lot connection for storm drainage system (■ = stormwater inlet grate, O = manhole/junction box, — = 20-in. exfiltration pipe, □ = gravity well)

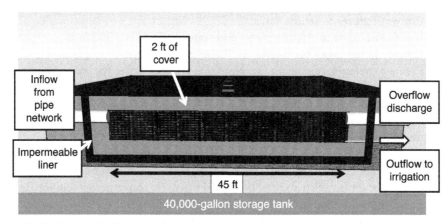

Figure 12.25 Proposed cistern system

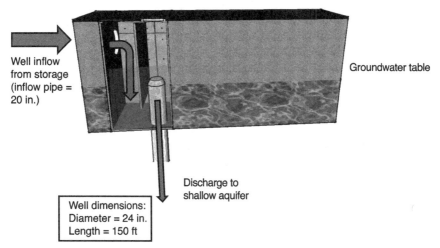

Figure 12.26 Gravity well concept for cistern overflow

12.3 Potable Water Systems

The potable water distribution portion of the project must conform with all local, state, and federal ordinances and building codes. The following is the basic procedure for water system design:

1. Identify connection points to the main potable water distribution system.
2. Create a spreadsheet for plumbing fixtures (toilets, sinks, showers, water fountains, etc.).
3. Identify the number of fixture units associated with each fixture in step 2 based on fixture unit conversions in the building code.

4. Convert fixtures to fixture units.
5. Use fixture units to size the meter and water main to the building.
6. Identify initial losses for meters and backflow preventer.
7. Create a preliminary piping plan. Note that the cold water system is designed first, as it requires the largest demand. The hot water simply parallels the cold system, except at toilets and urinals.
8. Enter the data into a water distribution piping model such as EPANET, WaterCAD, or other similar software tool.
9. Conduct modeling runs until all nodes have greater than 25 psi residuals. Note that for large buildings, a means to create reasonable demand must be sought or the pipe size will be grossly oversized and the model will never converge to a solution. In some jurisdictions, mechanical engineers perform this task, but the civil engineer must provide the appropriate size pipe to the building before the calculations can be done. Furthermore, the design engineer must ensure adequate clearance for the piping systems within the building structure.
10. Note that structures over three stories tall may require the installation of booster pumps to supply pressure for the upper floors.
11. Identify fire flow needs, if applicable.
12. Identify appropriate utility connection details from the local utility.

A couple of examples will help illustrate this design process. The first example is the assisted living facility discussed earlier in this chapter. The challenge is designing the potable water system for a building that has 206 bathrooms. The first step is to determine where connections can be made to the existing water main from the local water utility network. The public utility's pipelines normally are located in the road right-of-way or an easement adjacent to the property. Figure 12.27 shows a map of the water mains surrounding the site. There are adjacent water mains running along the east and north sides of the site. The new water piping system will connect at the northwest corner and east side of the site, shown as circles in Figure 12.27. Next, in designing the potable water system, the load factor of the building fixtures must be calculated using the applicable plumbing code section (Table 12.10). Depending on the type of fixture, many plumbing codes will designate a fixture unit load factor that is multiplied to determine the actual load factor for design purposes. The load factor is used along with data from Table 12.10 to determine the diameter of the water pipes running throughout the building and the recommended meter size at the main pipe entering the building (see Table 12.11).

The pipe network throughout the building must be selected using the same method. The plumbing code will state the minimum pipe size required at the connections depending on the fixture type (Table 12.12). At 1,287 fixture units, this project would require 3-in.-diameter pipes. However, if a 3-in. pipe were modeled for this case, it would fail at the demands of all these fixtures. That is because not all fixtures are in use at the same time. Smaller buildings will have more fixtures on (as a percentage) at any given time. Larger buildings, while exerting more demand, will have diminished demand per fixture. The codes usually will address this issue. Because it is not reasonable to assume that all water-using fixtures would be used at the same time, Hunter's method of estimating loads in plumbing systems

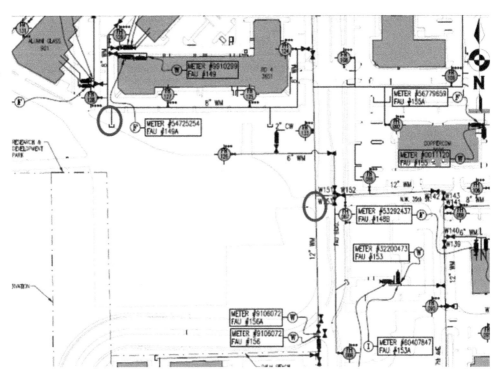

Figure 12.27 Existing water piping system surrounding the assisted living facility site (source: FAU)

Table 12.10 Excerpt of drainage fixture units for fixtures and groups (FBCP 2007, Table 709.1)

Fixture type	Drainage fixture unit value as load factor	Minimum size of trap (in.)
Bidet	1	$1\,1/4$
Combination sink and tray	2	$1\,1/2$
Dental lavatory	1	$1\,1/4$
Dishwashing machine, domestic	2	$1\,1/2$
Drinking fountain	$1/2$	$1\,1/4$
Emergency floor drain	0	2
Kitchen sink, domestic	2	$1\,1/2$
Kitchen sink, domestic with food waste grinder and/or dishwasher	2	$1\,1/2$
Lavatory	1	$1\,1/4$
Shower, flow rate		
5.7 gpm or less	2	$1\,1/2$
Greater than 5.7 gpm to 12.3 gpm	3	2
Greater than 12.3 gpm to 25.8 gpm	5	3
Greater than 25.8 gpm to 55.6 gpm	6	4
Service sink	2	$1\,1/2$
Sink	2	$1\,1/2$
Wash sink (circular or multiple), each set of faucets	2	$1\,1/2$

Table 12.11 Fixture unit conversion table for the assisted living facility example

Fixture type	Quantity	Pipe size (in.)	Fixture unit load	No. fixture units
		Water supply fixture units		
Bathroom group	206	$3/8$	6	1,236
Service sink	12	$1/2$	2	24
Washers	6	$1/2$	2	12
Sink (kitchen)	4	$1/2$	2	8
Water fountain	14	$3/8$	0.5	7
Total				1,287

Table 12.12 Minimum water service sizes for lines and meters based on fixture units (FBCP 2007, Table 603.1), applicable for both copper and plastic water piping

Number of fixture units flush tank water closet[a]	Diameter of water pipe (in.)[b]	Recommended meter size (in.)[c]	Approximate pressure loss for meter + 100-ft pipe (psi)[d]	Number of fixture units flush valve water closet[a]
18	$3/4$	$5/8$	30	—
19–55	1	1	30	—
—	1	1	30	9
56–85	$1\,1/4$	1	30	—
—	$1\,1/4$	1	30	10–20
86–225	$1\,1/2$	$1\,1/2$	30	—
—	$1\,1/2$	$1\,1/2$	30	21–77
226–350	2	$1\,1/2$	30	—
—	2	$1\,1/2$	30	78–175
351–550	2	2	30	—
—	2	2	30	176–315
551–640	$2\,1/2$	2	30	—
—	$2\,1/2$	2	30	316–392
641–1,340	3	3	22	—
—	3	3	22	393–940

[a] See Table 709.1 for fixture unit values.
[b] Minimum water service should be $3/4$ in. to control valve.
[c] All secondary submeters and backflow assemblies should be at least the same size as the line in which they are installed.
[d] Table based on minimum water main pressure of 50 psi.

uses a probability model and the fixture unit load number to generate a curve (see Figure 12.28) to estimate a more realistic design value for the demand of a system. The Hunter's conversion allows adjustment of the demands at each node proportionately prior to modeling. The adjustment will provide a more realistic estimate of the internal water distribution system demand and highlight critical pressure points where piping may need to be increased in size. In the assisted living facility example, using the total fixture unit load calculated (1,287 fixture units), the nomograph in Figure 12.28 indicates that a total of 244 gpm should be used as the design load of the piping network.

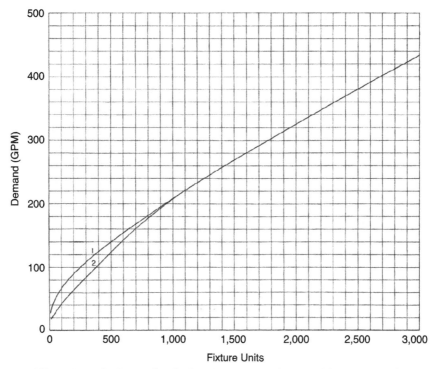

Figure 12.28 Hunter's method curve (1 = flush unit, 2 = tank flush system) (source: http://fire.nist.gov/bfrlpubs/build40/PDF/b40002.pdf)

A preliminary layout of the plumbing network, including all appropriate connections and nodes, is prepared using one of many software packages that model the hydraulic behavior of water distribution piping systems of a building or structure. These software packages can perform a simulation of the water movement within a pressurized pipe network by tracking the flow of water in each pipe and the pressure at each node. The objective of the calculations is to determine if the pressure from the potable water main is strong enough to deliver water to all of the fixtures in the building, such as the toilets, sinks, urinals, and water fountains. The building pipe network is input into the software package using the pipe sizes and modified loads in the system (from Hunter's method), taking into account losses from the meters and backflow devices and assuming an average available water main pressure. The models typically use the Hazen-Williams formula to calculate friction loss, pressures, and velocities at all locations of the network.

Ductile iron pipe (DIP) was chosen for installation (as per local standards) at the site water main loop, and type K copper was selected for use within the building. All exterior site piping must be installed with a minimum of 36-in. cover according to local standards (note that the bury depth will be deeper in areas that have frost concerns). The water main will enter both buildings at the west end, as shown in Figure 12.29. The supply branch to the first and second floors will be installed in the first-floor ceiling, while the branch supplying the

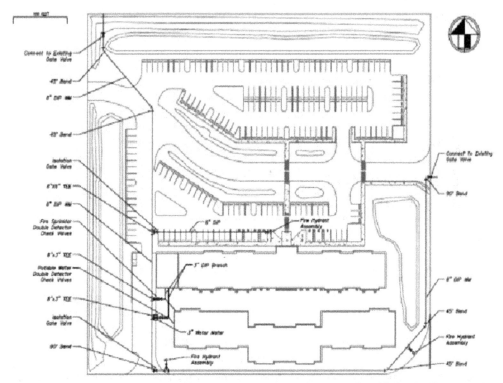

Figure 12.29 Final preliminary potable water system design for the assisted living facility example showing pipe sizing and material

third floor will be installed in the ceiling of the second floor. Figure 12.30 provides an example of an output from EPANET 2 showing that the minimum pressure conditions (25 psi) are met at all nodes under normal operating conditions. Note that manipulating pipe sizes typically will be required to get the model to converge to a working solution. The resulting pipe sizes were 4-in.-diameter DIP at the site loop, 3-in.-diameter type K copper from the meter into the building, 2-in.-diameter copper branches to each floor, and ¾-in. copper to supply individual fixtures (see Figure 12.30). The final preliminary potable water system design is shown in Figure 12.31.

The second example project is a proposed hotel development. The existing city water lines run along the west and east side of the site (see Figure 12.32), and the site already has three existing fire hydrants. The water main connection at the southeast corner of the site will be the primary water source for all fire systems in the building. The basic role of any water system design is to determine the number of fixtures in the building (toilets, sinks, showers, drinking fountains, etc.) and convert them to equivalent fixture units. Tables have been created to size piping and meters based on the estimated demand and fixture unit counts. The simplest way to determine fixture units is to create a spreadsheet with the fixture type and the conversion to fixture units and then multiply, as shown in Table 12.13.

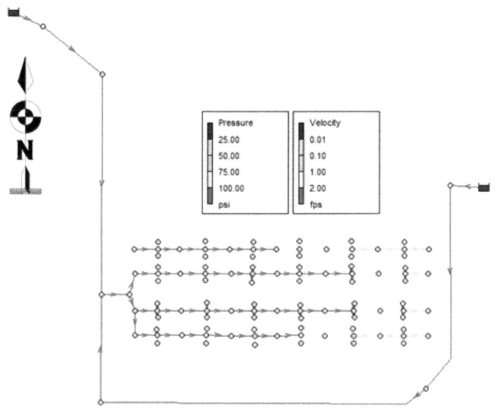

Figure 12.30 Model simulation of preliminary assisted living facility water network using EPANET 2

Prior to designing any of the pipe sizes or demands required, fixture locations were placed throughout the building based on the services and the scope requirements of the rooms. All hotel guest rooms contain a shower, a sink, and a toilet tank. The number and location of all the fixtures are listed in Table 12.13. The fixtures were then converted to fixture units by using Table 12.10. Hot water supply is only located where there are showers and hand wash locations, so sizing the hot and cold water the same way, based on the cold water size, ensures both systems will work properly. Flows are adjusted using Hunter's method.

Table 12.13 shows that there are 617 fixture units in the building. Using Table 12.12, the number of fixture units determines the size of the pipes needed to supply water to the building. Assuming that each room has a flush tank (as opposed to a valve system), the recommended pipe size is 2.5 in. and the meter size is 2 in. Because 2.5-in. pipe is uncommon, the next larger pipe size is recommended (3-in. pipe). For the design process, it is important to specify that the bathrooms in the guest rooms of the hotel will be counted as the "bathroom group (1.6-gal-per-flush water closet)." As noted previously, each fixture has a minimum flow rate and minimum pipe size using the data in Table 12.14.

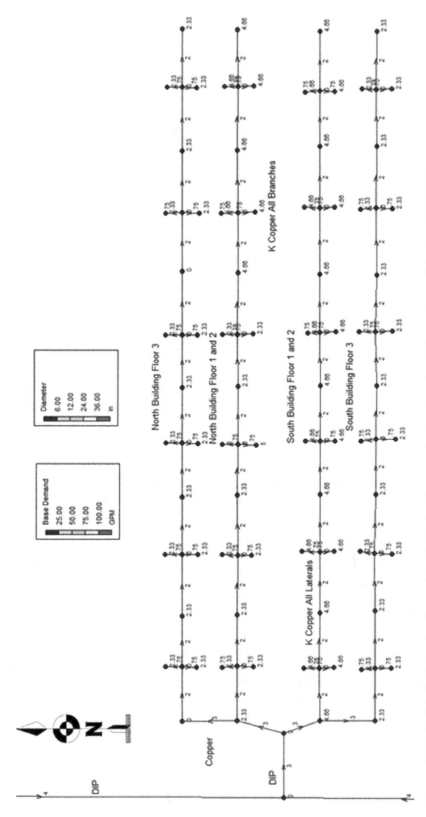

Figure 12.31 Model simulation for the assisted living facility example showing pipe sizing and material using EPANET 2

Figure 12.32 Location of existing water utility lines for the hotel property example

Table 12.13 Water distribution fixture units per floor for the hotel property example

Fixtures	Floor								Total fixtures	Fixture units	Fixtures in entire bldg
	1	2	3	4	5	6	7	8			
Automatic clothes washer, residential	0	0	0	2	2	2	2	2	10	2	20
Drinking fountain	2	2	0	0	0	0	0	0	4	0.5	2
Lavatory	13	10	0	0	0	0	0	0	23	1	23
Urinal (1 gal per flush or less)	2	2	0	0	0	0	0	0	4	2	8
Shower (flow rate 5.7 gpm or less)	0	0	13	13	13	13	13	13	78	2	156
Restaurant kitchen sink	0	2	0	0	0	0	0	0	2	2	4
Bathroom group (1.6-gal-per-flush water closet)	1	1	13	13	13	13	13	13	80	5	400
Dishwasher machine	0	2	0	0	0	0	0	0	2	2	4
Total fixture units											**617**

Table 12.14 Water distribution system design criteria per fixture supply at the outlet (FBCP 2010, Table 604.5)

Fixture supply outlet service	Flow rate (gpm)	Flow pressure (psi)	Minimum pipe size (in.)
Automatic clothes washer, residential	2	8	1/2
Drinking fountain	0.75	8	3/8
Lavatory	2	8	3/8
Urinal (1 gal per flush or less)	12	25	1/2
Shower (flow rate 5.7 gpm or less)	3	8	1/2
Restaurant kitchen sink	3	8	1/2
Bathroom group (1.6-gal-per-flush water closet)	7	20	1/2

The use of a hydraulic modeling program makes it easy to determine if the 2-in. meter creates too much head loss in the preliminary design layout of the system. The code will state the minimum pressures for every fixture. Software can help identify where issues with head loss and small piping impact the hydraulic network. It also can help identify if all nodes meet minimum pressure requirements (Figure 12.33). Common materials used for small underground cold water piping include schedule 80 polyvinyl chloride (PVC) or copper or in larger sizes C900 PVC or ductile iron. Fire flow services are exclusively ductile iron. The domestic water piping inside of the building should be copper pipe and fittings, based on local and state building codes, which often will specify the grade, either type K or M.

The third example project is the same mixed-use building discussed earlier. It has three stories. Table 12.15 outlines the fixture units. For this project, an EPANET simulation of

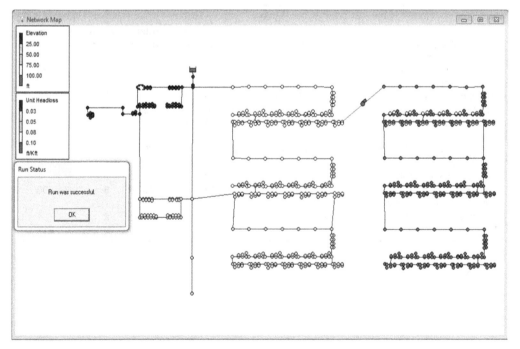

Figure 12.33 Screenshot of EPANET 2 model simulation showing node pressures for the hotel property example

Table 12.15 Spreadsheet for water distribution fixture units for the mixed-use building example

Fixture unit type	Fixture quantity	Fixture load factor	Fixture units in entire building
Water fountain	12	0.5	6
Shower	4	2	8
Water closet, public	16	4	64
Water closet, private	13	3	39
Urinal	8	2	16
Lavatory	16	1	16
Office sink	5	2	10
Utility sink	3	2	6
Kitchen sink	4	4	16
Total fixture units			**181**

the building clean water plumbing system had to be performed to verify that sufficient potable water is available under flow and pressure conditions throughout the fixtures. Referring to Table 12.12, the minimum diameter of water pipe is 2 in., the meter size is 2 in., and the approximate pressure loss is 30 psi. Using the information provided from the local utility, it was determined that the pressure coming into the building from the 6-in.-diameter DIP water main with a roughness coefficient of 140 is 50 psi. From there, the plumbing was designed to go into the second floor, as that is where all of the plumbing is located. The pipe from the main to the building is a copper pipe with a diameter of 3 in. and a roughness coefficient of 150. The main was put into the ceiling of the second floor and dropped between the walls. A tee valve was placed, which allows the incoming water supply to be split between the north and south half. Upon reaching these points, the pipes are reduced to the size required in the design. Figure 12.34 shows the final design for the building.

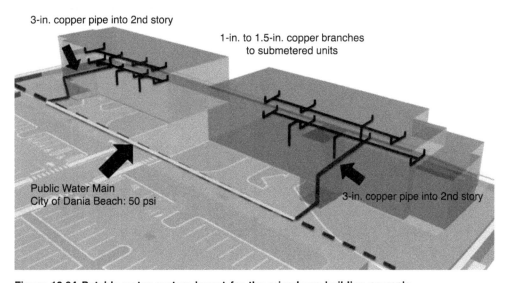

Figure 12.34 Potable water system layout for the mixed-use building example

12.4 Sanitary Sewer Systems

The sanitary sewer portion of the project must conform with all local, state, and federal ordinances and building codes. It uses the same basic fixture conversions as the water system, but sanitary sewer piping is gravity pipe. As a result, a grade is needed for design, which is a concern where there are long runs inside multistory buildings. The following is the basic procedure for sewer system design:

1. Identify connection points to the sewer system. A pump station may be required.
2. Create a spreadsheet for plumbing fixtures (toilets, sinks, showers, etc.), just as was done for the water system.
3. Identify the fixture units associated with each plumbing fixture.
4. Convert fixtures to fixture units.
5. Use fixture units to size the sewer main.
6. Create a preliminary piping plan.
7. Identify where changes in direction or length may require the placement of cleanouts or manholes.
8. Identify appropriate utility connection details from the local utility.

To illustrate the design process, a series of examples will be provided, similar to the potable water system examples presented earlier. Starting with the assisted living facility, the engineer must locate the existing sanitary sewer collection system to find the nearest available sewer connection. Sanitary sewers are designed to flow by gravity whenever possible or by pressurized systems if necessary. When connecting to gravity sewers, proximity and depth of the sewer line as well as pipe size are the parameters of greatest importance. The distance from the building to the gravity sewer line connection must be calculated, and the difference in elevation between the sewer exiting the building and the invert elevation of the sanitary sewer receiving the water must be determined, keeping in mind that there are minimum slope requirements and minimum cover requirements specified in the code. Figure 12.35 shows

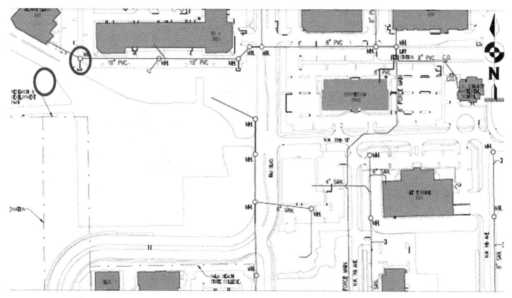

Figure 12.35 Existing sanitary pipe system surrounding the assisted living facility site

the existing sanitary piping system surrounding the assisted living facility site. The new sanitary piping system will connect at the northwest corner of the site at an existing sanitary manhole (shown as circles in Figure 12.35). This existing manhole is connected to a sanitary pipe system that runs east from the road.

Next, to determine the sanitary sewer pipe sizes, the fixture unit loads are used to select the appropriate pipe diameter from Table 12.16. The sanitary sewer pipe will enter each building as shown in Figure 12.36. This pipe will run vertically to each floor, with branches running east and west to service all fixtures. With this piping layout, each floor is divided into four zones, which equates to about 54 fixture units per zone (from a total of 1,287, divided by two buildings and three floors, for about 200 fixture units per floor, divided by four zones). Using Table 12.16, it was determined that 4-in.-diameter piping is needed at

Table 12.16 Maximum number of drainage fixture units connected to any portion of the building drain or the building sewer, including branches of the building drain [FBCP 2007, Table 710.1(1)]

Diameter of pipe (in.)	Slope per foot			
	1/16 in.	1/8 in.	1/4 in.	1/2 in.
1 1/4	—	—	1	1
1 1/2	—	—	3	3
2	—	—	21	26
2 1/2	—	—	24	31
3	—	36	42	50
4	—	180	216	250
5	—	390	480	575
6	—	700	840	1,000
8	1,400	1,600	1,920	2,300
10	2,500	2,900	3,500	4,200
12	3,900	4,600	5,600	6,700
15	7,000	8,300	10,000	12,000

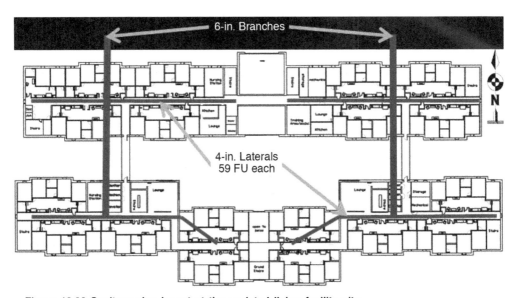

Figure 12.36 Sanitary pipe layout at the assisted living facility site

these laterals installed at a ⅛-in. slope per linear foot. Because there are two main pipes feeding the buildings, each pipe will need to carry 644 fixture units, which will require 6-in.-diameter piping installed at a ⅛-in. slope per linear foot. The main sanitary piping carries the full 1,287 fixture unit load, which requires an 8-in.-diameter pipe installed at a 1/16-in. slope per linear foot.

The local and state standards set a maximum distance between manholes of 400 ft. This layout has a distance of 1,225 ft (see Figure 12.37). The longest path of the overall sanitary piping is 850 linear feet, and the elevation of the first-floor slab is at 13.60 ft (see Figure 12.37). Assuming a minimum of 3-ft cover and using the slopes dictated by Table 12.16, the main sanitary pipe will be installed at the existing manhole at an invert height of 6.33 ft. The sanitary pipe will cross under the potable water pipeline near the northwest corner of the

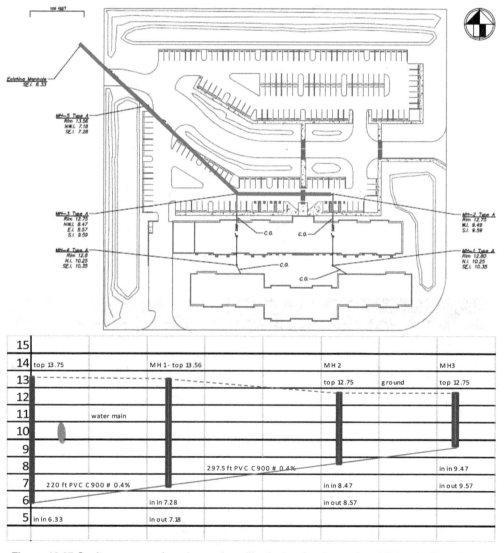

Figure 12.37 Sanitary sewer pipe plan and profile design for the assisted living facility site

property (see Figure 12.31). Local standards require a separation between these pipes to be at least 18 in. vertically. In this case, the sanitary piping will be several feet below the water pipe at this location. A manhole detail is provided in Figure 12.38.

For the hotel example, the same tables are used to determine the fixture units for water as for sewer systems. The water and sewer typically are installed in parallel because toilets,

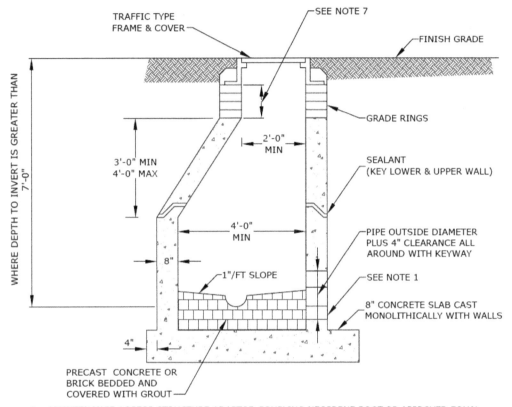

1. MAINTENANCE ACCESS STRUCTURE ADAPTOR COUPLING NEOPRENE BOOT OR APPROVED EQUAL ARE REQUIRED FOR ALL PIPE MATERIAL OR AS APPROVED BY WWS.

2. MAINTENANCE ACCESS STRUCTURE WALLS TO BE SEAL COATED INSIDE AND OUTSIDE WITH 16 MIL. THICKNESS OF COAL TAR EPOXY. THE 1st COAT IS RED AND THE 2nd COAT IS BLACK.

3. LIFT HOLES THROUGH PRECAST SECTIONS PERMITTED PER OSHA REQUIREMENTS.

4. ALL OPENINGS SHALL BE SEALED WITH WATERPROOF EXPANDING GROUT. SEE FIG. 322.

5. A FLOW CHANNEL SHALL BE CONSTRUCTED INSIDE MAINTENANCE ACCESS STRUCTURE TO DIRECT INFLUENT INTO FLOW STREAM.

6. ALL CONCRETE SHALL BE TYPE II CEMENT, MEETING LATEST ASTM REQUIREMENTS AND PROVIDED WITH LABORATORY CERTIFICATION ON PRECAST STRUCTURES.

7. THE CHIMNEY AREA SHALL BE MINIMUM OF 4" AND A MAXIMUM OF 12" IN HEIGHT. A MINIMUM OF 3 GRADE RINGS SHALL BE INSTALLED. SET IN 2 STRIPS OF SEALANT/ADHESIVE COMPOUND ON EACH SEALING FACE.

8. SET MAINTENANCE ACCESS STRUCTURE FRAME ON 2 STRIPS OF SEALANT PLUS A BED OF PORTLAND CEMENT AND SILICA SAND. BRING MORTAR UP OVER FRAME.

9. APPLY MORTAR COATING TO INSIDE AND OUTSIDE OF CHIMNEY. BRING MORTAR UP AND OVER FRAME.

Figure 12.38 Manhole detail for the assisted living facility site (source: City of Boca Raton, FL)

sinks, showers, etc. need water before any wastewater is generated (Figure 12.39). Table 12.13 indicated that for the 617 total fixture units in the building, a 6-in. vertical service line is required (refer to Table 12.16).

The same design procedure was used for the mixed-use facility. In this case, the building can be separated into two sections, each with over 300 fixture units. Using this number and a ⅛-in. slope per linear foot, two 6-in. connections would be satisfactory. Because the connection is in the street, no further construction is needed off-site.

The materials for sanitary piping typically are PVC schedule 40, but others may be used depending on the specific situation, as indicated in Tables 12.17 and 12.18. In the street, the pipelines are a minimum of 8-in. diameter. All gravity sewers should include a plan and profile sheet (see Figures 12.40 and 12.41 for examples). Manholes are to be placed at no more than 400 ft apart, at every junction and every change in slope. Minimum cover is typically 30 in. to the top of the pipe.

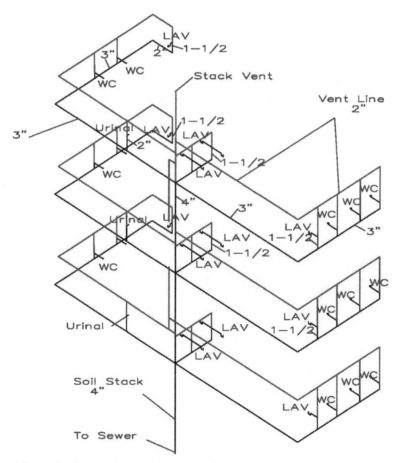

Figure 12.39 Example of hot and cold piping in parallel with sewer system included in perspective drawing for the hotel example

Table 12.17 Typical pipe material options (FBC 2010, Table 702.3)

Material	Standard
Acrylonitrile-butadiene-styrene (ABS) plastic pipe	ASTM D2661, ASTM D2751, ASTM F628
Asbestos-cement pipe	ASTM C428
Cast-iron pipe	ASTM A74, ASTM A888, CISPI 301
Coextruded composite ABS DWV schedule 40 IPS pipe (solid)	ASTM F1488
Coextruded composite ABS DWV schedule 40 IPS pipe (cellular core)	ASTM F1488
Coextruded composite PVC DWV schedule 40 IPS pipe (solid)	ASTM F1488
Coextruded composite PVC DWV schedule 40 IPS pipe (cellular core)	ASTM F891, ASTM F1488
Coextruded composite PVC IPS-DR, PS140, PS200, DWV	ASTM F1488
Coextruded composite ABS sewer and drain DR-PS in PS35, PS50, PS100, PS140, PS200	ASTM F1488
Coextruded composite PVC sewer and drain DR-PS in PS35, PS50, PS100, PS140, PS200	ASTM F1488
Coextruded PVC sewer and drain PS25, PS50, PS100 (cellular core)	ASTM F891
Concrete pipe	ASTM C14, ASTM C76, CSA A257.1M, CSA A257.2M
Copper or copper-alloy tubing (type K or L)	ASTM B75, ASTM B88, ASTM B251
Polyethylene (PE) plastic pipe (SDR-PR)	ASTM F714
PVC plastic pipe (type DWV, SDR26, SDR35, SDR41, PS50, or PS100)	ASTM D2665, ASTM D2949, ASTM D3034, CSA B182.2, CSA B182.4
Stainless steel drainage systems, types 304 and 316L	ASME A112.3.1
Vitrified clay pipe	ASTM C4, ASTM C700

Table 12.18 Typical pipe fitting options

Material	Standard
Acrylonitrile-butadiene-styrene (ABS) plastic pipe	ASTM D2661, ASTM D3311, CSA B181.1
Cast iron	ASME B16.4, ASME B16.12, ASTM A74, ASTM A888, CISPI 301
Coextruded composite ABS DWV schedule 40 IPS pipe (solid or cellular core)	ASTM D2661, ASTM D3311, ASTM F628
Coextruded composite PVC DWV schedule 40 IPS-DR, PS140, PS200 (solid or cellular core)	ASTM D2665, ASTM D3311, ASTM F891
Coextruded composite ABS sewer and drain DR-PS in PS35, PS50, PS100, PS140, PS200	ASTM D2751
Coextruded composite PVC sewer and drain DR-PS in PS35, PS50, PS100, PS140, PS200	ASTM D3034
Copper or copper alloy	ASME B16.15, ASME B16.18, ASME B16.22, ASME B16.23, ASME B16.26, ASME B16.29
Glass	ASTM C1053
Gray iron and ductile iron	AWWA C110
Malleable iron	ASME B16.3
Polyolefin	ASTM F1412, CSA B181.3
PVC plastic	ASTM D2665, ASTM D3311, ASTM F1866
Stainless steel drainage systems, types 304 and 316L	ASME A112.3.1
Steel	ASME B16.9, ASME B16.11, ASME B16.28

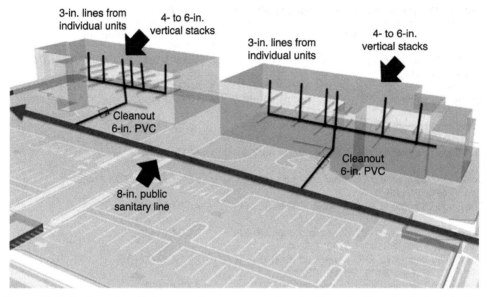

Figure 12.40 Design of preliminary sanitary sewer system for the mixed-use project example

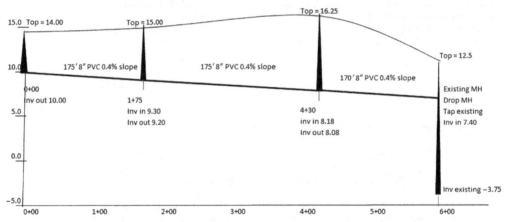

Figure 12.41 Typical sanitary sewer design profile view; note the exaggerated vertical scale

12.5 Heating, Ventilation, and Air Conditioning

With regard to the HVAC system, the design engineer needs to consider some key issues to help design these important elements of a building. First, because the HVAC system typically is installed in the ceiling, there must be enough space below the structural elements to allow the ducts to fit. If the air compressor is on the roof, there must be enough strength in the roof structure to support the weight and vibration of the unit. From an energy perspective, which drives the sizing of electrical components, the engineer must determine the air conditioning tonnage required to cool or heat the building. Note that heating loads typically are higher than cooling loads. The size of the HVAC system depends on the square footage of the building. A general rule of thumb used by HVAC contractors is for every 400 ft^2, 1 ton would be needed in a commercial building. Glass windows or walls, lighting, computers, power demands, etc. will create heat gains that must be cooled. Air conditioning designs consist of an air handler, an air compressor, and the ductwork. The following outlines the standard design procedure for an air conditioning system:

1. Determine the square footage of the floor area to be heated or cooled.
2. Identify extenuating issues, such as large volumes associated with the building (e.g., high ceilings).
3. Identify improvements beyond standard building construction that may increase system efficiency.
4. Run energy modeling to properly size the ductwork and mechanical systems.
5. Specify the make and model of the air compressor and air handler selected.
6. Identify local factors, such as percentages of outside air ventilation, air exchange rates, and the need for positive pressure versus negative pressure in the building. Keep in mind that some fresh air is needed to expunge carbon dioxide buildup in the structure.
7. Locate the air handlers in the building.

The same process is used for heating systems, except that the air compressor may be replaced by boilers, furnaces, or other heating systems. Note that heat is a greater draw on power than cooling systems in most climates. HVAC system design tends to be fairly site specific. Different codes may have requirements for introducing outside air to flush out carbon dioxide or carbon monoxide buildup. In addition, some high-performance buildings may use natural outside ventilation or air recirculation or energy recovery ventilators to earn LEED points.

An example is helpful. Assume a building has an area of almost 24,000 ft^2. The floor is divided into two sections; the north half is 12,700 ft^2 and the south half 11,000 ft^2. Using the 400-ft^2/ton rule of thumb, it is estimated that the north half would require about 32 tons and the south half about 28 tons. For residential houses, duplexes, individual condominiums, and similar structures, the rule of thumb may be sufficient, or even required by local codes, but this tonnage may be relatively inefficient for larger commercial buildings, where the cost of running the HVAC system is significant. Because efficiency reduces costs, it is important to verify the actual needs for larger buildings, which will tend to vary depending on the design elements inside the building. As a result, whenever possible, the size of the air conditioning unit should be verified using an HVAC software package. Most HVAC companies have user-friendly software to help engineers identify the correct air conditioning and air handling units for buildings, including the duct size for each space and function.

The design engineer has several options. Multiple air handling units within a building allow the use of smaller ducts but require more mechanical space and generally are less efficient. Small systems rarely have high Seasonal Energy Efficiency Ratio (SEER) ratings, which means smaller systems will be less efficient. For example, an ENERGY STAR®-qualified single-package central air conditioning unit would have a SEER rating higher than 14 (ENERGY STAR 2014). Large air compressors linked to large air handlers located in the building require large ducts. Space available for these large duct sizes becomes relevant in such cases. To select the air handlers needed for the condensing unit, designers often assume that 1 cubic foot per minute (cfm) of airflow is needed per square foot of building area (Stein et al. 2005). For the example in question, the north half would then require 12,700 cfm and the south half would require 11,000 cfm.

The next task is to create an energy model for the building. Note that HVAC may only be part of the load, so full modeling is suggested. The energy model will estimate the annual energy consumption and the corresponding cost for the energy. There are many software packages, but eQUEST®, which estimates the annual energy consumption of a building, is also set up to easily obtain the necessary data for LEED documentation. The intent is to show how much energy can be reduced from the baseline model to the design model. The baseline model is the design of the building assuming the minimum requirements, such as the minimum amount of insulation needed in the walls and roof.

For accurate results, the size of each room as well as the exposed area are required. Each room must detail the quantity and type of windows and doors. The wall and roof sections also are required for the program to know how well the building is insulated. Energy consumption can then be minimized by implementing energy-saving practices, such as increasing insulation in the walls and roof, changing the windows to low-emissivity glass filled with argon gas, changing the color of the building from a dark color to a light color, changing the building orientation, and changing the lightbulbs from incandescent to LED.

LEED requires that a baseline model be simulated in four rotations. One rotation is the same as the design model. The other three are each rotated 90° from each other (e.g., 90°, 180°, and 270° from the original orientation). At the end of the model simulation, output tables are created that are used for LEED documentation. A summary of the results for the building example is provided in Table 12.19. The baseline design is the average of all four rotations. The total energy consumption of the design was reduced by about 52% by changing the items mentioned earlier. Figures 12.42 and 12.43 show ducts and returns.

Another example is helpful here. For the assisted living facility discussed earlier, Carrier's Hourly Analysis Program (HAP) version 4.6 was utilized as the design method. Prior to inputting building information, zones had to be designated for specific air handling units. Because the proposed building layout runs in excess of 400 ft in length, the designers decided to break up the two buildings into the following zones:

◆ Zone 1: First-floor west wing
◆ Zone 2: Lobby 1
◆ Zone 3: Lobby 2
◆ Zone 4: Building 2 activities area
◆ Zone 5: Administrative offices
◆ Zone 6: Each pod or four assisted living facility units

Separating the building into various zones allowed for the design of individual air handling units in each zone. The zones are representative of areas that occur multiple times in the facility. For example, zone 6 occurs 49 times throughout the facility, so only a single unit had to be modeled.

Table 12.19 Example of energy modeling results using eQUEST

Category	Baseline design	Proposed design	Percent reduction
Energy use (kWh)	464,778	222,133	52.2
Annual cost ($)	$59,374	$28,510	52.0
kBTU/ft^2	71.49	34.17	47.8

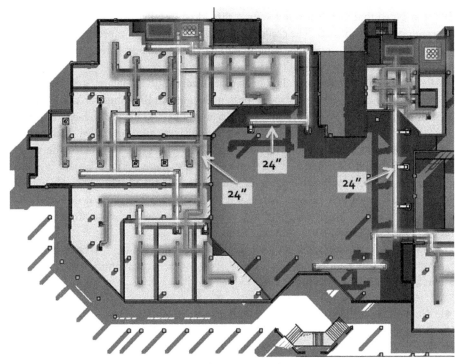

Figure 12.42 Example of the first-floor duct system in a building

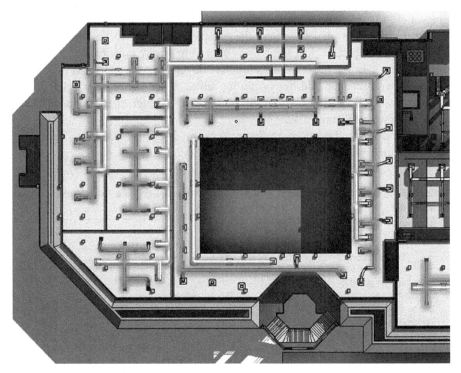

Figure 12.43 Example of the second-floor duct system in a building

It was decided that a chilled water system should be installed due to the large size of the cooling system for this project. The chilled water system replaces compressors at the building (they are now located at the chiller building). The system was designed based on the duty requirements of the complete facility using chiller units.

HAP version 4.6 requires that all zones are labeled and that the area and ceiling height for a zone are input in order to properly design the air handling unit for that zone. Table 12.20 summarizes the approximate square footage and ceiling height of each zone used in the HAP software to simulate a system design. The software automatically calculates system requirements such as velocity pressure, flow, and tonnage. The zones yielded the results in Table 12.20. It should be noted that there are instances where some judgment is required. In this example, the ceiling heights of zones 2 and 3 are 22 and 32 ft, respectively, which are not typical and cannot be modeled correctly with this software because it does not allow a floor level to exceed 12.0 ft. As a result, the two zones were modeled as double the square footage to address the increased volume of the rooms.

Based on this information, the designers selected the following units to be utilized in their respective zones (note that Trane® is only one of many excellent manufacturers that supply high-performance HVAC products):

◆ Zone 1   Trane Performance Climate Changer™ Air Handler   Unit 22
◆ Zone 4   Trane Performance Climate Changer Air Handler   Unit 6
◆ Zone 5   Trane Performance Climate Changer Air Handler   Unit 6
◆ Zone 6   Trane Hyperion™ XR GAM5A0B36

The chilled water system was sized based on the complete building footprint of both buildings. Both the ceiling height and square footage under air of both buildings were used to create the simulation model. Because the complete area of each building was approximately 160,000 ft², a parallel dual system was utilized to service individual air handler units in both buildings. The manufacturer's preliminary design software calculated that the size of the chilled water system should be two 96-ton units. A Trane Series R™ Helical 96-Ton Rotary Water-Cooled Chiller was selected, and a chilled water distribution network was designed as shown in Figure 12.44.

Using the airflow and pressure estimates from the design simulation and assuming an equally distributed volume of air, the sizes of the ducts can be determined. Using duct

Table 12.20 Zone data for model input and system output design data for the assisted living facility example

Zone	Description	Net floor area (ft²)	Ceiling height (ft)	Airflow requirements (cfm)	Cooling load (tons)
1	First-floor west wing	14,490	10	10,788	27.4
2	Lobby 1	2,522	22	Separate calculation	Separate calculation
3	Lobby 2	2,240	32	Separate calculation	Separate calculation
4	Building 2 activities area	1,587	10	2,610	5.4
5	Administrative offices	2,552	10	2,332	4.5
6	Pod	1,479	10	1,269	2.8

Figure 12.44 Chilled water distribution loop design for the assisted living facility example

calculator software called DuctSizer 6.4 by McQuay Air Conditioning, the typical duct supply and return layout was calculated for zone 6. The main supply duct will require a diameter of 6 in., and the branches will be 4-in. diameter. The pods utilize a louver return system located in the air handling unit cabinet. Every room in zone 6 has a 4-in. return line that terminates on the other side of the adjacent living room (see Figure 12.45).

An energy model for this project was obtained using eQUEST 3.65. This energy model analyzed the electric power consumption of the building throughout a calendar year. The proposed chilled water system was estimated to consume 776,000 kWh in a year compared to a baseline or typical system that uses 1,522,000 kWh annually, a savings of over 50% per year. The base and proposed energy models are shown in Figures 12.46 and 12.47.

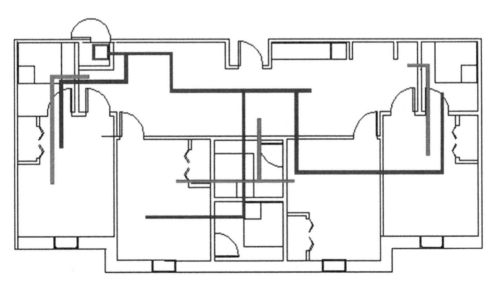

Figure 12.45 Zone 6 air distribution system layout for the assisted living facility suites (four rooms and central living area)

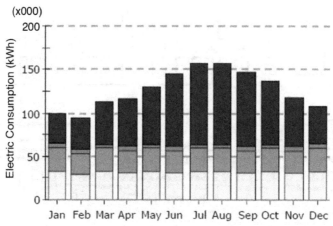

Figure 12.46 Electric consumption of typical buildings, as defined by eQUEST software

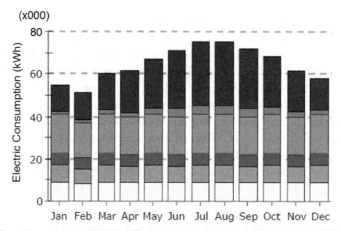

Figure 12.47 Electric consumption of chilled water system design for the assisted living facility example, as defined by eQUEST software

12.6 Parking Considerations

Parking spaces normally are allocated as a part of a local zoning code, although they may be supplemented by statewide or regional building codes. The zoning codes apply to many different development types, so it should be no surprise that the number of parking spaces required for any specific development type might vary. The following design protocol should be followed:

1. Identify the appropriate land use.
2. Determine the square footage, number of seats and/or barstools, or other features used in the code to determine the number of parking spaces.
3. Determine the number of parking spaces from steps 1 and 2. Often, the biggest challenge is to determine the land use and then calculate the number of spaces required from the local code.

4. Determine the appropriate size of spaces for the local jurisdiction. General engineering guidelines for parking lot design apply, but the local code will determine the allowable size of the spaces and often the aisle sizing as well. The most common sizes for parking spaces are 9 ft × 18 ft, 10 ft × 18 ft, 9 ft × 19 ft, and 10 ft × 20 ft.
5. Determine the number of disabled (accessible) spaces needed from the local or state codes. This may vary with land use.
6. Determine the layout options. When possible, 90° parking with 24-ft aisles should be used. Other configurations are available in local transportation department manuals.
7. Determine if tree islands are required, and if so, at what intervals.
8. Determine applicable radii on curbs, which often are 25 ft but may vary in low-speed parking lots. Most curbs should be type D.
9. Determine if bumpers are required for parking spaces.
10. Design the parking lot drainage system (refer to Section 12.2).

Sometimes the available land space will not accommodate the parking needs on one level, in which case a parking structure will be required. A parking structure should be integrated into the design when applicable as opposed to being a stand-alone structure. Many downtown buildings utilize this concept.

For example, if the local code specifies 1 parking space for every 250 ft² of retail space, then a new store with 100,000 ft² under roof would require 400 spaces. Note that a portion of those spaces will have to be disabled (accessible) spaces and must be located closest to the building entrance, preferably without having to cross a driving lane.

For a mixed-use building, often there are local codes, such as the following:

◆ 1 space for every 250 ft² of retail space
◆ 1 space for every 35 ft² of restaurant space
◆ 1 space for every 2 barstools
◆ 1 space for every 100 ft² of office space
◆ 1.7 spaces per residential unit
◆ 2 disabled (accessible) spaces for every 100 up to 300 spaces and 1 space for every 100 spaces thereafter

For example, suppose a proposed mixed-use development has 25,000 ft² of retail space, 7,000 ft² of restaurant space plus a bar with 80 seats, 10,000 ft² of office space, and 100 condo units. Using the local codes specified above, the parking requirements are as follows:

◆ 25,000 ft² × 1 space for every 250 ft² of retail space = 100 spaces
◆ 7,000 ft² × 1 space for every 35 ft² of restaurant space = 200 spaces
◆ 80 seats × 1 space for every 2 barstools = 40 spaces
◆ 10,000 ft² × 1 space for every 100 ft² of office space = 100 spaces
◆ 100 units × 1.7 spaces per residential unit = 170 spaces

The total is 610 spaces for this development. The number of disabled (accessible) spaces will be defined locally (check local codes) as a percent of the 610 spaces. Some local codes may provide an offset because the offices are occupied during the day but the residential units are not and vice versa at night, but this is more common in urban areas.

Once the number of spaces has been determined, the goal is to lay them out utilizing the least amount of space possible. Available options for parking spaces include 90°, 75°, 60°, and 45° angles from the pathway of cars. The angle also impacts the aisle width; 90° parking requires the most aisle space (24 ft minimum), whereas angled parking requires less aisle width but restricts the direction of traffic flow.

Parking area is impervious and requires significant stormwater attenuation. Therefore, minimizing paved areas is an important consideration. Parking decks can help, but they come at a high cost. The geometry of the site and topography also may limit options. Most parking lots are restricted as to slope, especially where disabled (accessible) spaces are located; 2% is a common limitation.

12.7 Transportation

Transportation is an essential part of modern life. Many important features of human settlement, including roads, bridges, and other transportation routes that people take for granted, are the products of transportation engineering. In many communities, growth is most noticeable around regional shopping centers and along major arterial roads that form commercial corridors because development typically is approved one project at a time with little consideration given to cumulative impacts in the form of deterioration of level of service on roadways and increasing delays. Because of ultimate roadway expansion limitations and the desire of communities to maintain high-quality mobility, a proposal to intensify development and increase traffic flow is evaluated prior to approval to accomplish one or more of the following:

1. Determine if the proposed development can be accommodated within the existing or planned transportation system.
2. Identify improvements beyond those already anticipated that can accommodate the proposed development.
3. Determine how a development of significant size can be designed so as to not adversely affect traffic operations or safety near the site.
4. Identify intersections that might be affected.
5. Identify if the level of service will be degraded (longer times between lights, longer delays, etc.).
6. Create a simulation of the traffic and intersections.
7. Add future expected demands.
8. Add development demands.
9. Determine impacts.
10. Create scenarios to minimize impacts.
11. Design useful improvements to improve traffic flow and reduce delays.

Some communities require transportation impact studies for nearly every development project, while others vary according to local policies and conditions. A transportation impact study is a means for a developer to provide planners and local officials with an objective assessment of both the anticipated impacts and needs of a proposed development. The trend has been toward increasing requirements as a basis for both assessing the extent of transportation impacts and negotiating mitigation.

Because each community is different, the options are less straightforward than in the prior examples; nevertheless, an example is useful. The example in this case is a government complex. The designers performed a transportation analysis using simulation software to determine the actual conditions and the future impacts that this complex will bring to the area in order to predict possible problems and generate plans to correct those impacts.

First, the principal intersections were analyzed and simulated using computer software, which allowed the designers to predict the current and future level of service. A traffic analysis should include pedestrians, trucks, buses, and trains as applicable, in addition to cars. For the government complex example, the designers created a simulation model of four key intersections. The software used in this project was Aimsun® 8.0. By creating a simulation, current delay times can be estimated for any intersection. Simulation programs also are used to predict the best way to improve a congested intersection by making changes to the inputs, running the simulation, and comparing the results. This is a very important tool for traffic engineers because it can save money and time versus intersection improvement after the fact. It also gives an engineer a realistic view of an intersection, thereby optimizing efficiency and safety.

The process includes the following steps:

1. The first step is to select aerial photographs to initiate the drawing. It is very important to accurately measure the width and length of the images to prevent errors.
2. The pictures are imported into the software package, and after the pictures are properly scaled, the approach lanes, through lanes, and sections are created.
3. The turning lanes are added in the appropriate direction from southbound, northbound, westbound, and eastbound.
4. The signals are incorporated based on their orientation direction.
5. The signal timing is calibrated in the master control plan.
6. Using traffic data obtained from the local traffic agency, the access point percentages and the number of cars are determined.
7. Key times of concern (typically rush hour) are identified for the worst-case scenario analysis.
8. Traffic including pedestrians, cars, trucks, bicycles, buses, and trains as applicable is added to the model.
9. The simulation is run and adjustments are made.

For the government complex example, the simulations were run during a 2-hr period from 8:00 A.M. to 10:00 A.M. during morning rush hour. Different parameters were analyzed, such as flow rate, density, and speed. Figure 12.48 shows the intersection in Aimsun 8.0. Table 12.21 shows the light timing schedule. Once the geometry of the networks was defined, vehicles were added to the model. Table 12.22 is an example of traffic count data used for this analysis.

Figure 12.48 Intersection nearest to the government complex example

Table 12.21 Traffic engineering timing data for intersection nearest to the government complex example

City of Boca Raton Traffic Engineering
Intersection Timing Sheet

Name NW 2 AVE & NW 2 ST	**Mod. No.** 2	**Initial Operation Date** 12/26/07
Intersection 76	**Mod. Date** 04/09/07	

Part I				**Actuated Timing Information**			

Noncoordinated Operation

Phase	1	2	3	4	5	6	7	8
Approach		NB		EB		SB		WB
Initial		14		12		14		12
Passage		2.0		2.0		2.0		2.0
Max1		50.0		30.0		50.0		30.0
Max2								
Yellow		4.5		4.0		4.5		4.0
All red		1.5		1.5		1.5		1.5
Walk		4.0		4.0		4.0		4.0
Ped clear		12.0		20.0		10.0		20.0
Recall		Min				Min		
Flash		Out		In		Out		In
C.N.A.								
Dual entry		x		x		x		x
Walk rest								
Flash pattern								
LT type		Perm		Perm		Perm		Perm
Detect type								

Table 12.22 Palm Beach County 2013 historic traffic growth table

Sta	Road	From	To	Lanes	Daily traffic volumes					2013 daily			2013 A.M. peak hour			2013 P.M. peak hour		
					2008	2009	2010	2011	2012	Date	Vol	Gr	2-way	NB/EB	SB/WB	2-way	NB/EB	SB/WB
4677	2nd Ave NW	Lake Worth Rd	Congress Ave	2	5844	5636	5594	5610	5493	2/13/2013	6047	2.63%	415	260	171	533	240	308
4679	2nd Ave NW	Congress Ave	Boutwell Rd	2	4640	4329	4127	4268	4158	1/14/2013	4176	0.39%	316	196	123	448	209	250
6843	2nd St	Boca Raton Blvd	NE 5th Ave	2	5878	5521	4993	4991	5376	1/23/2013	7037	12.12%	614	276	338	694	304	390
6890	4th Ave (NW)	4 Diagonal NW	Palmetto Park Rd	2	6851	7477	8306	8597	9366	1/30/2013	11507	11.48%	1259	789	489	1057	406	685
6826	4th Ave (NW)	Glades Rd	4 Diagonal NW	2	8055	8138	8558	8640	8728	1/23/2013	9065	1.94%	1016	682	334	851	370	496
6820	4th Ave (NW)	20th St NW	Glades Rd	2	9148	9317	8853	8631	8769	3/20/2013	9817	3.51%	991	463	528	890	379	514
6842	4th Ave (SW)	Palmetto Park Rd	Camino Real	2	5020	5284	6006	6270	5750	1/30/2013	4606	-8.47%	404	196	208	578	273	305
6894	5th Ave NE	Federal Hwy	S of Federal Hwy	2	6034	5804	5971	6518	6575	1/23/2013	7001	5.45%	552	288	264	652	322	330
6892	5th Ave NW	Spanish River Blvd	NW 20th St	2	6879	6537	6390	6665	7024	1/23/2013	7613	6.01%	970	504	469	636	350	286
4631	6th Ave S	Congress Ave	Sunset Ave	4D	24681	26889	28972	26345	29595	1/16/2013	29188	0.25%	2482	1425	1057	2499	1063	1454
4205	6th Ave S	Sunset Ave	I-95	4D	26665	32908	37165	34338	34273	3/11/2013	33837	-3.08%	2759	1452	1398	2844	1252	1697
4307	6th Ave S	I-95	SR 805 (Dixie Hwy)	4D	21646	25776	26868	26405	26931	1/16/2013	27394	0.65%	1953	901	1117	2242	1255	1006
6902	7th Ave (NW)	NW 12th Ave	NW 13th St	2	8315	7450	7424	8112	7866	3/20/2013	7672	1.10%	615	320	320	596	305	310
6840	9th Ave (SW)	Palmetto Park Rd	SW 13th St	2	2664	2230	2232	2343	2108	1/16/2013	2062	-2.61%	308	207	101	195	92	103
6883	10th Ave	Glades Rd	NW 15th St	2	7324	6880	6615	6153	8803	1/16/2013	9570	13.10%	825	533	357	930	481	456
4643	10th Ave N	Jog Rd	Haverhill Rd	4D	15411	15331	14782	14139	13882	3/13/2013	14539	-0.55%	1084	509	576	1217	578	639
4601	10th Ave N	Haverhill Rd	Military Tr	5	22360	25742	21819	21229	21682	3/13/2013	21808	-0.02%	1710	920	790	1732	768	964
4603	10th Ave N	Military Tr	Kirk Rd	5	26933	26808	26002	25842	26602	2/13/2013	27381	1.71%	1660	928	732	2057	1001	1084

Figures 12.49 to 12.52 show the analysis for one of the intersections. Figure 12.49 shows the flow rate per hour. The delays at the light are shown in Figure 12.50. The delays worsen (Figure 12.51) and the level of service degrades (Table 12.23) with the new development. Figure 12.52 shows the final results. The potential solutions are site specific: turn lanes, turning pockets, longer turn pockets, added lanes, changes to light timing, etc. The requirements will be different depending on the local conditions, and in some cases, the predevelopment traffic impacts are limited. As a result, often local governments will secure impact fees to fund more comprehensive solutions, a topic that is beyond the scope of this book.

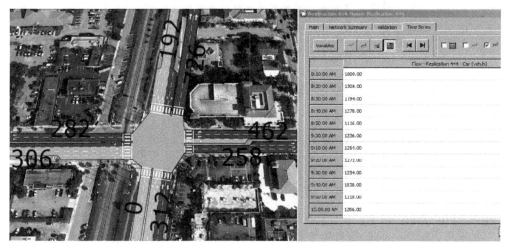

Figure 12.49 Flow rate per hour for traffic in intersection nearest to the government complex example

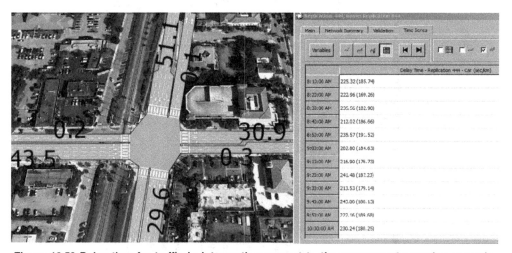

Figure 12.50 Delay time for traffic in intersection nearest to the government complex example

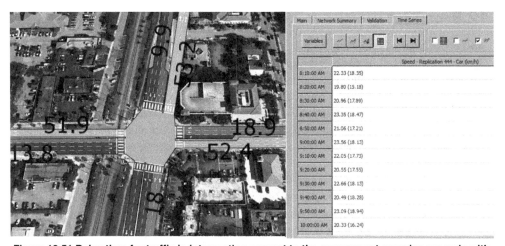

Figure 12.51 Delay time for traffic in intersection nearest to the government complex example with new development included

Table 12.23 Level of service for each roadway near the government complex example

Intersection	Level of service before	Level of service after
1. 2nd Avenue and 2nd Street	C	D
2. Dixie Highway and 2nd Street	D	D
3. Palmetto Park Road and 2nd Avenue	D	E
4. Palmetto Park Road and Dixie Highway	D	E

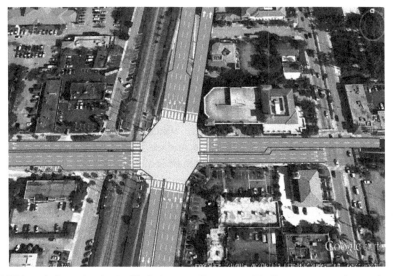

Figure 12.52 Level of service for traffic in intersection nearest to the government complex example after modeling adjustments

12.8 Landscaping

Most communities require landscaping for new developments. Some codes are extensive and others are less so, but in all cases the appropriate plants and planning need to be considered. While landscaping typically is the purview of landscape architects, engineers need some basic understanding of the concept. Before landscaping ideas can become a plan, the limitations and assets of the site must be understood. For example, if protected plant species are found on a site, they must be dealt with properly and may impact the size and shape of the buildings on the site. However, no design will compensate for poor growing conditions or improper plant selection.

Figure 12.53 is an example of a landscaping plan. To start such a plan, a boundary survey must be conducted to catalogue the existing vegetation on the site. This should be done by drawing a base map that shows all structures, sidewalks, driveways, fire hydrants, etc. The base map also should include the location of meters, utility boxes, telephone poles, etc., as these are expensive and/or complicated to move and may require easement access. The location of overhead and underground utilities should be included on this base map, along with water bodies, easements, and setbacks. The outline of all structures should include the location of window openings, doors, HVAC systems, and spigots.

Some other issues to consider include climate, wind, utilities, light exposure, water movement, and wildlife. Climate includes rainfall and temperature ranges. Plants grown in their preferred temperature zone will survive cold winters and hot summers and will require less water during the dry season. Plants typically are sold with a label that indicates their cold hardiness zone. Also, local codes sometimes establish lists of allowable plant species, or the local agricultural cooperative or regulatory agency will have lists of preferred species to use (more on this later). Microclimate is a term that relates to a small area affected by the surroundings. A small pocket area may have a microclimate that permits a plant to survive in an otherwise harsh environment. Examples are a courtyard that stays warmer than nearby areas or a wet area near a downspout. Any microclimates should be noted on the base map.

With respect to wind, in open areas with a long fetch, a strong wind can knock over or dry out plants that are not well established. In coastal areas, offshore winds can bring salt spray, which can injure plants that are not salt tolerant. Note that wind direction naturally shifts with the seasons, and it is difficult to determine the wind direction and speed just by observation without meteorological instruments.

Plant roots will seek water and wastewater utility pipes and wrap themselves around shallow underground electric, telephone, cable, and gas lines. It might be necessary to move these utilities before landscaping. All lines should be located on the base map after contacting the local utilities. Most jurisdictions operate a free utilities location service program to protect their infrastructure investments.

All plants require a certain amount of light to grow and flourish. However, it is common to have a mix of light conditions on a site. Plants may require full sun, partial sun, or shade. Matching the light conditions of the site with the light requirements of the plants selected will ensure healthy growth and reduce the need for supplemental irrigation. For example, shade plants grown in full sun require additional water. The base map should indicate which direction is north. In the northern hemisphere, the sun will always be in the southern half of the sky, even in the middle of summer. The engineer should track the movement of the sun throughout the day and make notes on the base map as to how many hours of sunlight each area receives. If observations cannot be made seasonally, imagine light availability with

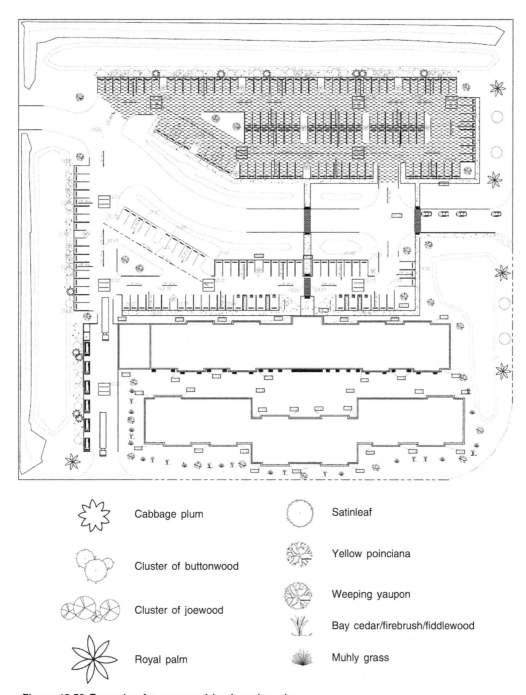

Cabbage plum		Satinleaf	
Cluster of buttonwood		Yellow poinciana	
Cluster of joewood		Weeping yaupon	
Royal palm		Bay cedar/firebrush/fiddlewood	
		Muhly grass	

Figure 12.53 Example of a proposed landscaping plan

the sun closer to the horizon, as it would be in the winter for example, when even a small structure will cast a large shadow.

Water moves either into the soil or across the soil surface in a sheet flow. Soil water availability can be a limitation or an asset. For dry areas, redirected water from sidewalks, gutters, or drains may help, although drought-tolerant plants may be a better solution. If poor drainage is an issue, the solution may be to fix the soil before planting. Because drainage and soil moisture likely will vary throughout the site, downspout locations and drain lines, if they exist, should be noted on the base map instead of just relying on soil borings. The engineer should check to see if water from the roof is running away from the foundation of existing buildings and try to observe the site during a rain event and make notes of areas where water ponds or moves rapidly. Finally, plant selection can attract wildlife in the form of birds and butterflies, but larger animals like deer or raccoons, drawn to the new habitat to feed, can become a nuisance.

Minimal irrigation is desirable, particularly when reducing water use for a more sustainable landscape is important. Other options for reducing potable water use for irrigation include xeriscaping, reclaimed water irrigation, and drip irrigation. All new growth should be native species that can survive with minimal to no water (drought tolerant). A tree schedule should be developed regarding plant selection and location on the site. In the example of a proposed landscaping plan in Figure 12.53, the plants were chosen based on their color, maintenance requirements, growth patterns, and native location, and a guide for new tree and vegetation preferences from the local land development code was used.

12.9 References

ENERGY STAR (2014). Air-Source Heat Pumps and Central Air Conditioners Key Product Criteria, <http://www.energystar.gov/index.cfm?c=airsrc_heat.pr_crit_as_heat_pumps>.

FBC (2010). Florida Building Code, Florida Department of Business and Professional Regulation, Tallahassee, FL, <http://ecodes.cyberregs.com/cgi-exe/cpage.dll?pg=x&rp=/nonindx/ST/fl/st/b200v10/index.htm&sid=20140722035821932211&aph=0&cid=iccf&uid=iccf0002&clrA=005596&clrV=005596&clrX=005596&ref=/nonindx/st/FL/index.htm> (accessed July 21, 2014).

FBCP (2007). Florida Building Code, Plumbing Section, Florida Department of Business and Professional Regulation, Tallahassee, FL, <http://ecodes.cyberregs.com/cgi-exe/cpage.dll?pg=x&rp=/nonindx/ST/fl/st/b900v07/index.htm&sid=20140722035716693206&aph=0&cid=iccf&uid=icsc0418&clrA=005596&clrV=005596&clrX=005596&aph=0&qy=joint+reinforcement&hlc=FFFF00&srchm=1&ref=/nonindx/ST/fl/index.htm> (accessed July 21, 2004).

FDOT (2014). <http://www.dot.state.fl.us/rddesign/DS/12/IDx/00232.pdf> (accessed July 21, 2014).

SFWMD (2014). Environmental Resource Permit Information Manual, South Florida Water Management District, West Palm Beach, FL, <http://www.sfwmd.gov/portal/page/portal/xrepository/sfwmd_repository_pdf/erp_swerp_manual.pdf> (accessed July 21, 2014).

Stein, B., Reynolds, J.S., Grondzik, W.T., and Kwok, A.G. (2005). *Mechanical and Electrical Equipment for Buildings*, 10th Ed., John Wiley & Sons, Hoboken, NJ.

USDA (1986). Urban Hydrology for Small Watersheds, Technical Release 55, 2nd Ed., Natural Resources Conservation Service, Conservation Engineering Division, U.S. Department of Agriculture, Washington, DC.

◆ 13 ◆

Structural Design Concepts

The structure of a building can be defined as an assembly of the essential components of the building that exist for the purpose of controlling its stability and geometrical shape. The most critical function of a building is to be able to resist any applied loads and transfer them safely to the ground without causing any type of failure or collapse. The reality is that the structural system of a building is more complex than just avoiding collapse because it also encompasses functional and aesthetic characteristics, which are provided by the owner, engineer, and architect of record.

A structural system must be engineered to satisfy requirements provided by the owner, architect, and, most importantly, applicable building codes based upon the geographic location of the project. In general terms, there are three characteristics that are intrinsic to building structural engineering: stability, strength/stiffness, and economy. *Stability* is necessary to maintain its designed shape without excessive deformations over a period of time. *Strength* means that the materials of choice are able to resist the predicted stresses imposed by the loads which the structural members will experience. It is very important to provide these members with a *factor of safety* so that the allowable stresses remain well below their respective failure stresses. *Stiffness* is how much a structure deflects as a response to an applied load. *Economy* goes beyond the dollar value associated with the cost of building materials. Construction economy is a subject that involves many factors, such as raw materials, fabrication of components, erection of the structure, and long-term maintenance.

When conceiving a building, the initial geometric form and structural planning process involves the shape of the building, the location of load-carrying members, the selection of materials, and the span lengths. During the development of the floor plans for a project, the engineer should start evaluating the initial structural planning and distribution of load-bearing walls and columns in order to minimize potential structural conflicts. The final structural design involves a complete analysis of all the parts and the sizes of the structural systems and drafting the structural design details for construction (Shaeffer 2002).

In general, codes outline the overall requirements for structural loading, although local or state building codes may designate different specifications. The engineer must become acquainted with the codes in the local jurisdiction and design to the most conservative standards. The engineer works with two types of codes: (1) general building codes and (2) design codes. *General building codes* specify the "minimum" design loads for a given structure. *Design codes* provide technical standards that are used to establish the requirements for actual structural design. The most common codes used in practice are:

1. *Minimum Design Loads for Buildings and Other Structures,* American Society of Civil Engineers (ASCE 7-10)
2. *International Building Code,* International Code Council (IBC 2012)
3. *Building Code Requirements for Reinforced Concrete,* American Concrete Institute (ACI 318-11)
4. *Steel Construction Manual,* 14th edition, American Institute of Steel Construction (AISC)

The structural design must withstand two types of loads that act on a structure. The first is categorized as gravity loads, and the second results from other natural forces. As the term suggests, *gravity loads* are the type of forces that a structure experiences in the vertically downward direction. Gravity loads can be subdivided into dead loads and live loads. *Dead loads* are those that include the self-weight of structural members and immovable objects of the structure. For example, the walls, floors, beams, mechanical equipment, and electrical equipment are considered to be dead loads. *Live loads* include movable objects, not permanently attached within or on top of the structure. Examples of live loads include people, furniture, and moveable equipment. *Other natural forces* refers to the lateral action that wind and earthquake loads can induce on a building structure. *Wind loads* vary in intensity because they depend on many criteria, such as wind speed, topography, building height, and location. Earthquakes generate very large *lateral forces* at the base, which transfer to the upper levels of the building due to ground motion. Due to the occurrence of these natural forces, further analysis and design criteria established by stringent building codes must be evaluated when properly designing for a structure.

Every member in a structure potentially has different loads. One issue the engineer must resolve is the number of different members permissible in the structure. Remember that contractors with numerous personnel on-site will construct the building. Having fewer different members makes constructability easier and lowers the risk of errors during the construction process. As a result, most engineers begin by determining the *critical members* in the building—those with the greatest moment and/or vertical loads. This is why most engineers try to lay out a grid pattern for columns and most buildings are square or rectangular. Designing for the critical members can reduce the potential for confusion. Of course, the cost needs to be considered as well—and critical members will be the biggest contributor.

The process of structural planning and design requires not only imagination and conceptual thinking, but also a sound knowledge of building codes and construction means and methods. The structural design also should include decisions on the availability of local materials, contractor capabilities, and the experience of the engineer. For example, an engineer who is not familiar with prestressed concrete design would be foolish to design a prestressed concrete structure. Likewise, if the local market does not include steel fabricators, masonry block may be the most effective (and perhaps least costly) solution given the circumstances.

The options available for structural systems that are common to every engineer's education include steel, reinforced concrete, and masonry. For smaller structures, the initial structural concepts typically include the use of reinforced concrete and concrete masonry units (CMUs) for the construction of load-bearing walls. CMU often is selected for structural walls on low-rise commercial buildings because when properly reinforced it can offer great resistance to horizontal and vertical loads. In addition, it is a durable construction material, offers excellent fire resistance, and can be installed easily by local laborers, which makes it an economical option. Furthermore, walls constructed with this material require low maintenance and can successfully be exposed to different temperature changes without causing any detrimental effect to the structure. Figure 13.1 shows a section of a CMU wall with the location of the reinforcement in the masonry wall.

Many buildings that use CMU wall construction also include steel for trusses and joists, which also are easy and convenient to install. Open web steel joists typically are used for roof framing and other floor framing because they can span long distances and support relatively large loads, such as those applied when a concrete slab is cast-in-place on top of cold-formed steel roof panels (see Figure 13.2). Both come in standard sizes and configurations and are readily available as prefabricated members. Specifications for steel joists are found online or in standard steel manuals.

In wind-prone areas, reinforced concrete roof slabs can provide sufficient weight to overcome uplift from wind loads when installed with tie-downs. The procedure for wind load design is defined in ASCE 7-10 (or latest edition). The key issues that define the wind loads

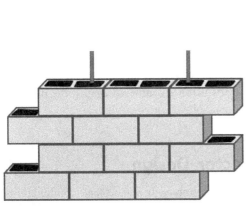

Figure 13.1 Reinforcing rods in concrete block wall composed of CMU

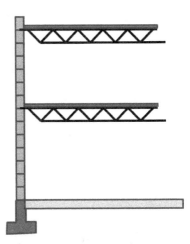

Figure 13.2 Example of concrete block wall with roof and second floor tied to wall concrete slab cast-in-place on top of cold-formed steel roof panels

are the profile height of the building and geographic location. The building profile is defined by the structural concepts and the need to include adequate space for heating, ventilation, and air conditioning (HVAC); water and sewer piping; and electrical and other services without conflicts with the framing. If there is a need to narrow the distances between floors yet still include HVAC, piping, etc., open web joists may frustrate this objective. Posttensioning or prestressed concrete construction may be better options, although neither subject receives much attention in undergraduate student education.

The support system also is an area for consideration. In large open buildings, columns will be difficult to avoid. In smaller buildings, interior walls can be designed to be load bearing by using concrete blocks or similar materials. Using a continuous wall distributes the loads and eases the foundation design. Building interior CMU walls also brings other benefits to the design, such as higher fire ratings, which are required between classrooms and hallway walls in educational facilities.

The design of most structures is controlled by specifications that are developed by various organizations and represent the industry's best practices. Municipal and state governments concerned with the safety of the public have established building codes with minimum requirements for construction of various types of structures within their jurisdictions. These codes can vary considerably, which can cause some confusion among architects and engineers (McCormac 2007). In designing the structural components, most building codes refer the design engineer to ASCE 7-10 for minimum design loads. An engineer should always view minimum design standards with a degree of skepticism. While these standards are prepared to guide design engineers in most situations, there may be a scenario for which the specified loads may not be adequate or where there is a clear need to build to a higher standard. The structural engineer should evaluate the minimum specified loads to determine whether they are adequate for the design of the new system and make appropriate adaptations to the design. Remember that codes are minimum standards. Engineers are encouraged to expand their expertise to include new ideas and materials. The engineering design industry is constantly evolving, so it is critical to use the most recent, up-to-date specifications for design purposes.

In designing the structural systems, all basic load combinations must be accounted for. There are two common design methods: (1) allowable stress design (ASD) and (2) load and resistance factor design (LRFD). The ASD method of computation is based on the stresses, whereas LRFD is based on the forces and moments capacity (load oriented). LRFD was developed in the 1980s and is based on the *limited state* philosophy. Limited state is a term used to describe the condition where the structure fails to perform as intended. The design of structures in this chapter will focus on the LRFD technique. In general, this method uses a statistical approach in determining factored loads that are compared against the ultimate strengths of structural members.

13.1 Load and Resistance Factor Design

The required strength of a member for LRFD is determined from the load combinations given in the applicable building code. For example, the loads of main concern in the south Florida region are as follows: dead loads, live loads, roof live loads, rain loads, and wind loads.

In another location, the critical issues may include snow loads and earthquake loads in addition to dead loads, live loads, and roof live loads. After determining each independent load, the ultimate design load, also referred to as the factored load, is calculated by using the worst-case scenario of the following seven load combinations given in ASCE 7-10, Section 2.3.2:

$$U = 1.4D \tag{13.1}$$

$$U = 1.2D + 1.6L + 0.5(L_r, \, S, \, \text{or} \, R) \tag{13.2}$$

$$U = 1.2D + 1.6(L_r, \, S, \, \text{or} \, R) + (L \, \text{or} \, 0.5W) \tag{13.3}$$

$$U = 1.2D + 1.0W + L + 0.5(L_r, \, S, \, \text{or} \, R) \tag{13.4}$$

$$U = 1.2D + 1.0E + L + 0.2S \tag{13.5}$$

$$U = 0.9D + 1.0W \tag{13.6}$$

$$U = 0.9D + 1.0E \tag{13.7}$$

where

U = ultimate design or factored load
D = dead load
L = live load
L_r = roof live load
S = snow load
R = rain load
W = wind load
E = earthquake load

The equation for the ultimate load must be solved using the worst-case scenario for each type of independent load from the seven load combinations above. These equations will need to be adjusted if there is soil or water on one or both sides of the proposed walls. A reinforced concrete design text should be consulted for more information.

13.2 Types of Loads

13.2.1 Dead Loads

The dead loads that must be supported by a particular structure include the weight of the structure itself plus all of the weights of items that are permanently attached to the structure, including the self-weight of the walls, roofs, ceilings, stairways, and permanently attached equipment. Some examples of these dead load items include HVAC components, plumbing fixtures, electrical cables, and green roofs. The dead loads acting on a structure are deter-

Table 13.1 Typical dead loads for selected building materials based on units weights (McCormac 2007)

Building material	Unit
Reinforced concrete	150 pcf (lb/ft^3)
Structural steel	490 pcf
Movable steel partitions	4 psf (lb/ft^2)
Hardwood flooring ($7/8$ in.)	4 psf
Mechanical duct allowance	4 psf

mined by inspection of the architectural, mechanical, and electrical drawings. Manufacturers' specifications and standard handbooks can be used to determine the weight of these items. The approximate weights of common materials are summarized in Table 13.1.

13.2.2 Live Loads

Live loads can vary in magnitude and position with time. They are caused by the occupants and intended function of the building. It is not unusual for a structure to have multiple functions and therefore to have different live loads to account for. Examples of common live loads are listed in Table 13.2. The engineer must be conservative in the design when choosing the loads for analysis. Using the heaviest load will translate to overdesigning the structure, but will provide insurance that the building will meet its performance criteria. For example, in a school or office building, a conservative 100-psf uniform load value could be used for performing calculations.

There are other types of live loads, such as highway bridge loads, which are caused by traffic and include loads due to vibration. Heavy trucks are mainly responsible for the analysis and design of these types of structures. More details for this type of loading on bridges can be found in *AASHTO LRFD Bridge Design Specifications*, which is published by the American Association of State Highway and Transportation Officials. Moving vehicles may have an unexpected trajectory over a bridge and may impose unanticipated loads on the deck; thus, an impact factor (I), which is derived from experimental studies, is used. Live loads on railroad bridges are specified in *Specifications for Steel Railway Bridges*, which is published by the American Railway Engineering Association. Live loads on curves, for both bridges and railroads, create horizontal loads caused by applied weight that must be accounted for (see Figure 13.3).

Table 13.2 Examples of uniformly distributed live loads (FBC 2010)

Occupancy or use	Uniform loads
Office use	50 psf (lb/ft^2)
Lobbies	100 psf (lb/ft^2)
Roofs	20 psf (lb/ft^2)
First-floor corridors	100 psf (lb/ft^2)
Hotel rooms	40 psf (lb/ft^2)
Stairs and exits	100 psf (lb/ft^2)
Stories (upper floors)	75 psf (lb/ft^2)

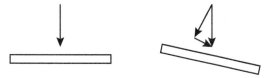

Figure 13.3 Load on flat surface (all gravity) (left) versus forces acting on surface on inclined plane (right)

13.2.3 Wind Loads

A procedure for estimating the wind pressures applied to buildings is presented in ASCE 7-10, Chapters 26 to 31. Several factors are involved in accounting for the effects of wind speed, including building shape and orientation, terrain profile, and importance of the building to human life and welfare, among other factors. The procedure might seem complicated, but its application is greatly simplified by using the appropriate tables presented in ASCE 7-10. When computing the wind loads for any structure, there are two things to check for: the main wind-force resisting systems (MWFRS) and components and cladding (C&C). MWFRS is defined as an assemblage of structural elements assigned to provide support and stability for the overall structure, whereas C&C refers to the elements of the building envelope that do not qualify as part of the MWFRS (ASCE 7-10). Design pressures for MWFRS are used to check the connections between components, such as hurricane straps on the roof that are used to overcome the uplift and overturning moments on the walls and foundation due to wind pressure. The C&C systems are used to check the individual members that make up the structure, without considering the structure as a whole.

Wind forces act as pressures on vertical surfaces facing the wind direction (windward). These may be positive pressures on the windward wall or negative pressures (suction) on the leeward walls (those walls facing in the opposite direction of the prevailing wind). Both pressure and suction may be applicable on sloping windward surfaces, and suction or uplift may be applicable on flat surfaces and on leeward vertical and sloping surfaces due to the creation of negative pressures. If the building shape is outside the scope of the specification, such as a building in the shape of an H or Y, ASCE 7-10 advises the engineer to conduct physical wind tunnel studies (McCormac and Csernak 2012).

A concise seven-step design process from ASCE 7-10 is used to calculate wind load (summarized in Table 13.3):

1. Determine the risk category of the building. These categories are listed in ASCE 7-10, Table 1.5-1. Several are as follows:
   a. Category I—Low risk to human life; nonoccupied buildings fit here.
   b. Category II—Buildings otherwise not assigned.
   c. Category III—Buildings that could cause a substantial impact to human life; occupied structures like multifamily residences, condos, and office buildings are included here.
   d. Category IV—Essential facilities like hospitals, schools, water treatment plants, police/fire/emergency medical services facilities, etc., plus any manufacturing facilities that deal with or manufacture hazardous materials or that generate hazardous waste products.

2. Determine the basic wind speed risk category based on the figures in ASCE 7-10. An example from the Florida Building Code is provided in Figure 13.4.
3. Determine the wind load parameters. These are the wind directionality factor (K_d) (ASCE 7-10, Table 26.6-1). The typical value for K_d is 0.85, but chimneys are 0.95.
4. Determine the exposure category. Exposure B is for buildings under 30 ft high in urban areas within 2,500 ft of the structure. This is normal for most urban areas. Open country and grasslands do not meet the Exposure B class. Exposure D is for areas within a mile of open water. Exposure C is everywhere else. The exposure classes are defined by a series of parameters in ASCE 7-10, Table 26.9-1.
5. Determine the topographic factor (K_{zt}) (ASCE 7-10, Table 26.8-1). The topographic factor is related to the flatness of the ground; it ranges from 0 to 1 for flat areas like south Florida and includes three other factors that involve topography of the area (see ASCE 7-10).
6. Determine the enclosure classification, which is related to the exterior surface and the type of resistance that exterior glass and doors have to wind. For example, in hurricane-prone areas, Category II to IV buildings must have wind-resistant windows.
7. Determine the gust effects, which are short-term wind bursts.
8. Determine the velocity pressure.
9. Determine the internal pressure coefficient (GC_{pi}) from ASCE 7-10, Figure 28.4-1. This value will be 0 for open buildings, +/–0.55 for partially enclosed buildings, and +/–0.18 for fully enclosed buildings, where + or – reflects leeward or windward direction.
10. Calculate wind pressure. For example, an office building in Miami-Dade County, FL might have a risk category of II for buildings with an exposure category of B. The engineer will be designing for 170-mph wind speed based on the location of the building and the wind velocity curves in Figure 13.4. Wind pressure (p) for an enclosed and partially enclosed rigid building is calculated using the following equation:

$$p = qGC_p - q_i(GC_{pi}) \qquad (13.8)$$

where

$$
\begin{aligned}
p &= \text{pressure (psf)} \\
GC_{pi} &= \text{internal pressure coefficient} \\
GC_p &= \text{external pressure coefficient} \\
q \text{ and } q_i &= \text{windward and leeward wall coefficient for both internal and external} \\
&\quad\ \text{pressure (see ASCE 7-10 for application)}
\end{aligned}
$$

Figure 13.5 shows a spreadsheet created to outline this procedure from ASCE 7-10 for an example project. The calculated maximum wind load is shown to be 81.5 psf on a wall. Of interest is the negative pressure of –56.4 psf on the roof, which must be overcome to keep the roof in place. Refer to ASCE 7-10 for the tables and graphs to determine all the relevant factors.

Table 13.3 ASCE 7-10 design for wind load for a structure

Seven-Step Procedure for Wind Loads

Step 1 Determine risk category of building or other structure (see Table 1.5-1)
Step 2 Determine the basic wind speed (V) for applicable risk category (see Figure 26.5-1A, B, or C)
Step 3 Determine wind load parameters:
  ◆ Wind directionality factor (K_d) (see Section 26.6 and Table 26.6-1)
  ◆ Exposure category B, C, or D (see Section 26.7)
  ◆ Topographic factor (K_{zt}) (see Section 26.8 and Figure 26.8-1)
  ◆ Gust effect factor (G) (see Section 26.9)
Step 4 Determine velocity pressure exposure coefficient (K_z or K_h) (see Table 29.2-1)
Step 5 Determine velocity pressure (q_z or q_h) (Equation 29.3-1)
Step 6 Determine force coefficient (C_f):
  ◆ Solid freestanding signs or solid freestanding walls (see Figure 29.4-1)
  ◆ Chimneys, tanks, rooftop equipment (see Figure 29.5-1)
  ◆ Open signs, lattice frameworks (see Figure 29.5-2)
  ◆ Trussed towers (see Figure 29.4-3)
Step 7 Calculate wind force (F):
  ◆ Equation 29.4-1 for signs and walls
  ◆ Equation 29-6-1 and Equation 29.6-2 for rooftop structures and equipment
  ◆ Equation 29.5-1 for other structures

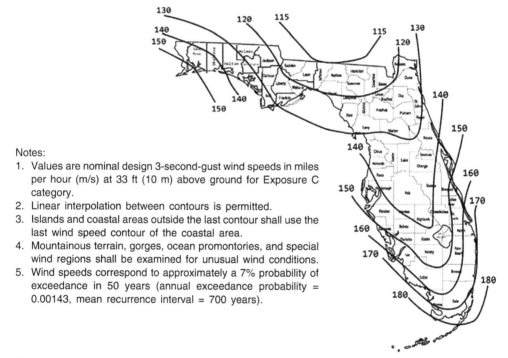

Notes:
1. Values are nominal design 3-second-gust wind speeds in miles per hour (m/s) at 33 ft (10 m) above ground for Exposure C category.
2. Linear interpolation between contours is permitted.
3. Islands and coastal areas outside the last contour shall use the last wind speed contour of the coastal area.
4. Mountainous terrain, gorges, ocean promontories, and special wind regions shall be examined for unusual wind conditions.
5. Wind speeds correspond to approximately a 7% probability of exceedance in 50 years (annual exceedance probability = 0.00143, mean recurrence interval = 700 years).

Figure 13.4 Minimum wind loads for Florida (FBC 2010)

ABC ENGINEERING, INC.	MWFRS Wind Loads ASCE 7-10 Enclosed & Partially Enclosed Buildings of All Heights	Job No: 11012 Designer: JCB Checker: RJK Date: 2/9/2014
	Notes: Grinding Building (+/- Z Direction)	

Basic Parameters

Risk Category	IV	Table 1.5-1
Basic Wind Speed, V	220 mph	Figure 26.5-1A
Wind Directionality Factor, K_d	0.85	Table 26.6-1
Exposure Category	B	Section 26.7
Topographic Factor, K_{zt}	1.00	Section 26.8
Gust Effect Factor, G or G_f	0.850	Section 26.9
Enclosure Classification	Enclosed	Section 26.10
Internal Pressure Coefficient, GC_{pi}	+/- 0.18	Table 26.11-1
Terrain Exposure Constant, α	7.0	Table 26.9-1
Terrain Exposure Constant, z_g	1,200 ft	Table 26.9-1

Wall Pressure Coefficients

Windward Wall Width, B	425 ft	
Side Wall Width, L	490 ft	
L/B Ratio	1.15	
Windward Wall Coefficient, C_p	0.80	Figure 27.4-1
Leeward Wall Coefficient, C_p	-0.50	Figure 27.4-1
Side Wall Coefficient, C_p	-0.70	Figure 27.4-1

Roof Pressure Coefficients

Roof Slope, θ	1.1°	
Median Roof Height, h	30 ft	
Velocity Pressure Exposure Coef., K_h	0.70	Table 27.3-1
Velocity Pressure, q_h	73.8 psf	Equation 27.3-1
h/L Ratio	0.06	
Windward Roof Area	0 ft²	
Roof Area Within 15 ft of WW Edge	6,375 ft²	

Location	Min/Max	Horiz Distance From Windward Edge				
		0 ft	15 ft	30 ft	60 ft	
Windward Roof Coefficient	Min	-0.90	-0.90	-0.50	-0.30	Figure 27.4-1
Normal to Ridge, C_p	Max	-0.18	-0.18	-0.18	-0.18	
Leeward Roof Coefficient	Min	-0.90	-0.90	-0.50	-0.30	
Normal to Ridge, C_p	Max	-0.18	-0.18	-0.18	-0.18	
Roof Coefficient	Min	-0.90	-0.90	-0.50	-0.30	
Parallel to Ridge, C_p	Max	-0.18	-0.18	-0.18	-0.18	

Structure Pressure Summary (Add Internal Pressure q_hGC_{pi} or q_iGC_{pi} as Necessary)

Height, z	K_z	q_i	Walls				Roof			Internal	
							Normal to Ridge		Parallel		
			WW	LW	WW + LW	Side	WW	LW	to Ridge	Positive	Negative
0 ft	0.57	60.5 psf	41.2 psf		72.5 psf					13.3 psf	
3 ft	0.57	60.5 psf	41.2 psf		72.5 psf		Min:	Min:	Min:	13.3 psf	
6 ft	0.57	60.5 psf	41.2 psf		72.5 psf		-56.4 psf	-56.4 psf	-56.4 psf	13.3 psf	
9 ft	0.57	60.5 psf	41.2 psf		72.5 psf					13.3 psf	
12 ft	0.57	60.5 psf	41.2 psf		72.5 psf					13.3 psf	
15 ft	0.57	60.5 psf	41.2 psf	-31.4 psf	72.5 psf	-43.9 psf				13.3 psf	-13.3 psf
18 ft	0.61	63.8 psf	43.4 psf		74.7 psf					13.3 psf	
21 ft	0.63	66.6 psf	45.3 psf		76.7 psf		Max:	Max:	Max:	13.3 psf	
24 ft	0.66	69.2 psf	47.1 psf		78.4 psf		-11.3 psf	-11.3 psf	-11.3 psf	13.3 psf	
27 ft	0.68	71.6 psf	48.7 psf		80.0 psf					13.3 psf	
30 ft	0.70	73.8 psf	50.2 psf		81.5 psf					13.3 psf	

Figure 13.5 Wind load calculation from Excel® spreadsheet

13.2.4 Roof Loads

The roof can only be designed after the roof load is determined. Uplift as a result of wind loads in high-wind areas and/or snow or rain loads are almost always the critical factors depending on location. Winds loads will be discussed first.

13.2.4.1 Wind Loads

There are two construction means to accomplish an acceptable roof design in high-velocity wind zones: (1) some form of tie-downs or straps or (2) the use of the weight of the roof slab itself. A combination of both also can be used. To come up with an adequate roof load, the slab thickness needs to be considered first. Concrete has a density of 145 to 150 pcf. Using the unit weight of concrete with the maximum wind load pressure acting on the roof, the preliminary thickness of the slab can be determined using the following formula:

$$\text{Slab thickness} = \frac{\text{Wind load}}{\text{Concrete density}} \tag{13.9}$$

The self-weight of the slab (dead load) is then incorporated in the roof load. The dead load of the slab can be computed by multiplying the thickness of the slab by the concrete unit weight. Of course, adding weight to the roof means that the members supporting that roof must be increased significantly, which adds to the cost.

13.2.4.2 Rain Loads

Rain loads should not be a controlling factor in designing a roof. The intent of most codes is to get water off the roof as fast as possible, so it does not add weight and because standing water on a roof inevitably will leak into the building envelope. As a result, codes for rain-prone areas generally require two different roof drainage systems: primary and secondary. Each portion of a roof should be designed to sustain the load of all rainwater that will accumulate on it if the primary drainage system for that portion is blocked, plus the uniform load caused by water that rises above the inlet of the secondary drainage system (e.g., scuppers) at its design flow. If water on a flat roof accumulates faster than it runs off, ponding will occur. An acceptable design must eliminate ponding. Referencing ASCE 7-10, the rain load (R) experienced by a roof can be computed as:

$$R = 5.2(d_s + d_h) \tag{13.10}$$

where

    d_s = the static head of the water when the primary drainage system is blocked (in.)
    d_h = the additional depth of water (in.)

For example, if the values for d_s and d_h are 5 and 3 in., respectively, the rain load can be found to be 41.6 psf. The roof must be able to support this load.

    In addition, the flow off the roof needs to be calculated. Using the tributary area for the scupper and the rainfall intensity of a design storm, the flow rate that a particular drain can support can be calculated from the following formula:

$$Q = 0.0104Ai \tag{13.11}$$

where

Q = the flow off of the roof (gpm)

A = the area served by each scupper (ft^2)

i = the location rainfall intensity (in./hr) gathered from local codes

13.2.4.3 Snow Loads

In some geographical parts of the United States, roof loading can be quite severe due to the effect of the accumulation of snow. Like wind load, designs for snow loads are specified based on a zone map that reports the 50-year recurrence intervals of a "risky" snow depth (see Figures 13.6 and 13.7). It must be noted that no single code can exactly predict all of the

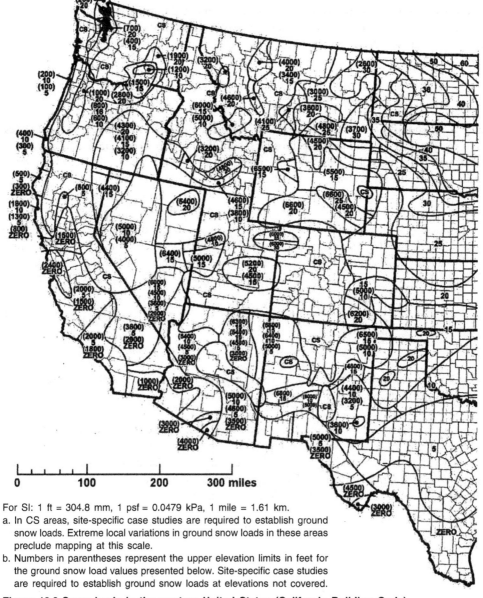

For SI: 1 ft = 304.8 mm, 1 psf = 0.0479 kPa, 1 mile = 1.61 km.

a. In CS areas, site-specific case studies are required to establish ground snow loads. Extreme local variations in ground snow loads in these areas preclude mapping at this scale.

b. Numbers in parentheses represent the upper elevation limits in feet for the ground snow load values presented below. Site-specific case studies are required to establish ground snow loads at elevations not covered.

Figure 13.6 Snow loads in the western United States (California Building Code)

implications that snow loading may impose on a structure. For a flat roof with a slope less than 5%, the snow load on the roof (p_f) can be computed by using the following empirical formula:

$$p_f = 0.7C_eC_tIp_g \tag{13.12}$$

where C_e is the exposure factor dependent upon the terrain and is based on exposure to winds. The value ranges from 0.8 to 1.2 (see ASCE 7-10, Table 7-2). C_t is the thermal factor dependent of the average temperature in the interior of the building (e.g., C_t = 1.0 for a normally heated structure, but ranges from 1.3 for open buildings to 0.75 for heated green-

For SI:
1 ft = 304.8 mm
1 psf = 0.0479 kPa

Figure 13.7 Snow loads in the eastern United States (California Building Code)

houses). I is the importance factor related to the building category (ASCE 7-10, Table 1.5-1) and ranges from 0.8 for a Category I structure to 1.2 for essential structures. p_g is the ground snow load (psf).

13.2.5 Earthquake Loads

Structures located in earthquake-prone areas must be designed to withstand loadings due to interaction with the ground motion and its response characteristics. The loadings that a structure will experience are influenced by the lateral resistance of the structure and its horizontal displacement caused by the ground motion. The magnitude of the forces is directly related to the mass, stiffness amount, and type of ground acceleration. During an earthquake, the ground vibrates in the horizontal and vertical axes. The horizontal acceleration creates shear forces in the columns. The effect of a structure's displacement, velocity, and acceleration can be determined and characterized as an earthquake response. The loadings on the structure can then be calculated using advanced theories of structural dynamics, performed using commercially available software. This type of analysis must be performed for large structures such as multistory buildings in earthquake-prone areas (Hibbeler 2009).

A static analysis can be performed for small structures to provide adequate design for structural integrity. This can be done by approximating the dynamic loads caused by an earthquake and applying them as externally applied static forces in the lateral direction of the structure. The design process involves computing the base shear acting on the structure and designing the members to effectively counteract this load. The engineer also must consider the moment that the structure may experience due to earthquake loadings. ASCE 7-10 also includes a brief guideline on how this principle can be adapted for proper design. Keep in mind that the Pacific Coast is not the only place where there are earthquakes. Figure 13.8 shows where the U.S. Geological Survey identifies seismic activity in the United States.

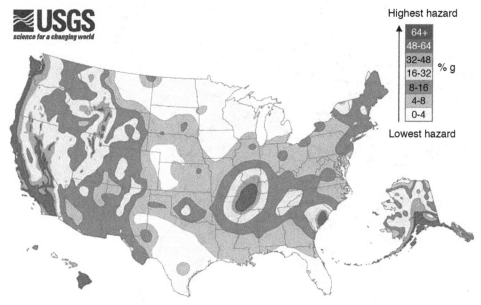

Figure 13.8 U.S. Geological Survey seismic activity map

Earthquake-prone areas must deal with lateral loads as well as vertical ones. The loads move in multiple directions. Due to the complexity of the subject, this topic is best left to other civil engineering texts.

13.2.6 Other Loads

The effect of blast loads, temperature changes, and differential settlement of the foundation also can apply stresses on a structure. In certain scenarios, these other loadings may dictate the design. Although these types of loads or how to design for them will not be discussed in this chapter, they are noted to broaden the perspective of the structural engineer as to the wider range of analyses required to be performed.

13.3 Structural Design Concepts

The roof or floor typically is supported by joists that rest on walls (Figure 13.2) or columns that rest on the building foundation. Each acts to support the load, while transferring it to adjacent members. Ultimately, the building load is to be effectively transferred to the foundation and the soil below it. Finding the critical loads to properly design specific members for the types of forces that will be experienced during building occupancy is the job of the designated structural engineer. While the engineer has to deal with many variables, there are some basic guidelines to follow:

◆ Determine the type of construction proposed (steel, concrete, reinforced concrete, or other material); it should be something with which the engineer is familiar.
◆ Define the tributary areas of all beams, columns, and joists.
◆ Determine the means to construct the floor slabs.
◆ Calculate the loads on beams and define reinforcement for tension, shear, and other needs.
◆ Calculate the loads on columns.
◆ Determine loads to be transferred to the foundation.

In the following sections, this process is described in more detail, beginning with identification of the critical loads for the structure.

13.3.1 Concept of Tributary Area

In designing a structure, there is a very important concept known as the *tributary area* (A_T), which is used as a simplified approach for the determination of the load acting at a particular structural element. This method is used to visualize and compute the load that the element must carry for proper structural design based on equilibrium analysis. An equally important concept is the *influence area* (A_I). This area is different from the tributary area in the sense that it reflects the area over which any applied load would have an influence on the member under investigation. Based on ASCE 7-10:

$$A_I = K_{LL}A_T \tag{13.13}$$

Table 13.4 Sample live load element factors (K_{LL}) (ASCE 7-10)

Element	K_{LL} factor
Interior columns	4
Edge columns without cantilever slabs	3
Edge beams without cantilever slabs	2
Interior beams	2
All other members, including cantilever beams, one-way slabs, and two-way slabs	1

where K_{LL} stands for the live load element factor. Table 13.4 gives sample values of the live load element factor.

With respect to tributary areas, Figures 13.9 to 13.11 illustrate the point. Figure 13.9 shows an example for the tributary area of the largest slab. The slab area is 60 ft × 20 ft, holding 1,200 ft². If the slab live load is 100 psf, the load is 120 kips. Figure 13.10 shows the critical beam. The tributary area coincidentally also is 20 ft × 60 ft. Figure 13.11 shows the critical column. The tributary for the column coincidentally also is 1,200 ft², but must be multiplied by the number of floors and the roof load must be added.

To account for the size of the influence area and provide a more realistic value of the actual live load acting on the structure, a live load reduction factor can be used for influence areas in excess of 400 ft² (Geschwindner 2012). The live load may be decreased to the following reduction equation:

$$L = L_o \left(0.25 + \frac{15}{\sqrt{A_I}} \right) \qquad (13.14)$$

where

L = newly reduced live load (psf)
L_o = code-specified nominal live load (psf)
A_I = influence area (ft²)

For the tributary areas in the examples shown in Figures 13.9 to 13.11, the live load reduction would be:

$$L = L_o \left(0.25 + \frac{15}{\sqrt{A_I}} \right) = 100 \left(0.25 + \frac{15}{\sqrt{1,200}} \right) = 68.3 \text{ psf}$$

Limitations on the use of this reduction of live load must be checked before applying it (see ASCE 7-10 or latest edition or local code).

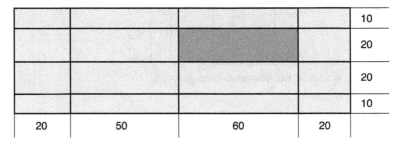

Figure 13.9 Tributary area of slab

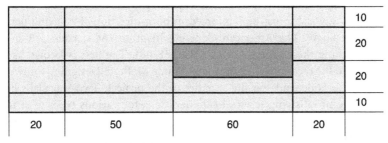

Figure 13.10 Tributary area of critical beam

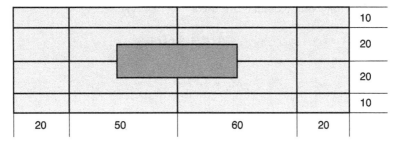

Figure 13.11 Tributary area of critical column

13.3.2 One-Way and Two-Way Slabs

A deck that is supported in such a way that it delivers its load to supporting members by a single directional bending action is commonly known as a *one-way slab* (Figure 13.12 left). This is a structural floor where the ratio of the long span (L_2) to the short span (L_1) is greater than or equal to 2.0 ($L_2/L_1 \geq 2.0$). For practical purposes, if a rectangular slab with an aspect ratio greater than 2.0 is supported by beams on all four sides, the slab will behave as a one-way slab by bending primarily in the long direction, so it acts like a beam. As the length of L_1 becomes smaller, the beams provide a larger stiffness to carry the load. In reinforced concrete, a one-way slab is assumed to be a rectangular beam with a large ratio of width to

Figure 13.12 One-way versus two-way slab configuration

depth, typically designed for a 1-ft-wide section; the slab is assumed to be connected by these 1-ft strips of beams. One-way slabs are calculated like beams (Figure 13.13).

If the ratio of the long to short span is less than 2.0 ($L_2/L_1 < 2.0$), the slab will behave as a *two-way slab* (refer to Figure 13.12 right), which will tend to bend in a double directional action. This bending will resemble a satellite-dish-shaped surface. Because of this bending behavior, the two-way slab must be reinforced in both directions by having perpendicular bars layered accordingly. The American Concrete Institute (ACI) code 318 specifies two methods for designing these types of slabs (ACI 318-08). The first is known as the Direct Design Method and the second is the Equivalent Frame Method. Designing for two-way slabs using these techniques depends on the specifics of the project. Two-way slabs have shown to be very effective in satisfying service load conditions for various types of structures, including but not limited to warehouses, parking structures, and industrial buildings.

A one-way slab design requires several steps:

◆ Determine the design moment for the slab based on the self-weight, live and dead loads. The following formulas are used:

$$w_u = 1.2(D) + 1.6(L) \qquad (13.15)$$

$$M_u = \frac{w_u l^2}{8} \qquad (13.16)$$

where

w_u = total load (lb or kip)
D = dead load (lb or kip)
L = live load (lb or kip)

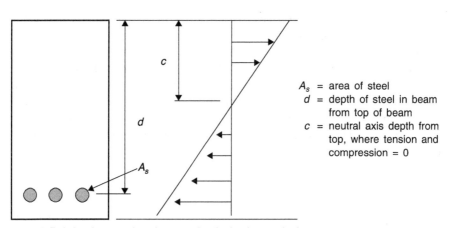

Figure 13.13 Reinforcing steel and stress loads for beam design

M_u = ultimate moment of the slab (convert to in.-lb)

l = clear span (ft)

◆ Determine the depth of the compression block. The following formulas are used:

$$z = 0.9d \qquad (13.17)$$

$$M_n = A_s f_y z \qquad (13.18)$$

$$a = \frac{(A_s)(f_y)}{0.85(f_c)} \qquad (13.19)$$

where

z = distance from pin to centroid of applied load (lever arm) (in. vertical)

d = depth to rebar in slab

M_n = moment about the center for compression for the tensile force (converted to in.-lb)

A_s = area of steel per foot of width

f_y = tensile strength of steel (psi)

f_c = compressive strength of concrete (psi)

a = area of reinforcing steel

◆ Recalculate the area of steel (A_s) to define the steel reinforcing to be used (see rebar size in Table 13.5):

$$\frac{M_n}{0.9} = A_s(f_y)\left(d - \frac{a}{2}\right) \qquad (13.20)$$

Table 13.5 Rebar size

		U.S. rebar size chart					
Imperial bar size	"Soft" metric size	Mass per unit length		Nominal diameter		Nominal area	
		lb/ft	kg/m	in.	mm	in.²	mm²
#2	#6	0.167	0.249	0.250 = ¼	6.350	0.05	32
#3	#10	0.376	0.561	0.375 = ⅜	9.525	0.11	71
#4	#13	0.668	0.996	0.500 = ⁴⁄₈	12.700	0.20	129
#5	#16	1.043	1.556	0.625 = ⅝	15.875	0.31	200
#6	#19	1.502	2.240	0.750 = ⁶⁄₈	19.050	0.44	284
#7	#22	2.044	3.049	0.875 = ⅞	22.225	0.60	387
#8	#25	2.670	3.982	1.000	25.400	0.79	509
#9	#29	3.400	5.071	1.128	28.650	1.00	645
#10	#32	4.303	6.418	1.270	32.260	1.27	819
#11	#36	5.313	7.924	1.410	35.810	1.56	1,006
#14	#43	7.650	11.410	1.693	43.000	2.25	1,452
#18	#57	13.600	20.284	2.257	57.300	4.00	2,581
#18J		14.600	21.775	2.337	59.400	4.29	2,678

◆ Check tension control (c_t):

$$c_t = \frac{A_s}{0.8} \gg 0.005 \tag{13.21}$$

◆ Check tensile strain (ϵ_t):

$$\epsilon_t = 0.003 \, \frac{(d - c)}{c} \gg 0.005 \tag{13.22}$$

where

c = neutral axis depth (in.)

◆ Define the minimum reinforcement ratio (ρ), where the largest value of ρ is the controlling case:

$$\rho_{\text{actual}} = \frac{a}{(w)(c)} \tag{13.23}$$

$$\rho_{\min} = \frac{200}{f_y} \tag{13.24}$$

$$\rho_{\min} = \frac{3\sqrt{f_c}}{f_y} \gg 0.0035 \tag{13.25}$$

where

w = width of section (in.)

◆ Check shrinkage and temperature reinforcement:

$$A_s = 0.0018bh \tag{13.26}$$

where

b = width of section (in.)
h = depth to rebar (in.)

Example 1. One-Way Slab Design of Flexure Reinforcement

The load taken by the slab has to be obtained for a 12-in. strip of slab and a clear span of 4.5 ft, as illustrated in Figure 13.14.

Step 1. Determine the design moment for the slab:

$$\text{Self-weight of 12-in. strip} = 150 \times \frac{3 \times 12}{144} = 37.5 \text{ lb/ft}$$

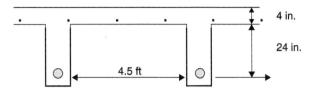

Figure 13.14 Reinforcing steel for slab design using a 22-in. strip of slab and a clear span of 4.5 ft (assume 1.5-in. cover over rebar on beams and 1 in. on slab)

$$D_L = 54 \text{ psf} \times 1 \text{ ft} = 54 \text{ lb/ft}$$

$$L_L = 100 \text{ psf} \times 1 \text{ ft} = 100 \text{ lb/ft}$$

$$w_u = 1.2(54 + 37.5) + 1.6(100) = 269.8 \text{ lb/ft}$$

$$M_n = \frac{w_u l^2}{8} = \frac{(269.8)(4.5)^2}{8} = 682.9 \text{ ft-lb} \cong \mathbf{8{,}195 \text{ in.-lb}}$$

Step 2. Determine the depth of the compression block, assuming:

$$z = 0.9d = 0.9 \times 3 = 2.7 \text{ in.}$$

$$M_n = A_s f_y z$$

$$8{,}195 = A_s(60{,}000)(2.7)$$

$$A_s = 0.051 \text{ in.}^2/12\text{-in. strip}$$

$$a = \frac{(0.051)(60{,}000)}{0.85(5{,}000)(12)} = \mathbf{0.06 \text{ in.}}$$

Step 3. Recalculate the area of steel:

$$\frac{8{,}195}{0.9} = A_s(60{,}000)\left(3 - \frac{0.06}{2}\right)$$

$$A_s = 0.046 \text{ in.}^2$$

Step 4. Check tension control:

$$c = \frac{0.046}{0.8} = 0.06 \text{ in.}$$

$$\epsilon_t = 0.003\left(\frac{2 - 0.06}{0.06}\right) = 0.1 \gg 0.005$$

$$\therefore \text{ Section is tension controlled}$$

Step 5. Check minimum reinforcement:

$$\rho_{actual} = \frac{0.06}{(12)(2)} = 0.0025$$

$$\rho_{min} = \frac{200}{f_y} = 0.0033$$

$$\rho_{min} = \frac{3\sqrt{5,000}}{60,000} = 0.0035 \text{ (controls)}$$

$$0.0046 \gg 0.0035$$

OK! Accept design

Step 6. Check shrinkage and temperature reinforcement:

$$A_s = 0.0018bh = (0.0018)(12)(3) = 0.0648 \text{ in.}^2$$

$$\boxed{\therefore \text{ Use #3 sp. at 24 in. center to center}}$$

The same basic process works for beams.

13.3.3 Beams and Girders

Beams are considered to be mainly "prismatic" horizontal members used to carry gravity loads. Beams are primarily designed to resist bending moment loads; moreover, a beam also must be designed against internal shear failure due to large loads that may act on it. In some circumstances, torsional loads also may act on a beam. Special attention must be paid when designing for concrete beams because concrete is strong in compression but relatively weak in tension. Therefore, steel reinforcing rods are used in the cross sections of concrete beams to be able to effectively accommodate the areas where tensile stresses occur. Reinforced concrete cross sections are effective in resisting the applied moments in the top and bottom flanges, while the web is effective in resisting the applied shear forces. A reinforced concrete text should be consulted for specific solutions for concrete beams. The process will include obtaining the tributary area, live and dead loads, and moments.

Steel beams are much easier to specify. Steel manufacturers have created standard shapes, sizes, and loading for steel members. As a result, if the engineer knows the loads and moments, the steel member can be selected from a steel manual relatively easily. The only real issues involve connections, fire protection, and moments. The relevant moment equations are:

$$\text{Negative moment: } -M_{AB} = -0.107w_u l^2 \qquad (13.27)$$

$$\text{Positive moment:} \quad +M_{AB} = +0.0772 w_u l^2 \tag{13.28}$$

where

w_u = the applied load from $1.2D + 1.6L$
l = length
AB = the two ends of the beam

Example 2. Design of Continuous Steel Beam for Flexure by LRFD

The beam is continuous over four spans of 24 ft (see Figure 13.15). This beam supports a live load (L) of 2.5 kips/ft and a dead load (D) of 1.8 kips/ft. Use A992 steel.

Solution:

$$w_u = 1.2D + 1.6L = 6.16 \text{ kips/ft}$$

From the beam, shear, moment, and deflection diagrams in the AISC *Steel Construction Manual*, Table 3-23, Case 42, the critical span is the exterior span AB:

Negative moment: $-M_{AB} = -0.107 \; w_u l^2 = -0.107 \times 6.16 \times 24^2 = 380$ ft-kips

Positive moment: $+M_{AB} = +0.0772 \; w_u l^2 = +0.0772 \times 6.16 \times 24^2 = 274$ ft-kips

Determine the plastic section modulus (Z_{req}):

$$Z_{\text{req}} = \frac{M_u}{\phi F_y} = \frac{(380 \times 12)}{(0.9 \times 50)} = 101.3 \text{ in.}^3$$

where

M_u = the larger of the absolute value of $-M_{AB}$ and $+M_{AB}$
F_y = the modulus of steel (50 ksi)

Select W-shape from the AISC *Steel Construction Manual*, Table 3-2: **W16×57**

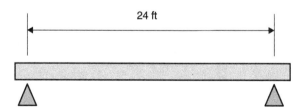

24 ft

Figure 13.15 Example of beam sitting on column supports

13.3.4 Columns

Columns are structural members that are positioned vertically and carry axial compressive loads. They can have different shapes depending on what type of material they are made out of. In steel design, I- or W-shaped cross sections or square sections typically are used. In reinforced concrete design, circular and square cross sections are the most common types. Concrete columns also are reinforced with steel reinforcing rods in the longitudinal direction to resist tensile stress and in the planar direction to resist shear stress. With today's advances in computational technology, columns also can be designed as composite members made up of steel sections encased in concrete. Columns that experience both axial compression and bending moments are commonly referred to as *beam columns* (Hibbeler 2011). The main function of columns is to provide support for the load-carrying beams and slabs. The columns resist the loads from the upper floors and transfer them to the columns below and then ultimately to the foundation (Figure 13.16). As can be seen in Figure 13.16, the roof and all floors above the roof must be accounted for with the column. Hence, in large buildings, the columns carry very large loads. If a critical column collapses due to improper design, this may lead to a progressive failure of the floors above it, which in turn may result in a complete collapse of the entire structure.

Generally speaking, columns can be designed as *short/nonslender columns* or as *long/slender columns*. Columns can fail because of improper material design, such as yielding of the steel at the tension face, initial crushing of concrete at the compression face, or through buckling. Columns that fail due to initial material failure are referred to as *short columns*. As the height of the column increases, the column starts transitioning from a short to a long column. The probability of failure of a long column due to buckling increases due to its slenderness ratio property (kl_u/r). The slenderness ratio is the effective length of the column (kl_u) to the radius of gyration (r). The height (l_u) is the unsupported length of the column; k is a factor dependent on the end support of the column and whether it is braced or unbraced (Nawy 2009). Referring to the ACI criteria for a concrete column, an unbraced column can be defined as a *short column* if its slenderness ratio is less than or equal to 22.

The means to determine column loads is the same as for beams or slabs: find the tributary area and apply the loads, recognizing that each floor and the roof add loads. Reinforced concrete columns require significant calculations and are best covered in a reinforced con-

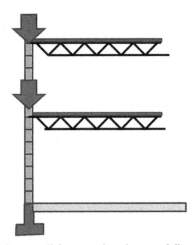

Figure 13.16 Load transfers down wall from roof and second floor

crete text. Steel is easier to work with, although the slenderness issue is common to both steel and concrete columns. Fire protection is a critical issue with steel columns, but selecting a column is straightforward, as follows:

◆ Step 1. Determine the maximum required strength using LRFD load combinations from Equations 13.1 to 13.7.
◆ Step 2. Select the least-weight shape in the AISC *Steel Construction Manual*, Table 4-1 by scanning for the LRFD values than can support the calculated load.

Example 3. Design of Steel Columns

The column shown in Figure 13.17 must resist the following loads in the appropriate combinations: dead load = 60 kips, live load = 175 kips, and rain load = 180 kips. Using A992 steel, assume that the live load comes from a distributed load less than 100 psf (due to the live load reduction).

Step 1. Determine the maximum required strength using LRFD load combinations from Equations 13.1 to 13.4:

$$1.4D = 1.4 \times 60 = 84 \text{ kips}$$

$$1.2D + 1.6L = 1.2 \times 60 + 1.6 \times 175 = 352 \text{ kips}$$

$$1.2D + 0.5L + 1.0W = 1.2 \times 60 + 0.5 \times 175 + 1.0 \times 180 = 339.5 \text{ kips}$$

$$0.9D + 1.0W = 0.9 \times 60 + 1.0 \times 180 = 234 \text{ kips}$$

Therefore the column must carry w_u = 352 kips

Based on the support conditions, the effective length factor (k) about the x- and y-axes can be taken as 1.0; therefore, KL = 15 ft.

Step 2. Select the least-weight shape in the AISC *Steel Construction Manual*, Table 4-1 by scanning for the LRFD values that can support the calculated load (w_n is the weight the beam can support from the table):

Select W8×48 with ϕw_n of 367 kips > w_u

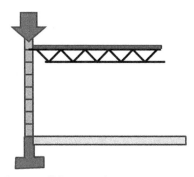

Figure 13.17 Load transfers down wall from roof

13.3.5 Walls

There are three types of walls used for construction purposes: (1) nonload-bearing walls, (2) load-bearing walls, and (3) shear walls. *Nonload-bearing walls* are those that support their own weight and very little lateral forces (McCormac and Brown 2009). Basement walls, retaining walls, and façade-type walls are examples of nonload-bearing walls. The ACI code provides specific limitations on the design of these types of walls in Section 14.6. For example, the thickness of a nonload-bearing wall may not be smaller than 4 in. or $\frac{1}{30}$ times the smallest distance between members that provide lateral support. Refer to the ACI code for further limitations of the design of these walls.

Load-bearing walls support vertical loads and lateral moments. Because these walls are stiff, they are able to provide valuable resistance when a structure is subject to earthquake and wind loads. ACI code 318 provides two design methods for load-bearing walls: (1) empirical design method and (2) rational design method. The empirical method is used if the resultant of all factored loads is within an eccentricity equal to or less than $\frac{1}{6}$ the thickness of the wall. In other words, the empirical design method is adequate for relatively concentric loads being applied to walls of short height. The rational design approach is similar to column design, regardless of whether the eccentricity is smaller or larger than $\frac{1}{6}$ the thickness of the wall. The Portland Cement Association (PCA), ACI, and others provide design tools for load-bearing walls. Slender walls such as tilt-up walls can be designed appropriately using PCA references.

Shear walls are used to provide adequate stiffness for tall buildings to effectively resist horizontal forces such as side sway due to wind and earthquake loads. If these buildings are not properly designed for lateral loads, they may experience high stresses and relatively large side sway. The strength of the shear wall comes from it being an adequate member to resist flexural forces. Common shear wall systems are placed in tall buildings around service cores, stairwells, and elevator shafts. Shear walls generally provide a stiff box-type structure. Multiple shear walls can be used in a tall building, connected together by link beams or by shear coupling walls, which are structural bracing members constructed of steel or reinforced concrete. Both will augment building stiffness. Shear walls also may be used as load-bearing walls. The concept of a shear wall is similar to a vertical cantilever beam, which is subject to bending moment and shear forces. Shear is predominantly more important in short walls, but moment may govern the design for walls with greater heights (McCormac and Brown 2009). A shear wall will require tensile reinforcement on both ends of the wall since the horizontal force can act from either direction, thus causing respective tensile stresses. Chapter 11 of ACI code 318 provides the required formulas for proper analysis and design for shear walls. The procedure for shear walls is as follows:

◆ Step 1. Determine if the wall thickness is satisfactory from the factored shear force (V_u):

$$V_u = \phi 10 \sqrt{f_c'} \, bd \tag{13.29}$$

where

V_u = ultimate shear force (kip)
ϕ = strength reduction factor = 0.75 typically
f_c' = concrete yield strength (psi)
d = the effective depth (in.), which is $0.8l_w$
b = the thickness of the wall (in.)

◆ Step 2. Calculate the factored overturning moment (M_u) and weight of the wall at critical section c_s. The critical section for shear occurs at the smaller of $l_w/2$ and $h/2$, where h = height of the wall (ft):

$$M_u = 1.4V_u(b - c_s) \tag{13.30}$$

$$N_u = (0.15)\left(\frac{b}{12}\right)(l_w)(h - c_s) \tag{13.31}$$

where

N_u = the factored axial load (kip) (positive for compression, negative for tension), including creep and shrinkage, and will be zero under some cases
l_w = the critical wall section length (ft)

◆ Step 3. Compute and design for shear capacity provided by the concrete (V_c) for the wall:

$$V_c = 3.3\lambda \sqrt{f_c'}\, hd + \frac{N_u d}{4l_w} \tag{13.32}$$

where λ is a factor for adjusting for lightweight concrete. Assume one for regular concrete.

◆ Step 4. Check V_c and take the smaller of Equations 13.32 and 13.33:

$$V_c = \left[0.6\lambda\sqrt{f_c'} + \frac{l_w\left(1.25\lambda\sqrt{f_c'} + 0.2N_u/l_w h\right)}{M_u\Big/V_u - l_w\Big/2} \right] hd \tag{13.33}$$

◆ Step 5. Check if shear reinforcing is required. If $V_u > \phi V_c$, it is:

$$\frac{\phi V_c}{2} < V_u \tag{13.34}$$

◆ Step 6. Design for horizontal shear reinforcement. Required stirrup spacing for the condition $V_u > \phi V_c$ is given by:

$$\frac{A_v}{s} = \frac{V_u - \phi V_c}{\phi f_y d} \tag{13.35}$$

where

A_v = the area of shear reinforcement within a distance s (in.)
f_y = the yield strength of steel (psi)

and d is converted to inches.

◆ Step 7. Find the maximum vertical spacing of horizontal stirrups based on the smaller of:

$$\frac{l_w}{5} = 18 \times \left(\frac{12}{5}\right) = 48 \text{ in.} \qquad \text{or } 3b = 3 \times 8 = 24 \text{ or } 18 \text{ in.}$$

◆ Step 8. Check the steel ratio (ρ_t):

$$\rho_t = \frac{A_v}{A_g} \tag{13.36}$$

where

A_v = the vertical steel area (in.^{2})
A_g = the area of the blocks (in.^{2})

◆ Step 9. Design for vertical shear reinforcement:

$$\min \rho_l = 0.0025 + 0.5\left(2.5 - \frac{h}{l_w}\right)(\rho_t - 0.025) \tag{13.37}$$

where

ρ_l = the longitudinal steel ratio

◆ Step 10. Find the maximum vertical spacing of horizontal stirrups based on the smallest of:

$$\frac{l_w}{3} = 12 \times \left(\frac{12}{3}\right) = 48 \text{ in.} \qquad \text{or } 3b = 3 \times 8 = 24 \text{ or } 18 \text{ in.}$$

◆ Step 11. Check the design for vertical reinforcement:

$$\frac{M_u}{\phi b d^2} \tag{13.38}$$

From tabular reference sheets A.9 to A.14 for given combinations of strengths for steel and concrete (McCormac and Brown 2009), find ρ and then verify the area of steel:

$$A_s = \rho bd \qquad\qquad (13.39)$$

Example 4. Concrete Shear Wall Design

The height of the wall (H) = 15 ft, length of the wall (l_w) = 30 ft, thickness (b) = 8 in., and V_u = 250 kips (see Figure 13.17).

Step 1. Determine if the wall thickness (V_u) is satisfactory:

$$V_u = \phi 10\sqrt{f_c'}\, bd \qquad \text{where } d = 0.8l_w$$

$$V_u = 0.75 \times 10 \times \sqrt{3,000} \times 8 \times (0.8 \times 20 \times 12) = 630 \text{ kips} > 250 \text{ kips}$$

∴ **OK!**

Step 2. Calculate the factored overturning moment and weight of the wall at the critical section (which is half the height of the wall in question). The critical section for shear is determined as the smaller of $l_w/2$ and $H/2$:

$$\frac{l_w}{2} = 10 \qquad \frac{H}{2} = 7.5$$

Therefore, use 7.5:

$$M_u = 250 \times 7.5 = 1,875 \text{ ft-kips} = 22,500 \text{ in.-kips}$$

$$N_u = (0.15)\left(\frac{8}{12}\right)(20)(15 - 7.5) = 15 \text{ kips}$$

Step 3. Compute and design for shear capacity provided by the concrete (V_c) for the wall:

$$V_c = 3.3\lambda\sqrt{f_c'}\, hd + \frac{N_u d}{4l_w}$$

$$= 3.3 \times 1.0 \times \sqrt{3,000} \times 8 \times (20 \times 12 \times 0.8) + 15 \times \frac{0.8 \times 20}{4 \times 20} = 280 \text{ kips}$$

Step 4. Check V_c and take the smaller of Equations 13.32 and 13.33:

$$V_c = \left[0.6\lambda\sqrt{f_c'} + \frac{l_w(1.25\lambda)\sqrt{f_c'} + 0.2N_u/l_w h}{M_u\big/V_u - l_w\big/2}\right] hd$$

$$V_c = \left[0.6 \times 1.0 \times \frac{\sqrt{3{,}000 \text{ psi}}}{1{,}000 \text{ lb/kips}} \right.$$

$$\left. + \frac{(20 \text{ ft} \times 12 \text{ in./ft})(1.25)(1.0) \dfrac{\sqrt{3{,}000 \text{ psi}}}{1{,}000 \text{ lb/kips}} + 2.0 \times \dfrac{15 \text{ kips}}{20 \text{ ft} \times 8 \text{ in.} \times 12 \text{ in./ft}}}{1{,}875 \text{ ft-kip} \times 12 \text{ in./ft} \Big/ 250 \text{ kips} - (20 \text{ ft})/2} \right]$$

$\times$ 8 in. $\times$ 20 ft $\times$ 12 in./ft = 122 kips is the smaller number

Step 5. Check if shear reinforcing is required:

$$\frac{\phi V_c}{2} = \frac{0.75 \times 210}{2} = 78.5 \text{ kips} < 250 \text{ kips}$$

∴ Reinforcement required

Step 6. Design for horizontal shear reinforcement:

$$\frac{A_v}{s} = \frac{V_u \times \phi V_c}{\phi f_y d} = \frac{250 - 0.75 \times 212 \text{ kips}}{0.75 \times 60 \text{ ksi} \times 0.8 \times 20 \times 12} = 0.0105 \text{ in.}$$

Try two #3 bars with a cross-sectional area of 0.11 in.^{2}:

$$s = \frac{(2 \times 0.11)}{0.0105} = 21 \text{ in.}$$

Two layers of horizontal bars will be placed at the calculated spacing.

Step 7. Find the maximum vertical spacing of horizontal stirrups as the smaller of Step 6 or 18 in.:

$$\frac{l_w}{5} = 18 \times \left(\frac{12}{5} \right) = 48 \text{ in. or } 3b = 3 \times 8 = 24 \text{ or } 18 \text{ in.}$$

18 in. is the smaller value. ∴ Use #4 bars (A_s = 0.2 in.^{2} at 16 in. on center vertically, which will allow for a block wall with reinforcement).

Step 8. Check the steel ratio:

$$\rho_t = \frac{A_v}{A_g} = \frac{2 \times 0.2}{8 \times 16} = 0.003125 > 0.0025$$

∴ **OK!**

∴ Use #4 bars as horizontal stirrups at 16 in. on center vertically.

Step 9. Design for vertical shear reinforcement (ρ_l):

$$\min \rho_l = 0.0025 + 0.5 \left(2.5 - \frac{H}{l_w} \right) (\rho_t - 0.0025)$$

$$= 0.0025 + 0.5 \left(2.5 - \frac{180}{20 \times 0.8 \times 12} \right) (0.003125 - 0.0025) = 0.00299$$

Try #4 bars:

$$s = \frac{(2 \times 0.20)}{(8 \times 0.00299)} = 16.72 \text{ in.}$$

Two layers of horizontal bars will be placed at the calculated spacing.

Step 10. Find the maximum vertical spacing of horizontal stirrups:

$$\frac{l_w}{3} = 12 \times \left(\frac{12}{3} \right) = 48 \text{ in.} \qquad \text{or } 3h = 3 \times 8 = 24 \text{ or } 18 \text{ in.}$$

∴ Use #4 bars as vertical stirrups at 16 in. on center horizontally

Step 11. Check design for vertical reinforcement at the end of the wall:

$$M_u = 250 \times 15 = 3,750 \text{ ft-kips at the base of the wall}$$

$$\frac{M_u}{\phi bd^2} = \frac{12,000 \times 3,750}{0.75 \times 8 \times (0.8 \times 20 \times 12)^2} = 203 \text{ lb}$$

From the tabular reference sheet, $\rho = 0.01$:

$$A_s = \rho bd = 0.01 \times 8 \times (0.8 \times 20 \times 12) = 9.95 \text{ in.}^2$$

∴ Use 20 #9 bars on each end of wall

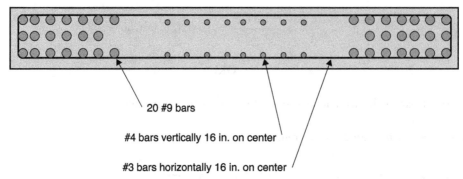

20 #9 bars

#4 bars vertically 16 in. on center

#3 bars horizontally 16 in. on center

Figure 13.18 Shear wall reinforcement

See Figure 13.18 for the final design. Note that as standard practice, with the exception of stirrups which come preformed, the potential for confusion on the part of the contractor's staff suggests the engineer should not specify different size bars that vary by one number. Hence, many engineers will specify steel based on even- or odd-sized steel rebar.

13.3.6 Lateral Load Analysis

Building frames are subject to not only gravity loads but also lateral loads such as wind forces. There are two simplified approaches that can be used as an approximate method to account for these lateral forces. The first method is known as the *portal frame method*, which was first introduced by Albert Smith in 1915 and can still be used as a satisfactory analysis for buildings up to 25 stories in height (McCormac and Brown 2009). The second method is known as the *cantilever method*, and it can be used to analyze buildings up to 35 stories in height. Taller buildings require more complicated methods that are beyond the scope of this book.

Using the portal frame method, the loads acting on a structure are assumed to be acting at the joints of the structural elements. Based on this assumption, the moments acting on the members will vary linearly, and the inflection points will be located at the members' midpoints. When using the portal frame method, the wind loads acting on a structure are assumed to be resisted entirely by the frame. To make use of this method, three assumptions must be made for each portal or girder that makes up the frame, because the entire frame is considered to be made up of independent portals that in turn make up the global structure. The first assumption is that the columns bend in such a way that the inflection point occurs at the midpoint of the column. The second assumption is that the girders bend in such a way that the point of inflection occurs at their centerlines. Finally, the horizontal shear acting among the columns is assumed to be twice the value for interior columns and the same value for exterior columns. Refer to Figure 13.19 for a clearer understanding of this concept. The lateral load (P) acting on the structure is computed from a wind analysis. The exterior columns experience a shear force (V) of one magnitude, and the interior columns experience double the shear force ($2V$). P_w represents the factored axial force imposed on the column due to lateral and gravity forces. M_w is the combined factored moment acting on the column due to the lateral loads and gravity loads. Note that the portal frame method will give each member its equivalent load acting on the structure due to the lateral forces; moreover, these loads must be further combined and factored with the loads acting on the members due to gravity forces. The analytical procedure for a portal frame analysis is accomplished as follows:

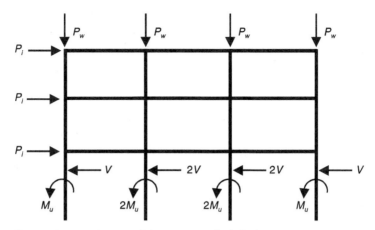

Figure 13.19 Frame diagram for portal frame method of design

1. The shear forces in the various levels are obtained for each column based on the lateral load (P_i) acting on the story (i).
2. The moments acting on the columns (top and bottom) equal the applied shear times half the column height.
3. At any joint in the frame, the sum of the moments acting on the columns equals the sum of the moments in the girders. The girder shears equal the girder moments divided by half of the girder lengths.
4. The axial forces in the columns are directly obtained by the shears acting on the girders that are transferred to the column.

Structures are three-dimensional objects. Doing three-dimensional analysis by hand calculation is tedious work. With today's technological advances, structural analysis software is available to accomplish this three-dimensional analysis task. The user of the software must be thoroughly familiar with it in order to obtain meaningful and realistic results, and in some jurisdictions the designer is held responsible for results generated by the software.

13.3.7 Serviceability

Serviceability of a concrete structure is determined by its deflection, cracking, corrosion of steel reinforcement, and surface deterioration. If the surface of concrete is exposed to hazardous chemicals, a special type of cement and additives should be used in the concrete mix design. In this section, the evaluation of cracking behavior of concrete beams and one-way slabs is introduced. The engineer should design all members by taking into consideration the effect of cracking and stiffness of the member, short- and long-term deflection, and how the cracked concrete element can still perform both aesthetically and mechanically, without loss of performance.

With present technological advances in the design of structural members, higher strength concretes with f'_c values greater than 12,000 psi and higher strength steel have led to the design of more slender and efficient members. The more slender a member, the more deflection becomes an important controlling criterion. The problems that may arise due to excessive deflections of members are (Nawy 2009):

1. Excessive deflection of a floor slab may cause dislocation in the partitions it supports.
2. Excessive deflection of a beam can damage a partition below, and excessive deflection of a lintel beam above a window opening could crack the glass panel.
3. In the case of open floors and roofs such as top garage floors, ponding of water might become an issue.

Simplified deflection control criteria proposed in the ACI code are presented in Table 13.6.

The load deflection of a reinforced concrete beam behaves in a trilinear fashion. The first stage, known as the *precracking* stage, occurs when the member is "crack-free." The second stage is known as the *postcracking* stage and occurs when the member develops acceptable controlled cracking. The third stage is the *postserviceability* cracking stage and refers to the moment the stress in the tension reinforcement steel starts yielding. The first and second stages have their own specified design criteria that the structural engineer must follow.

In general, time will increase the magnitude of deflection of a member. Therefore, the designer must evaluate the short- and long-term deflection in order to satisfy the maximum permissible criteria for the particular member in the structure. Time-dependent effects are produced by the superposition of the strains caused by creep, shrinkage, and temperature. Permissible deflections in a structural system are governed by the amount of deflection that can be sustained by the interacting structural components without the loss of building aesthetics, building use, and/or stresses to the member itself. The level of acceptability of deflection is defined in ACI code 318, which provides the designer with a minimum permissible ratio for computed deflections, as presented in Table 13.7. The designer must calculate the deflections for the members using computational procedures and then compare the calculated deflections to the maximum permissible deflections in the table. If the calculated deflections are greater than the deflection limit, the structural engineer must revisit the design and possibly increase the cross-sectional area of the member.

Table 13.6 Minimum thickness for beams and one-way slabs [ACI 318-11, Table 9.5(a)]

Member not supported or attached	Minimum member thickness[a]			
	Simply supported	One end continuous	Both ends continuous	Cantilever
Solid one-way slab	L/20	L/24	L/28	L/10
Beams	L/16	L/18.5	L/21	L/8
Ribbed slabs	L/17	L/18.5	L/21	L/8

[a] L is the span of the member.

Table 13.7 Maximum permitted deflections [ACI 318-11, Table 9.5(b)]

Type of member	Deflection	Max deflection[a]
Roofs not attached to structural members	Immediate deflection likely	L/180
Floors not attached to structural members	Immediate deflection likely	L/360
Roofs attached to structural members	Long-term deflection	L/480
Floors attached to structural members	Long-term deflection	L/240

[a] L is the span of the member.

13.3.8 Structural Detailing

Engineering drawings are drawings prepared by the engineer of record for the owner. The engineering drawings and structural specifications form a part of the contract documents, as discussed first in Chapter 4. Engineering drawings should contain an adequate set of notes and all other essential information in a form that can be quickly and correctly interpreted. Details are one of the requirements. The engineer is responsible for furnishing a clear set of design requirements to the team to ensure that the appropriate drawings include relevant details on things like connections, welds, bolting, rebar lapping, and any other components that the contractor will need to actually construct the building. The drawings also should show reinforcing bars and welded wire fabric. Without them, the contractor will not understand how the engineer intended for the connections to be made for the beams, joists, and columns. The engineer should refer to applicable building codes when preparing specifications or drawings and include the calculations for them. The engineer should then review the work to ensure that the details meet the standards (ACI 315).

Engineering drawings or specifications for structural elements like beams, girders, columns, walls, and foundations should show the type and grade of reinforcing steel, service load, concrete strength, concrete dimensions, class of tension splice or lap lengths, concrete cover for the reinforcement, required joints, and any other information needed for the preparation of reinforcement placing drawings. Orientation should be clear when orientation matters (e.g., when columns are not square), in which case a north-facing arrow should be placed on every sheet. The scale used also should be indicated on all engineering drawings, preferably under the title of each view. All lettering should be clear, legible, and for normal full-size prints not less than ⅛-in. high.

Labeling every door and window in a drawing set is a cumbersome task. To simplify this task, because many of the doors and windows are the same, a listing or schedule of doors and windows, beams and joints, columns, and other appurtenances often is included in the drawing set (Table 13.8). Schedules for beams and girders must contain:

Table 13.8 Schedules for beams and girders

Member	Number	Location	Reference to floor plan	Length (ft)
Beams				
W12×14	23	Outer east and west walls	A-C, F-H	17-6
W16×26	21	Internal/external	5-6, C-D	27-8
W14×22	8	Internal	D-D1	25-0
W16×31	2	Center	D1-E1	20-0
W12×19	16	Internal	E1-F	27-8
W8×10	21	North/south exterior wall	1-2, 6-7	10
Girders				
W16×31	4	Internal	C	25-0
W16×60	3	Internal	C, E1	27-8
W18×48	3	Internal	C, E1	27-8
W18×50	2	Internal	D	27-8
W18×35	2	Internal	F	27-8
W21×50	1	External edge	A	25
W16×31	3		A, F	25

Note: The list is incomplete for a project and is provided simply as an illustration.

1. Size of beam
2. Size of member
3. Number and size of straight and bent reinforcing bars (rebars)
4. Special notes on bending rebars
5. Number, size, and spacing of stirrups or stirrup-ties within the beam
6. Location of top steel
7. Any special information, such as two layers of steel required

For continuous beams, the number and spacing of top bars to be placed in reinforced concrete T-beams for crack control must be shown. Column design must show the size, number, and location of columns; grade and size of reinforcement; and all necessary details where column section or reinforcement changes. Splicing steel always must be clearly defined, showing arrangements of splices, whether butt or lapped, any staggering of rebars, and type of splicing required for butt splices. Orientation of reinforcement in two-way symmetrical columns must be shown when reinforcement is not two-way symmetrical (ACI 315). Two structural detailing examples are provided in Figures 13.20 and 13.21.

For steel, wood, and masonry structural detailing, the design engineer should refer to the proper specifications and requirements. Figure 13.19 showed an example of a structural framing plan for a building that must undergo a structural analysis investigation for adequate design. Note that members have certain drawing requirements in the plan view. Columns should be depicted to scale in the proper orientation and dimension (i.e., square

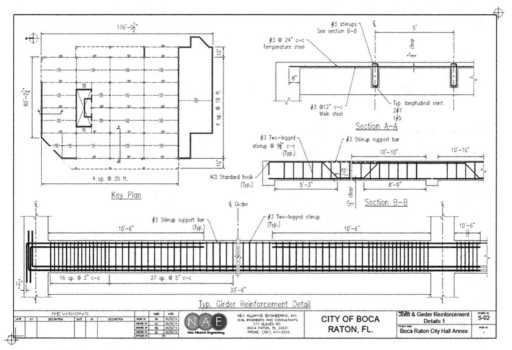

Figure 13.20 Example of steel placement in a girder

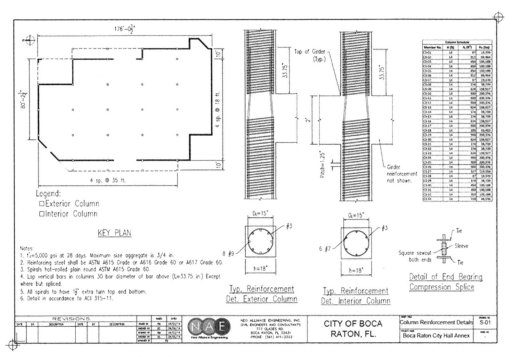

Figure 13.21 Example of steel placement in a column

columns should look square). Beams normally are shown as thicker lines between columns. Joints may be shown as thinner lines between beams. Concrete walls should be noted, typically with some sort of fill. Likewise, elevation views should represent the member geometry properly.

13.4 Foundation Design Concepts

Foundations prevent settlement and stability issues in a structure. The foundation is the lowest supporting layer of a structure. However, the foundation cannot be designed without knowing two critical items: (1) the way in which the loads will be transferred to the foundation and (2) the load-bearing capacity of the receiving soil. To determine the foundation requirements, the structural design must be completed. The first step is to construct load diagrams (Figure 13.22) so that the loads transferred to the foundation can be determined. Both moments and vertical loads should be depicted (recall Figure 3.19). Internal and external critical load diagrams should be available for every structural design. For the load-bearing capacity of the receiving soil, a standard penetration test boring log (see Figure 13.23) is required.

For practical purposes, there are basically two types of foundations: (1) shallow foundations and (2) deep foundations. *Shallow foundations* are the most common and the easiest

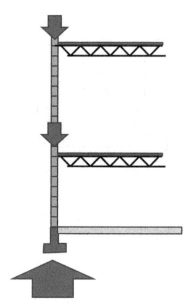

Figure 13.22 Load transfer down from the roof and second floor to the soil

to calculate. However, they are dependent upon accurate geotechnical data. The assumption with shallow foundations is that the loads are small enough to be spread over the native soil or new soil without having to dig too deep. Removal of large amounts of soil material is a costly venture.

When the soil will not support the building, *deep foundations* are used to transfer the load of a structure down through a weak upper layer of soil to a stronger subsoil or bedrock layer below. Information required for the design of an effective foundation includes: (1) the load that will be transmitted by the superstructure to the foundation system, (2) the requirements of the engineering code, (3) the behavior and stress-related deformability of soils that will support the foundation system, and (4) the geological conditions of the soil under consideration, which can be determined from a soil boring log, cone penetration test, or standard penetration test.

13.4.1 Shallow Foundations

Foundations come in different shapes and sizes for different uses. Options vary from deep, shallow, or continuous footers to footers that support tilt-up construction and curtain walls. Numerous tests outline each of these, and the variety of soils and rock conditions makes any extensive discussion here confusing. Therefore, although several examples of the more common foundation types will be outlined, engineers are directed to the appropriate codes and standards for the locale as well as texts on foundations. For example, for one- and two-story buildings constructed with concrete masonry walls, strip footers typically are used because the load is small and easily can be spread across a relatively small area (see Figures 13.24 and 13.25). For low-rise buildings that use columns to support tilt-up panels or curtain walls,

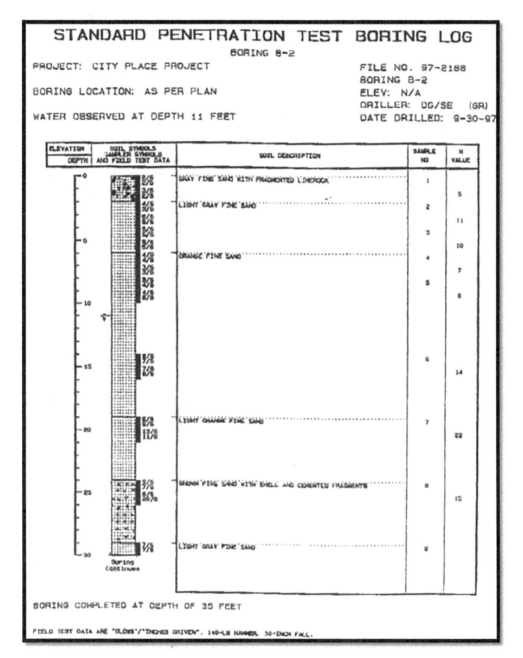

Figure 13.23 Soils and boring log

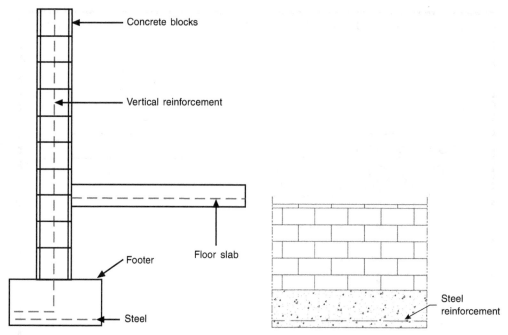

Figure 13.24 Standard CMU foundation for strip footer: side view

Figure 13.25 Standard CMU foundation for strip footer: front view

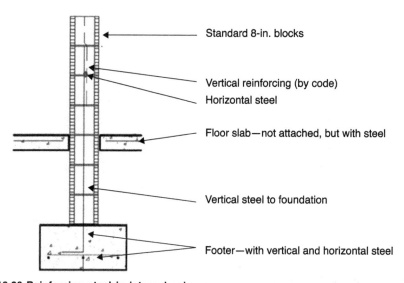

Figure 13.26 Reinforcing steel in internal column

shallow footers may be appropriate (see Figure 13.26). For taller buildings or in locations where the soil characteristics are a challenge, deeper foundations or piles will be more appropriate. The type of pile is important, and there may be local preference. In areas with hard rock, for example, driven piles may not be the standard due to potential vibrations that can crack walls and windows in other buildings.

Table 13.9 Examples for soil capacity (Nawy 2009)

Type of soil (Nawy 2009)	Bearing capacity (tons/ft^2)
Crystallized bedrock, granite, gneiss	100
Schist or slate	40
Sedimentary rock, sandstone, limestone	15
Gravel and gravel sand mixes (GW and GP soils)	
Dense	5
Moderate	4
Loose	3
Sand and gravely sand, well-graded SW soil	
Dense	3.75
Moderate	3
Loose	2.75
Sands and gravely sands, poorly graded (SP soils)	
Dense	3
Moderate	2.5
Loose	1.75
Silty gravels and gravel/sand/silt mixed (GM soils)	
Dense	2.5
Moderate	2
Loose	1.5
Silty sand	2
Clay mixtures (all)	2
ML and CL soils	1
CH and MH soils	1

The protocol to start all foundations is the same:

◆ Step 1. Determine the loads being transferred to the soil by the walls and columns.
◆ Step 2. Determine the load-bearing capacity of the soil (see Table 13.9).
◆ Step 3. Determine any required factors of safety.
◆ Step 4. Determine the steel required for bending, punching, and moments. Remember that the moments of the exterior walls are transferred to the foundation, and the steel reinforcing must account for this.

Once these issues have been addressed, the specifics of any particular foundation type can be determined. Keep in mind that the larger and taller the building, or the greater the loads and moments applied, the more complex the foundation design will be. Strip footers are the simplest.

13.4.2 Strip Footer Foundations

The design for a strip footer is straightforward; it requires only the width of the footer and the reinforcement. The procedure is as follows:

◆ Step 1. Determine the width of the footer (b) based on the load and the bearing capacity of the soil.
◆ Step 2. Design the reactive upward design pressure for lateral reinforcement:

$$p_a = \text{Actual pressure} = \frac{\text{Applied load}}{\text{Base width}} \tag{13.40}$$

$$p_u = \text{Ultimate reactive pressure} = \lambda \times \frac{\text{Applied load}}{\text{Width}} \tag{13.41}$$

where λ is related to the percent of the load applied by live loads and typically is 1.4 to 1.6.

◆ Step 3. Determine the lateral bending and shear:

$$d_e = d - d_r \tag{13.42}$$

where

d_e = effective depth
d = thickness of the footer
d_r = depth to the rebar

$$M_u = w_u \frac{\left[\left(\dfrac{b}{2} - \dfrac{d_e}{2}\right)^2\right]}{2} \tag{13.43}$$

where

M_u = ultimate moment to overcome (ft-kip)
b = width of the footer (ft)

◆ Step 4. Determine the area of steel per foot of width (A_s):

$$A_s = \frac{M_u}{bd^2} \tag{13.44}$$

◆ Step 5. Pick rebar and check rebar based on the number of bars that can be placed in the member (minimum 1.5 in. apart) and steel area required This is done by trial and error.
◆ Step 6. Check rebar.
◆ Step 7. Check shear stress (v_u):

$$V_u = w_u \left(\frac{b}{2} - \frac{d_e}{2}\right) \tag{13.45}$$

$$v_u = \frac{V_u}{(b_v d_e)} \tag{13.46}$$

$$V_c = 2\lambda b d_e f_c^{0.5} \tag{13.47}$$

where

V_u = shear load (lb or kip)
v_u = concrete shear stress
V_c = factored shear capacity (lb or kip)
b_v = lateral bending shear = $b/1.2$

If $v_u < V_c$, no shear reinforcement is required. If $v_u > V_c$, reinforcing is required.

Example 5. Strip Footer

A footer supports a 15-ft wall that transfers a total load of 10,000 lb/ft. The foundation is on a poorly graded sand and sandy gravel soil with moderate compaction. From Table 13.9, the soil supports 5,000 psf. The foundation is above the water table (no correction) and assume the footer is 2 ft thick to start.

Step 1.

$$A_f = \frac{\text{Load}}{p_n} = \frac{10,000 \text{ lb/ft}}{5,000 \text{ psf}} = 2 \text{ ft}$$

where p_n is the soil bearing capacity. Therefore, the footer must be 2 ft wide.

Step 2.

$$p_a = \text{Actual pressure} = \frac{10,000 \text{ lb/ft}}{2 \text{ ft}} = 5,000 \text{ psf} = 5 \text{ kips/ft}^2$$

$$p_u = 1.5 \times 5,000 \text{ psf} = 7,500 \text{ lf/ft}^2 = 7.5 \text{ kips/ft}^2$$

Step 3. Determine the lateral bending and shear:

$$d_e = 2 \text{ ft} - 3 \text{ in.} = \text{rebar cover} = 1.75 \text{ ft}$$

$$M_u = 7,500 \frac{\left[\left(\frac{2}{2} - \frac{1.75}{2}\right)^2\right]}{2} = 58.6 \text{ ft-lb/ft}$$

Note for students: underground is always a minimum of 3-in. cover.

Step 4. Determine the area of steel (A_s) (in square inches per foot of width) using Equation 13.44:

$$A_s = \frac{58.6 \text{ ft·lb}}{2 \text{ ft} \times (1.75 \text{ ft})^2} = \frac{9.6 \text{ lb}}{(\text{ft})^2} \cdot \frac{(\text{ft})^2}{(12 \text{ in.})^2} = 0.06 \text{ in.}^2$$

Step 5. Pick rebar as one #3 bar per foot.

Step 6. A #3 bar has an area of 0.11 in.^{2}:

$$\text{Load} = 0.11 \text{ in.}^2 \times \frac{[2 \times 12 \times (1.75 \times 12)^2]}{12} = 24.26 \text{ ft-lb/ft}$$

24.3 ft·lb/ft < 58.6 ft·lb/ft. Three #3 rebars will be needed to overcome the load.

Step 7. Check shear stress (v_u):

$$V_u = 7,500 \left(\frac{2}{2} - \frac{1.75}{2} \right) = 1,875 \text{ lb}$$

$$v_u = \frac{1,875}{(1.67 \times 1.75)} = 642 \text{ psf}$$

$$V_c = 2 \times 1.0 \times (2 \times 12) \times 20 \text{ in.} \times 3,000^{0.5} = 52,581 \text{ lb}$$

$v_u < V_c$, so no shear reinforcement is required.

13.4.3 Shallow Footer Foundations

When the receiving soil can easily handle the load of the building, a shallow foundation typically is an effective option. There are three methods to evaluate: (1) Terzaghi's method, (2) the general bearing capacity method, and (3) the layer method. According to Terzaghi's method, a foundation is shallow if its depth is less than or equal to its width (Terzaghi 1943). Meyerhof (1951, 1963) included a shape factor s-q with the depth term. The layer method looks at backflow in layers as opposed to a full unit. The latter investigations have suggested that foundations that measure three to four times their width may be defined as shallow foundations (Das 2007). For more details on any of these method, the reader is directed to consult a foundation engineering textbook.

Shallow footers are common under columns for buildings three to six stories in height. Certain soils types will permit even larger loads, but moments become more critical. The shallow footer is more complicated than the strip footer because the loads are greater and the column is a much smaller portion of the area of the footer. As a result, reinforcement and resistance to punching are important issues to consider. The protocol for designing a shallow footer is described as follows (check local codes for variations):

◆ Step 1. Determine the allowable bearing capacity of the soil based on test data and soil borings. Make sure the soil borings go several feet below the proposed footing depth to ensure that poor soils are not located immediately under the proposed foundation:

$$p_n = \text{Soil bearing capacity (psf)} - (\text{Depth to bottom of footing}$$
$$\times \text{ Soil weight} + 100 \text{ psf}) \tag{13.48}$$

◆ Step 2. Obtain the service loads (p_a) and the moments as translated to the foundations from the structural engineer. Hence, the structural and geotechnical engineers need to communicate regularly on the structural aspects (loads and moment) of the project.
◆ Step 3. Determine the area of the footer:

$$A_f = \frac{\text{Load}}{p_n} \tag{13.49}$$

◆ Step 4. Calculate the factored intensity loads (q_s) for the controlling load condition:

$$q_s = \frac{p_a}{A_f} \tag{13.50}$$

◆ Step 5. By trial and adjustment, find the effective depth to avoid punching shear. Nominal shear for punching is the smaller of:

$$V_c = \left(\frac{2 + 4}{\beta}\right) \lambda b_o d_e f_c^{0.5} \tag{13.51}$$

$$V_c = \left(\frac{2 + \alpha_s d_e}{b_o}\right) \lambda b_o d_e f_c^{0.5} \tag{13.52}$$

where

$$b_o = (w_c + d_e) \times 4 \tag{13.53}$$

α_s is 40 for interior columns, 30 for edge columns, and 20 for corner columns.
◆ Step 6. Evaluate one-way and two-way action using V_c and then calculate the factored moment (V_n) using ϕ; note that often this value is 0.9 (most of the variables were defined in the prior example for strip footers):

$$V_c = 2\gamma w_f d_e f_c^{0.5} \tag{13.54}$$

$$V_n = \frac{V_c}{\phi} \tag{13.55}$$

$$V_u = q_s \left(\frac{w_f}{2} - \frac{w_c}{2} - d_e\right) \times w_f \tag{13.56}$$

where

w_f = width of footer
w_c = width of column in same direction

Check the moment arm on the foundation:

$$\text{Moment arm} = \left(\frac{w_f}{2} - \frac{w_c}{2} \right) \tag{13.57}$$

$$M_u = q_s w_f \left(\frac{\text{Moment arm}^2}{2} \right) \tag{13.58}$$

$$M_n = \frac{M_u}{\phi} \tag{13.59}$$

For two-way action:

$$V_u = q_s \left[\frac{A_f - (w_c + d_e)}{2} \right] \tag{13.60}$$

$$V_n = \frac{V_c}{0.75} \tag{13.61}$$

◆ Step 7. Select appropriate reinforcement:

$$M_n = A_s f_y \left(\frac{d_e - a}{2} \right) \tag{13.62}$$

◆ Step 8. Determine the size and spacing of the flexural reinforcement in the long and short directions uniformly distributed across the footing:

$$\text{Check } f_c^{0.5} < 100$$

◆ Step 9. Verify that bond requirements are met.
◆ Step 10. Check stresses on column.
◆ Step 11. Determine the number and size of dowel bars (vertical steel), and design the reactive upward design pressure for lateral reinforcement.

Steps 9 to 11 are similar to reinforced concrete columns and beams. Refer to an appropriate reinforced concrete text and local codes for exact procedures.

Example 6. Size of Footer

In this brief example of sizing a shallow footer foundation, the footer has a load of 500,000 lb applied to it. The column is 16 in. × 16 in. (1.33 ft × 1.33 ft). Find the area of the concrete footer assuming 3,000-psi concrete. The unit weight of the densely packed gravel soil is 135 pcf. The footer slab thickness is 2 ft and lies below the 3-ft water table. The solution is found as follows:

Step 1.

$$p_n = 10{,}000 \text{ psf} - (5 \times 135 \text{ pcf} + 100 \text{ psf}) = 9{,}225 \text{ psf}$$

Step 2. Loads are given as 500,000 from a 16-in. × 16-in. column.

Step 3.

$$A_f = \frac{\text{Load}}{p_n} = \frac{500{,}000 \text{ lb}}{9{,}225 \text{ psf}} = 54 \text{ ft}^2$$

A 7.5-ft × 7.5-ft square footer will work.

Step 4.

$$q_s = \frac{500{,}000}{(7.5 \times 7.5)} = 8{,}888 \text{ psf} < p_n \qquad \text{OK}$$

Step 5. Nominal shear for punching is the smaller of:

$$b_o = (w_c + d_e) \times 4 = 144 \text{ in.}$$

$$\text{Ratio of column } w_c = 1.0 = \beta$$

α_s is 40 for interior columns:

$$V_c = \left(\frac{2+4}{\beta} \right) \lambda b_o d_e f_c^{0.5} = \left(\frac{2+4}{1} \right) \times 1.0 \times 144 \times 20 \times 3{,}000^{0.5} = 946{,}500 \text{ lb}$$

$$v_c = \left(\frac{2 + \alpha_s d_e}{b_o} \right) \lambda b_o d_e f_c^{0.5} = \left(\frac{2 + 40 \times 20}{144} \right) \times 1 \times 144 \times 20 \times f_c^{0.5} = 1{,}191{,}850 \text{ lb}$$

Step 6. Determine the effective thickness of the slab. Normally there is cover over the rebar. Assume that cover is 3 in. and the rebar itself is up to an inch; therefore, the effective thickness (d_e) is 1.67 ft:

$$V_c = 2 \times 1.0 \times (7.5 \times 12) \times 20 \text{ in.} \times 3{,}000^{0.5} = 197{,}180 \text{ lb}$$

$$V_n = \frac{99,800}{0.9} = 104,700 \text{ lb}$$

$$V_u = 8,888 \times \left(\frac{7.5}{2} - \frac{\frac{16}{12}}{2} - 1.67 \right) \times 7.5 = 94,222 \text{ lb}$$

Check two-way action:

$$V_u = 8,888 \times \left[7.5^2 - \frac{(16 + 20)}{12} \right] = 473,300 \text{ lb}$$

$$V_n = \frac{V_c}{0.75} = \frac{473,300 \text{ lb}}{0.75} = 631,000 \text{ lb}$$

b_o = Perimeter failure zone is at column for puncture = $(16 + 20) \times 4 = 144$ in.

Step 7. Select appropriate reinforcement. Start with calculating the moment art to obtain all variables. Check moment art on the foundation:

$$\text{Moment arm} = \left(\frac{7.5}{2} - \frac{\frac{14}{12}}{2} \right) = 3.25 \text{ ft}$$

$$M_u = q_s w_f \left(\frac{\text{Moment arm}^2}{2} \right) = 8,888 \times 7.5 \times 3.25^2 = 704,100 \text{ in.-lb}$$

$$M_n = \frac{M_u}{\phi} = 704,100 \text{ ft-lb} \times 0.75 = 938,800 \text{ in.-lb}$$

$$A_s = \frac{M_n}{f_y(0.9 d_e)} = \frac{938,800 \text{ in.-lb}}{(60,000)(0.9 \times 20)} = 0.87 \text{ in.}^2$$

$$a = \frac{A_s f_y}{0.85 w_f f_c} = \frac{0.87 \times 60,000}{0.85 \times (7.5 \times 12) \times 3,000} = 0.22 \text{ in.}$$

$$\text{Check } A_s = \frac{M_n}{f_y \left(d_e - \frac{a}{2} \right)} = \frac{938,800 \text{ in.-lb}}{(60,000)\left(20 - \frac{0.22}{2} \right)} = 0.79 \text{ in.}^2 \qquad \text{OK}$$

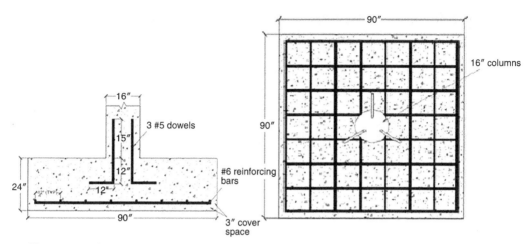

Figure 13.27 Reinforcing steel for footer

Step 8. Determine the size and spacing of the flexural reinforcement in the long and short directions uniformly distributed across the footing:

$$\text{Check } f_c^{0.5} < 100$$

$$(3,000^{0.5}) = 54.8 < 100 \quad \text{OK}$$

Step 9 to 11. From this point, the engineer will select the appropriate rebar size and spacing and perform the checks as needed to satisfy the result. One option is shown in Figure 13.27.

13.4.4 Pile Foundations

Piles will be the foundation solution for when loads are transferred from tall buildings, when heavy loads are transferred, when the soils may wash (such as in coastal areas), or when the subsurface is otherwise unstable. Typically precast or prestressed concrete will be used for the piles, but wood and steel also may be used in some applications if permitted by local codes. If the maximum capacity of single piles is not enough, group piles can be used, but the pile cap must be designed. Different configurations can be used once the required number of piles is computed. For pile foundations, the objective is to determine the diameter, length, and number of piles suitable to manage the applied load.

The process for the design of a pile foundation is essentially completed in two steps. The first step is to find the allowable side shear (R_s) and allowable end-bearing resistance (R_b) and then add up the two terms from this step to determine the total allowable force (R_T), which is divided by the load coming down the critical column to find the number of piles required. The procedure is as follows:

$$R_T = R_s + R_b \tag{13.63}$$

Calculating for resistance from friction (where the boring log was broken down into 10 segments):

$$R_s = D\pi\Sigma f_{s,i} L_i \tag{13.64}$$

$$f_{s,i} = k\sigma'_v \tan \varphi = \beta\sigma'_v \tag{13.65}$$

$$\beta = B - 0.135Z^{0.5} \text{ limited to} \rightarrow 0.25 < \beta < 1.2 \tag{13.66}$$

$$\sigma'_v = (\gamma_{sat} - \gamma_{water}) \times Z \tag{13.67}$$

where

D = diameter of the pile (in.)
$f_{s,i}$ = ultimate unit side shear resistance (ksi)
β = factor
Z = height from top surface to middle of pile segment (ft)
σ'_v = vertical effective stress (ksi)

Calculating for end-bearing resistance (at the end of each segment):

$$R_b = q_p \left(\frac{(\pi D^4)}{4} \right) \tag{13.68}$$

$$q_p = 0.6(N_{ave}) \tag{13.69}$$

where

q_p = ultimate unit end-bearing resistance (ksi)
N_{ave} = standard penetration test N value from the boring log (ksi)

Piles are designed with a safety factor of 2.5, with a specified pile diameter. Calculating the minimum reinforcement ratio (ρ_s) is determined in accordance with AASHTO code. The reinforcement for the piles is as shown by the formulas as follows:

$$P_s = 0.45 \left[\left(\frac{A_g}{A_c} \right) - 1 \right] \left(\frac{f'_c}{f_y} \right) \tag{13.70}$$

where

P_s = volume of spiral turn = area of bar $\times \pi \times$ diameter (in.)
A_c = cross-sectional area of concrete inside spiral steel = $0.25 \times \pi \times$ (core diameter in inches)^2
A_g = gross cross-sectional area of concrete of pile cross section (in.^{2})

f'_c = concrete compressive strength (ksi)

f_y = yield strength of steel (ksi)

Where one pile cannot carry the load, multiple piles with a pile cap can be used.

Piles under pile caps should be laid out symmetrically. The column should be placed at the geometric center of the pile cap in order to transfer load evenly to each pile. In general, pile spacing should be three times the pile diameter in order to transfer load effectively to soil. Figure 13.28 shows an example of this placement. The pile cap thickness normally is governed by deep beam shear. The forces are determined in a manner similar to a footer. Heavier pile loads are governed by direct shear. The critical section of punching shear stress is at a distance $d/2$ from the edge of the pile, where d is the effective depth of the pile cap. For a corner pile, the critical section normally extends to the corner edge of the pile cap because it gives less shear area.

The design procedure for pile caps is as follows:

◆ Step 1. Determine the loads and appropriate factor of safety (minimum 2.5).
◆ Step 2. Determine the pile construction method (driven, bored, auger cast, etc.).
◆ Step 3. Determine the material (steel, reinforced concrete, prestressed concrete, wood, etc.).
◆ Step 4. Estimate the number of piles needed based on individual pile loads.
◆ Step 5. Design the layout of the pile cap.
◆ Step 6. Calculate the factored pile load.
◆ Step 7. Assuming a pile cap thickness, calculate the factored moment and shear at the critical section.
◆ Step 8. Check direct shear.
◆ Step 9. Calculate the moment and shear at the face of the column, and check deep beam shear.
◆ Step 10. Check punching shear and edge distance.
◆ Step 11. Design flexural reinforcement.

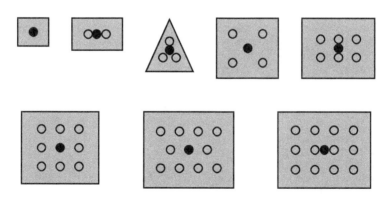

Pile diameter	12 in.	14 in.	16 in.	18 in.	20 in.	22 in.	24 in.
Pile spacing	3 ft 0 in.	3 ft 6 in.	4 ft 0 in.	4 ft 6 in.	5 ft 0 in.	5 ft 6 in.	6 ft 0 in.

Figure 13.28 Piles should be spaced at three times the pile diameter in order to effectively transfer load to soil (solid circle = column, open circle = pile, concentric circles = column and piles are in same centerline)

Piles can be calculated by hand when they are limited in number, but specialized software is available to help engineers design piles and pile caps. An example here would be cumbersome and difficult to follow. Engineers should reference a foundation text and consult online sources for pile design.

13.5 References

ACI 315-80 (1980). *ACI Detailing Manual 1980*, American Concrete Institute, Detroit, MI.

Das, B. (2007). *Principles of Foundation Engineering*, 8th Ed., Cengage.

FBC (2010). Florida Building Code, <http://ecodes.biz/ecodes_support/free-resources/2010Florida/Building/10FL/Building.html>.

Geschwindner, L.F. (2012). *Unified Design of Steel Structures*, 2nd Ed., John Wiley & Sons.

Hibbeler, R.C. (2009). *Engineering Mechanics: Combined Statics and Dynamics*, 12th Ed., Prentice Hall.

Hibbeler, R.C. (2011). *Structural Analysis*, 8th Ed., Prentice Hall.

McCormac, J.C. (2007). *Structural Analysis: Using Classical and Matrix Methods*, 4th Ed., John Wiley & Sons.

McCormac, J.C., and Brown, R.H. (2009). *Design of Reinforced Concrete*, 8th Ed., John Wiley & Sons.

McCormac, J.C., and Csernak, S.F. (2012). *Structural Steel Design*, 5th Ed., Prentice Hall.

Meyerhof, G.G. (1951). "The ultimate bearing capacity of foundations." *Geotechnique*, 2(4), 301–332.

Meyerhof, G.G. (1963). "Some recent research on the bearing capacity of foundations." *Canadian Geotechnical Journal*, 1(1), 16–26.

Nawy, E.G. (2009). *Reinforced Concrete: A Fundamental Approach*, 6th Ed., Prentice Hall

Shaeffer, R.E. (2002). *Elementary Structures for Architects and Builders*, 4th Ed., Prentice Hall.

Terzaghi, K. (1943). *Theoretical Soil Mechanics*, John Wiley & Sons.

Cost Estimating

For clients, among the most important issues are the time of construction, the utility of the project, quality, and of course costs. Remember that all clients have a budget, and as a project design moves forward, the engineer will refine the cost of the project while also trying to stay within budget. Decisions will be made by the client to add or subtract items, alter floor space, install equipment, etc. as the costs become better defined. The engineer's job is to design a functional project with that budget in mind; however, cost should not be the overarching issue—public health, safety, and function should be the priorities.

Estimating the cost of a project relies on the engineering team's ability to define all aspects of the budget, including the capital outlays, the operations and maintenance budget, the lifetime of the asset, its salvage value at the end of its life, and other items. The capital outlays are comprised of the up-front costs and the direct fixed asset costs associated with the design, materials, equipment, and construction (including labor). Labor costs for construction will include salaries, fringe benefits, overtime, consultants' fees, and indirect costs, which include overhead for supervision, security, telephones, debt service, and the like. Operations and maintenance costs include electricity, utilities, chemicals, repairs, and other items that are considered annual or recurring costs instead of capital outlays. For most projects, the operations and maintenance costs can comprise more than 80% of the total costs over the life of the building (Halpin 2006). As a result, the concept of life cycle analysis, which involves the evaluation of a project over its full life, will be introduced later in this chapter.

14.1 Purpose and Process

Every project will have up-front expenditures that will cover the initial capital cost plus the annual operations and maintenance cost after construction is over and occupancy begins. Therefore, a predictive process must be employed to quantify resources required by the project scope. Cost estimating also must address risks and uncertainties. The outputs of estimating are used primarily as inputs for budgeting, cost or value analysis, and decision making in project planning, design, and/or schedule control.

The process of developing an accurate cost estimate involves obtaining up-to-date quotes for big-ticket items and equipment, obtaining reliable estimates of the cost of materials and labor, procuring proposals from subcontractors or subconsultants, reviewing bidders' rates, and evaluating engineering alternatives by performing a benefit-cost analysis. Recent bids for similar projects are the best source of information for pricing the cost of a project. These bids can be obtained from engineers, local governments, and those who bid on projects regularly. Remember, however, that cost estimates like these are specific to the economy of the region at the time, so comparing the costs of a project in Jacksonville, FL with one in Chicago, IL or San Francisco, CA would likely be inaccurate. In addition, if an area is growing fast, the bids will be higher because the contractors have work, a common situation seen in the mid-2000s in south Florida, Las Vegas, NV, and certain other fast-growing places. Conversely, if the economy is in recession, then the bids may be low because the contractors are just trying to stay in business or trying to have enough cash to retain good employees. While the best source of information for costs is a similar project, it is unlikely that many similar projects can be found locally. In such cases, the engineer can use data from other regions, corrected for differences in the economic market, or use specialized software that is adjusted for time and region, like the program that RSMeans® has offered online (at a cost) for many years.

There are many unknowns *a priori*, which means the engineering team must have a strong understanding of the project scope. Because of these uncertainties, many project managers regard cost estimating as an art. If the project manager can predict the budget, including the required man-hours, material costs, contingencies, and overruns, then the project will make a profit as opposed to losing money if the budget is underestimated. In some instances, information may be available to one bidder due to prior experience with which other bidders may be unfamiliar. Trying to sort out the unknowns and retrieve that information is an essential part of the due diligence and planning process. Many engineers maintain extensive records of past bids or develop estimation manuals for cost estimating purposes. Young engineers should consider doing the same in their future careers.

14.2 Stages of Cost Estimating

As noted in Chapter 4, there are four stages of design. For many engineers, the approximate percent a project is complete at each stage is roughly as follows: (1) conceptual design (10% of total design), (2) predesign (35%), (3) preliminary design (70 to 80% complete), and (4) final design (100% complete). In each stage, a cost estimate will be provided to the owner. At the conceptual stage, very little of the actual design work has been accomplished, and many of the details of the project are not known, so this estimate will be less accurate. At this stage, the cost estimate will be an *order of magnitude estimate*, which is made without any detailed engineering data. Such estimates may use past experience but will be ±35% accurate within the scope of the project. Order of magnitude estimates can be used in evaluating project feasibility, establishing a preliminary budget for cost control during the design phase, screening alternatives, or during the planning and development phase of a project with little or no design completed. Using published cost indices will help improve the accuracy of these estimates. Some preferred indices include the following:

- General—*Engineering News-Record*, U.S. Department of Commerce
- Contractor prices from reputable construction management firms
- Valuation indices: American Appraisal®, Handy-Whitman Index
- Special purpose indices: U.S. Department of Labor Bureau of Labor Statistics, state highway department
- Forecasting services: Data Resources, Inc., Chase Econometrics
- Location indices: RSMeans, *Engineering News-Record*

For the predesign phase, an *approximate (rule of thumb) estimate* can be developed without the benefit of detailed engineering data. The estimator will use similar projects to obtain estimates, but the size of the building and accompanying information like parking must be known. The indices noted previously can be used. The accuracy is ±15 to 25%. A contingency normally is included to cover a percentage of any unanticipated cost overruns and any items added to the scope after the design phase.

Once preliminary design is complete, much of the equipment, construction techniques, and construction market will be known, so a *definitive (or detailed) estimate* can readily be prepared. The preliminary cost will be garnered from well-defined engineering data, vendor costs, quotes, unit prices, etc. The accuracy should be ±10%, but the construction market may create uncertainty in this estimate, so a contingency should be included.

Once the final project design is complete and the project is ready to bid, a *prebid estimate* will be developed. The prebid estimate should be within ±5% of the actual construction costs. A contingency is not necessary at this point. The best estimators are engineers who finish with an estimate just slightly higher than the actual low bid. The closer the prebid cost estimate is to the actual cost, the less work that will be required to do value engineering (external review of the plans to reduce costs through materials and/or construction methods), product replacement, scope reduction, or total redesign to make the budget work.

Nothing will disrupt a budget plan faster than unanticipated add-ons after cost estimating is complete or construction has already started. Such events cause scheduling delays, long lead time orders, construction conflicts in the field due to lack of integrated design or poor coordination of subcontractors, or just poor understanding of the project scope. Armed with a well-defined project scope, a work breakdown structure can be developed (see Figure 14.1), which is a road map of how the project will be constructed. If the scope is not well defined, then the work breakdown structure cannot be developed and the cost estimate will not be accurate.

The cost estimation process follows a step-by-step approach:

- Provide a complete *definition* of the work to be done.
- Construct the *work breakdown structure* and estimate the activities (time/cost).
- Establish *acceptable costs* for each activity in the work breakdown structure.
- Develop/construct a *logic network diagram.*
- Review *time/cost* with the respective functional managers.
- Review the *base costs* with the owner/client.
- Decide on a *course of action.*
- Adjust.
- Document the cost estimate in the project file.

Input from vendors, suppliers, and subcontractors is necessary for contractors to develop bids. Engineers should be familiar with these same entities and involve them in the design and cost estimating process to improve accuracy and efficiency.

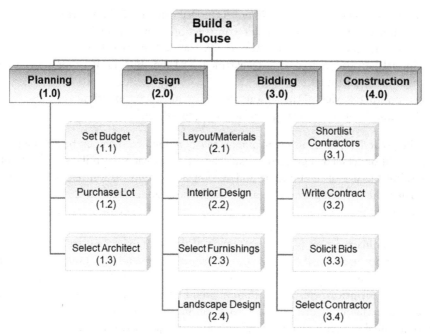

Figure 14.1 Example of a work breakdown structure

14.3 Bidding Process

How a bid is put together and how cost estimates are developed by the engineer are defined in part by how the engineer expects the project to be bid. Methods of bidding include:

◆ **Price per unit**—Requires specifically knowing all the components of the project. This method is often used for roads, water and sewer piping, stormwater piping, curbing, and the like.
◆ **Lump sum**—The quantities are not known precisely, but the components can be estimated. Buildings are often bid this way.
◆ **Cost plus**—The quantities and scope are not well defined. The idea is for the contractor to track actual costs with an overhead amount.

Barriers to good cost estimating include poor estimating techniques and/or standards, resulting in unrealistic budgets, misunderstanding the sequence of start and completion of activities and events, inadequate understanding of the work breakdown structure, and the lack of any management policy on reporting and control practices. Poor work definition at the lower levels of the organization and inadequate formal training or access to estimating tools can result in unnoticed, or often uncontrolled, increases in scope of effort. A lack of understanding of the scope or poor comparisons of actual and planned costs can lead to a failure of the cost estimating process.

The design process requires an understanding of how the contractor will construct the project to avoid unforeseen technical problems that create schedule delays, require overtime or idle time, or escalate material prices, all of which will increase the costs because contractors recognize the inherent risks associated with each of these inefficiencies, and the effect on their bottom line profits, and will adjust their prices based on their perception of that risk. Sharing the design concept with contractors during the initial engineering phase (integrated design process) will eliminate many construction barriers, leading to better cost estimates and a better bid process.

Other cost estimating problems can stem from a failure to understand customer requirements, misunderstanding the time needed to complete the construction, omissions in the plans, misinterpretation of information, and failure to assess and provide for contractor risks. Employing an integrated design process in which vendors and contractors review the design during the engineering phase will help to eliminate or mitigate the effects of these issues. Uncertainty is an unavoidable component of business decisions, but the integrated design process is a method for identifying, quantifying, and eventually minimizing uncertainties.

For example, a utility needs a building to house two large generators that will run the plant when the power is out. The utility hires a consultant to design the building and then produce a corresponding cost estimate. The engineer fast-tracks the design in order to get bids, with limited review by the utility. The engineer does not get input from contractors and vendors because, after all, "it is just a simple building to house generators." The bids come in, and the prices seem reasonable and are under the initial cost estimate. Construction proceeds very quickly. The building is finished, and then the contractor turns in the final pay request. A building inspection reveals that no generators are included. When asked about this omission, the contractor points out that there is no spec for generators in the bid documents because "it is just a simple building to house generators." And of course, it turns out that the openings are not large enough to slide the generators inside. The cost to remedy the situation adds $750,000 to the project (which is about 40% of the total budget). No wonder the bids were so low. This is a true story—but the names were omitted to protect the embarrassed.

14.4 Asset Management

Once the project is in operation, it becomes an asset of the owner. Because significant investments are made to secure the asset, it is expected to function properly, over a long lifetime. However, to ensure that the asset operates within the desired parameters, it must be suitably maintained. An inventory of parts must be on hand to facilitate operation or repairs, and some form of financial system should be in place to track expenditures. The ongoing operation, preventive maintenance, and rehabilitation of major equipment all are included as part of an operations system. Maintenance and operations management systems are designed with the intent to identify where there are potential infrastructure needs to be addressed as a part of a capital program, to be able to track costs, and to identify where excessive maintenance is occurring. Maintenance management systems also may be required to demonstrate compliance with regulations. With data management systems, the responsibilities will be transferred to other areas for the long term (i.e., once the capital project is completed, the maintenance management system should immediately initiate work orders for ongoing maintenance). Maintenance can add value to the asset of the entity that owns it, while ongoing depreciation will reduce asset value on an annual basis. Maintenance staff will look at the age and potential deterioration of the asset to determine maintenance needs. These needs will be gleaned from the work orders and other output from the maintenance system, which should indicate infrastructure that is nearing the end of its life or has excessive amounts of repairs. Tracking from maintenance programs and work orders will help the owner identify where priorities may need to change for infrastructure improvements. Needs assessments come out of the maintenance management system and indicate where upgrades are required to deal with changes in technology, address concerns with age, and improve the operation and maintenance of the system, generally by lowering costs by simplifying operations.

One of the primary issues with large assets is service life, which refers to how long the asset will remain in working condition. Keep in mind that most building projects are designed to last at least 50 years, and there are many examples well beyond that time frame, although the expected lifetimes for mechanical equipment are much shorter. Table 14.1 shows examples of the life expectancy of various assets. The engineer has significant input into this service life based on the materials and techniques specified in the design process. Figure 14.2 is an example of how the cost to maintain an asset increases as the condition of the asset decreases over its life. In truth, all assets decline with age and periodically require rehabilitation to improve reliability and serviceability (Figure 14.3), but an asset is never "like new." Figure 14.3 shows that while most finance experts depreciate infrastructure using straight-line depreciation (see Chapter 11), the reality is that the in-service condition generally stays significantly above that (Bloetscher 2011). Moreover, as a result of the deterioration occurring more slowly than anticipated with straight-line depreciation, the ability to maintain infrastructure condition improves if upgrades or rehabilitation efforts are done at the appropriate time. Waiting too long to perform preventative maintenance can increase the risk of catastrophic failure. For most entities, budgets, work orders, and capital construction projects are viewed as separate items, but in reality, they should be linked in order to be able to evaluate the risks.

Risk is defined as the probability that some adverse impact will occur. Risk is a difficult issue to address as a part of any set of criteria upon which a capital expenditure is based. Evaluating risk of failure and vulnerability of the asset is crucial to the decision-making process, however. To minimize the potential for failure, owners must assess the potential risks and the likelihood they might occur. When major engineering projects fail, there are significant risks. These risks include property damage, health impacts, lost wages, public safety, regulatory responses, moratoria, and wasted dollars. Therefore, the benefits of having an asset

Table 14.1 Examples of life expectancy of assets (Bloetscher 2011; Uddin et al. 2013)

Asset	Life expectancy (years)
Bridge decks	50
Bridges	100
Tunnels	>100
Nuclear power plants	60
Ductile iron pipe	80–100
PVC C900 pipe	50–100
Clay pipe	100
Concrete pipe	50–100
Road base	30
Road surface	12–15
Mechanical pumps in lift stations	15
HVAC	15
Refrigerator	10
Automobile	7–10

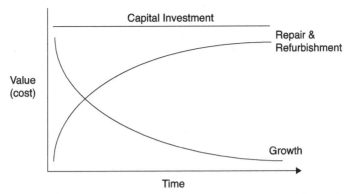

Figure 14.2 Comparison of the cost of maintenance versus the remaining value of the asset (Bloetscher 2011)

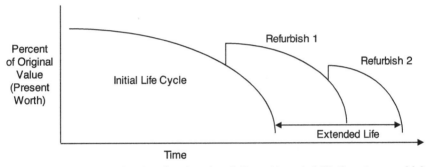

Figure 14.3 Life of an asset, showing deterioration, followed by rehabilitation stages, which never quite return the full value of the asset (Bloetscher 2011); note that refurbishment may occur later in life before full devaluation takes place, but refurbishment never really reaches the initial condition, and extension of the life of an asset decreases with each refurbishment

management system embedded within the organization are lower risks, improved system reliability, decreased operations and maintenance expenditures, and improved safety. Implicit in this concept is the idea of optimization of the design process, equipment selection, and materials choices to minimize long-term operating costs. Risk and deterioration lead to repair and replacement costs, and once these costs reach a certain point, they lead to a new capital project. Once an asset has reached a point where it is no longer economical to operate or has deteriorated to the point where replacement is more cost effective than repairs or where the infrastructure no longer serves its intended purpose or meets regulatory standards, the asset needs to be replaced (Bloetscher 2011). Together, all of these systems form a concept that is known as asset management, which is the financial understanding of the infrastructure that is constructed.

14.5 Life Cycle Analysis

Asset management is a business process and a decision-making framework that covers an extended time horizon, draws from economics as well as engineering, and considers a broad range of assets. Asset management can be used to develop data that allows the engineer to make better decisions about long-term asset life. This is termed *life cycle analysis* (LCA), which is used to evaluate the lifetime costs—design, construction, and operations, with rehabilitation and disposal included. LCA considers obsolescence, reliability, repairs, and the potential for catastrophic failure. When comparing LCA options, the concepts of engineering economics are used because life cycle costs easily can be brought back to present worth values by starting with capital investment. LCA includes acquisition (design, construction) and operations (implementation, construction, usage, phase-out). Figure 14.4 shows an example of two projects. The graphed line for option A shows a lower capital cost but higher operating costs across the life of the asset. Option B has a higher capital cost but lower operating costs. Because 80% of life cycle costs are operations, it is expected that option B will have a lower LCA than option A. The point where the two lines cross is the point where the two options are equal. Most developers and owners find that if this breakeven point is 3 to 7 years out, the added capital costs are worthwhile. If longer than that, the uncertainty about operations and maintenance creates too much potential for error.

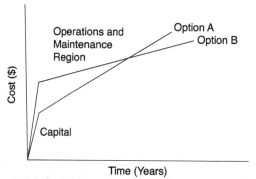

Figure 14.4 Example of two projects: option B has a higher capital cost but lower operations and maintenance costs, and option A has a lower capital cost but higher operations and maintenance costs over time; over the life of the assets, the project with the higher capital cost (option B) results in a better quality project, with lower operations and maintenance cost and thus a lower life cycle cost

14.6 References

Bloetscher, F. (2011). *Utility Management for Water and Wastewater Operators*, American Water Works Association, Denver, CO.

Halpin, D.W. (2006). *Construction Management*, 3rd Ed., John Wiley & Sons, Hoboken, NJ.

Uddin, W., Hudson, W.R., and Hass, R. (2013). *Public Infrastructure Asset Management*, McGraw Hill, New York

14.7 Assignments

1. Develop cost estimates for your project as you proceed. Create a graph of the costs as they have changed over time. What do you notice?
2. Develop a preliminary detailed cost estimate for your capstone project design.

◆ 15 ◆

Conclusion

Now that the comprehensive work of the previous chapters is nearing completion, the end of the capstone design course is nearing as well. The intent is to give students in their senior year a means to apply all the knowledge gained in the many engineering fundamentals classes taken as undergraduates in one multidisciplinary project with multiple realistic constraints. Hopefully, every student in the class has had the opportunity to apply the knowledge gained in the engineering curriculum to develop the skills necessary to become a competent engineering intern. In the "real world," projects almost always involve multiple disciplines, and this course is designed to engage students in integrating their prior coursework to fulfill a design project in a multidisciplinary team setting.

This capstone course is designed to be a transition from traditional classroom learning to the profession. Although there is still much to learn, the capstone experience prepares young engineering graduates for the expectations of the workplace. After passing the FE exam and graduating from an ABET-accredited institution, there will likely be a 4-year apprenticeship before sitting for the PE exam and becoming a licensed engineer. The license is the key milestone to move forward in this noble profession. As a young engineer, there will be many potential career opportunities and many opportunities to grow and develop. Engineers are in high demand because there are many types of engineering jobs available from which to choose, and the condition of today's infrastructure ensures that engineers will remain in high demand. In the long term, the growth and development of human civilization will rely on the ability to resolve challenges using science, technology, engineering, math, and creativity. In reality, all challenges are just the surface of the expansive need for competent engineering graduates.

At this point, students should have confidence in their public speaking, leadership, teamwork, and writing skills and technical abilities. At the conclusion of the course, students will have compiled an impressive notebook that includes the detailed design work for the group. Students can show prospective employers a comprehensive project from start to finish. Through the capstone course, students have developed valuable skills in procuring work as part of the response to a request for proposal or qualifications as well as due diligence in the Phase I Environmental Site Assessment process. A former student of one of the authors was offered a full-time job and promotion because he was able to demonstrate these skills.

The site plan and floor plans should be a source of pride for group members. They are the basis for the final design notebook. Site plans and floor plans in AutoCAD® format will be dimensioned properly, clearly readable, and meet all requirements for professional-quality engineering drawings. Elevations should clearly match the floor plans. Renderings of the proposed project are desirable and can be readily created to allow prospective employers or potential clients to experience the richness of the project in three dimensions.

The student notebooks should now include drawings, design details, and technical memoranda for all parts of the project. Specifically, structural, geotechnical, water, stormwater, drainage, sewer, and transportation technical memoranda should be included. HVAC, landscaping, LEED®, life cycle costs, and other ancillary technical memoranda enrich the project experience with little added effort. All technical memoranda should include AutoCAD drawings of the project that match the site and floor plans. Integrating these pieces and describing them is an important outcome of the capstone experience.

Proposals to enhance engineering design education have included the development of design expectations across the curriculum, team-based learning activities, and assessments to gauge student attainment of outcomes, but the key obstacle for undergraduate students is transitioning from traditional lecture-based coursework to more realistic, practice-oriented training. That was the goal of this class. All of these issues are intended to be key components of the classroom experience. It is precisely these components that are missing from the traditional engineering curriculum, which emphasizes the regurgitation of equations and repetition of standard problem sets, neither of which reflects the "real world."

A key component of the capstone course is to stimulate creativity and critical thinking to solve a real-world problem with multiple realistic constraints using appropriate engineering standards and codes by putting engineering skills into practice in the classroom. The skills sharpened in the capstone class serve as the foundation for a prosperous career. Professional engineering is a process. The first aspect of the process requires a full understanding of the problem and associated challenges (due diligence) as a means to define the problem in a context that can lead to the second aspect—the successful design of the solution.

Congratulations—you are now ready to walk across the graduation stage and start the next important step toward becoming a licensed professional engineer or land surveyor.

Index

A

AASHTO 342, 386
Abbreviations 79, 85, 125
ABET 1, 2, 39, 80, 397
Access 9, 56, 65, 117, 135, 151, 166, 169, 180, 187, 189, 191, 196, 197, 199, 200, 204, 205, 209, 211, 213, 217
Accessibility 6, 55, 56, 135, 152, 182, 213, 214
ACI 318 338, 354, 362, 370
ADA 213, 214, 219, 229
Adjusted cost 265
Advanced Buildings Core Performance Guide 152
Advanced Energy Design Guides 152
Alternative analysis 6, 10, 110, 131–141, 231
  defined 131
  evaluation criteria 35, 61, 131, 133, 134, 136
  results 136
  selection criteria 134–135, 136, 138
  selection matrix 110, 136–137
  sensitivity analysis 137–138
American Society of Civil Engineers 35
Annual worth 231, 234, 239–241, 259–260
Annuities 232, 234, 239, 245, 246–248
Answering questions 10
Architect/architecture 4, 41, 57, 65, 155, 190, 204, 212, 213, 337, 340, 342
ASCE 7 10 338, 339, 340, 341, 343, 344, 347, 349, 350

ASHRAE 152
Asset management 42, 393–395

B

Basis of design 7, 8, 63
  report 2, 63, 64, 95, 96–98
Bathrooms 148, 211, 212, 214, 215, 216, 280, 304, 309
Beams 99, 338, 351, 352, 358–359
Benefit-cost ratio 231, 232, 390
Bidding process 63, 74
Bids 390, 391, 392–393
  cost plus 392
  lump sum 392
  price per unit 392
Book value 265, 267
Branding 20–23
Breakeven analysis 258–259
Brownfields 151, 172
Building 45, 49
  access 217
  capital projects 53
  codes 2, 172, 204–205, 213, 218, 337, 338, 340, 347, 371, 374, 380, 382, 385
  departments 177
  design codes 338
  elevations 95, 100, 195, 205
  emissions 144
  footprint 194, 211, 215
  foundations 5, 7, 9, 65, 95, 96, 204, 373–388

high performance 2, 4, 8, 10, 143–154
materials 65, 134, 279–288
program 109, 189, 200, 211–213, 217, 223
structures 32, 99

C

Canons 35
Capital 231, 232, 257, 258, 264, 265
recovery 240
Capitalization 85, 124
Capital outlay 389, 393, 394, 396
Capstone course 1, 24, 26, 95, 101, 128
design 1, 2
objectives 3
CERCLA 157, 159, 172, 173
Charette 7, 190, 191, 215
Clean Air Act 160
Closing line 92
Codes 10, 58, 64, 65, 97, 98, 99, 102–103,
151, 189, 190, 211, 214
building 2, 172, 204–205, 213, 218
ethical 35–36
local 61, 191–193
Colons 92, 122
Columns 360–361
beam 360
long/slender 360
short/nonslender 360
Commas 79, 85, 121, 122
Communication 3, 6, 8, 10, 17, 20, 25, 26, 28,
58, 59, 60, 71
verbal 71
written 71
Compare options 134, 137
Compound interest 234
Conceptual design 64
Construction 2, 4, 17, 37, 38, 41, 49, 53, 54,
55, 57, 58, 62, 63, 65, 87, 95, 97, 98,
99, 113, 133, 135, 144–147, 150, 153,
184, 187, 190, 193, 195, 205
Construction documents 6, 66, 152
Construction materials 9, 41, 43, 63, 65, 66,
99, 116, 144–146, 149, 150, 152, 166,
204
Construction scheduling 66
Consultants 10, 17, 22, 55–56, 57, 58, 63, 64,
132
subconsultants 58, 60, 61
Consumer Price Index 260, 391
Context 134

Cost analysis 10, 96, 133
BCA 133
Cost estimates 390–392
approximate 391
definitive 391
detailed 391
order of magnitude 390
prebid 391
rule of thumb 391
Cost estimating 389–396
Cover letter, 93
Creativity 2, 3, 6, 14, 24
Credit interpretation ruling (CIR) 149
Creeds 35–36
Criteria 4, 6, 12, 100, 110, 121, 131, 133
selection 35, 61, 134–135, 136, 138
Critical loads 338
Corrosion 280
Cut-and-fill 205

D

Dead loads 338, 341–342
Decision making 8, 18, 24, 25, 26, 28, 33, 36, 54,
55, 61, 78, 87, 91, 94, 97, 99, 100, 126,
131, 132, 133, 140, 150, 189, 191, 211
Deflection 370
Depreciation 264–268, 394
straight line 265, 394
Design 3, 4, 5, 7, 9, 17, 21, 22, 35, 49, 54, 56,
58, 63–66, 71, 94, 95, 97, 98, 99, 101,
103, 133, 143, 144, 146, 166, 189, 190,
191, 196, 204, 205, 211, 213, 214, 222,
223, 229
capstone 1, 2, 4, 5, 7, 14, 217
conceptual 2, 9, 64
final 8, 10, 63, 65–66
high performance 2, 4, 8, 144, 145, 149
predesign 63, 64–65
preliminary 2, 8, 63, 65, 182, 198
process 2
steps in 63–66
Diction 75
Disbenefits 133
Dry cleaners 175
Due diligence 8, 36, 37, 46, 98, 161, 177, 390, 398

E

Earthquake loads 350–351
Easement 6, 175, 189, 193

ECHO 172
Economic benefits 133
Economy (of building) 337
Emails (writing) 21, 86, 88–90
Emergency Response Notification System 172
Energy 4, 10, 144, 145, 146, 149, 151–152, 153, 194, 215
ENERGY STAR® 146, 147, 152
Engineering 18, 143, 161
    career 10, 14, 17, 18, 49, 71, 103
    design 3, 5
    economics 6, 10
    ethics 6, 32–42
Engineering economics 231–278
Engineers 1, 2, 3, 17, 18, 32, 35, 53, 55, 65, 95, 97, 98, 99, 100, 131, 133, 157, 158, 187, 190, 198, 204, 211, 212, 213, 222
    licensure 39–42
    responsibility 32, 35, 36, 49, 54, 66, 93, 99, 214
    training of 1, 2, 4, 17, 18, 19, 20, 39, 74
EnviroFacts 172
Environment 1, 2, 24, 58, 60, 63, 64, 98, 143, 144, 145, 146, 147, 149, 150, 153, 154, 159, 172, 192
Environmental professional 159, 160–161, 178
Environmental site assessment 6, 8, 93, 109, 157, 189
    CERCLA 159
    controlled REC 159
    data gap 182
    de minimis 159, 181
    evaluation 181–182
    hazardous substance 160, 166
    historical REC 158
    interviews 180–181
    nonscope 182–187
    Phase I 159, 161, 177, 182, 187, 192
        site recon 161–168
        records review 169–180
    Phase II 187
    Phase III 187
    REC 158
    records review 169–180
EPANET 304, 308, 309, 312
eQUEST 322, 325
Ethics 3, 6, 40, 42
    engineers and 32–33, 36, 39
    history of 30–36
    origin of 33–34
    philosophers 35
Exfiltration 196, 197

F

Factor of safety 337
Failure 97
Fatal errors 112
FE Reference Handbook 10, 40
Figures (format) 78, 79, 80–85, 101
Final Design Notebook 8, 10
Fixture units 308, 309, 314
Floor plan 6, 8, 9, 66, 95, 100, 110, 194, 196–213
Footers
    shallow 380–385
    strip 377–380
Foundations 5, 7, 9, 65, 95, 96, 204, 373–388
    deep 374, 376
    piles 376, 385–388
        side shear 385
    shallow 373, 374–385
Future worth 232, 233, 236–239, 241–243, 244

G

GBCI® 149, 150
Gradient 233, 234, 243–246, 252, 254, 262
Gravity loads 338
Green (building) 143, 144–146
Groups 5, 6, 10, 22, 27, 28, 32, 33, 34, 85, 94, 191

H

HVAC 7, 9, 41, 65, 66, 95, 96, 99, 151, 152, 166, 187, 211, 214
    design of 320–326
Hyatt Kansas City 97–98

I

Incompetence 37
Indoor air quality 149, 152–153, 160, 184
Inflation 260–264
Influence area 351
Infrastructure 1, 17, 34, 49, 131, 132, 193, 194
Interest rate 233, 234–235, 249–252
    effective interest rate 233, 249–252, 260, 262
    annual worth and 259
Interview 4, 8, 10, 18, 21, 61, 93, 158, 160, 161, 172, 180–181, 182, 187
ISO 146–147

J

Job opportunities 17, 19–20

L

Landfills 145, 149, 152, 172
Landscaping 6, 9, 64, 79, 96, 99, 102, 190, 194, 195, 334–336
LEED® 4, 57, 89, 96, 100, 103, 110, 126, 144, 145, 146, 149, 150–153, 321, 322, 398
Letter of transmittal 93
Letters 85, 86
  transmittal 93
  writing 91–93
Licensure 20, 21, 31, 32, 36, 39–42
Life cycle 1, 144, 145, 231, 252, 254, 257, 258, 264, 265, 395–396
Live loads 338, 341, 342, 343–344, 378
Logic network diagram 391
LRFD 340–341
LUST program 159, 172

M

MARR 259
Meeting minutes 10, 86, 87, 103
Memoranda 9, 86, 91, 95, 96, 97, 98–100
Minimum program requirements, LEED 150
Misconduct 37

N

National Priorities List 172
Negligence 37
Net cash flow 258
netronline 172
Nominal interest rate 249

O

Only source of income 36
Opening statement 92
Owners 2, 17, 53, 54, 56, 58, 94, 144, 154, 172, 180
  what to look for 55–56

P

Parking 6, 9, 57, 64, 65, 189, 190, 191, 194–203, 204, 205, 211, 326–328
Parking space size 195
Payback period 258
Payments 232, 234, 235, 239, 240, 246–248, 249–252
  annuities 232, 239, 246

comparing 257–258
  number of 233, 249–252
  shifted 246–248
PCB 159
Phase I 159, 161, 177, 182, 187, 192
Phase II 187
Phase III 187
Philosophy 32, 34–35
Plagiarism 75
Potable water 9, 144, 161, 209, 303–313
PowerPoint® 109
Practically reviewable 169
Preliminary Design Briefing Report 9
Premium 144, 231
Present worth 133, 136, 137, 138, 231, 232, 233, 234, 235–236, 239, 243, 245, 246, 247, 248, 253, 255, 256, 257, 258, 260, 264, 396
Primacy 20, 77, 78, 88, 92, 93, 99, 100, 101, 107, 113
Process 2, 3, 5, 8, 24, 26, 27
  of becoming an engineer 39–40
  construction 145, 152
  design 63–66, 96–98, 131–141
  floor plan design 211–229
  LEED application 150
  review 58, 61–62
  selection 55–56, 61–63, 71
  site development 189–210
Project 1–5, 7, 8, 10, 19, 24, 28, 35, 36, 37, 38, 39, 43, 44, 46, 49, 55, 60, 66, 77, 78, 83, 93, 94, 95, 96, 189, 200, 205
  capital 53–55
  delivery 63–68
  design alternative 133–136
  LEED certified 144, 145, 146, 149, 150, 154
  manager 6, 17, 24, 56, 93, 108, 113, 122
  reports 96–103
  requirements 6, 96–98, 108–110, 131–133, 137
  schedule 25, 66–68, 132
  scope 24, 57–61, 96–98, 132, 133, 191
  technical memoranda 98–100
Progress reports 6, 9, 73, 86, 93–96
Proofreading 85–86, 90, 101, 107, 112, 124
Proposals 55, 56, 58, 61–63, 86
  evaluation 61–62
  presentation 103
  private sector process 63
Publicly available 169
Public sector 17, 19, 49, 56, 58, 147

Public speaking 71, 103–113
  questions 112–113
  visual aids 109–112
Punctuation 75, 85, 121–122

Q

Quality assurance/quality control (QA/QC)
    99
Questions 10, 11, 18, 26, 61, 64, 93, 98, 101,
    107, 110, 112–113, 133, 166, 180, 181

R

Railway bridges 342
Rate of return 259
Rating factor 136
RCRA 160, 172, 173
Reasonable cost 169
Reasonable time frame 169
Reasonably ascertainable information 169,
    172, 175, 176, 181
Recency 77–78, 92, 93, 100, 101, 107, 112,
    113, 116, 136
Recognized environmental condition 158
  historical 158
Recovery period 265
Redundancy 128–129
Request for proposal 56
Request for qualifications 56
Risk 389, 392, 393, 394, 395
Roadways 7, 32, 34, 49, 64, 65, 132, 159, 161,
    187, 189, 193, 194–209
Roof 7, 9, 65, 66, 95, 143, 280–287, 293, 298,
    320, 332, 336
  drains 95
  roof loads 346–351
    rain 347–348
    snow 348–350
    wind 347

S

Salutation line 92
Salvage value 265
Sampling 187
Sanitary sewer 6, 8, 9, 19, 20, 32, 49, 64, 65,
    95, 96, 99, 102, 132, 133, 145, 158,
    189, 191, 194, 196, 209, 280, 314–320
  design of 314–320
  profile 316, 320

Schedule 8, 25, 53, 54, 58, 60, 61, 62, 66, 73,
    94, 95, 100, 110, 184
Self-assessment 17–19
Semicolons 123
Serviceability/cracking 369–370
Setbacks 6, 189, 190, 193, 199
Shallow footers 380–385
Sick building syndrome 280
Simple interest 234
Site plan 2, 6, 8, 9, 57, 64, 65, 92, 95, 100,
    102, 110, 189–209, 217, 222, 223
Spelling 85, 88, 101, 112, 124–125
Stability 337
Standard business letter 91–92
Statement of interest and qualifications 56
Stiffness 337
Stormwater 6, 8, 9, 32, 38, 49, 57, 65, 96, 99,
    102, 151, 189, 194, 195, 198, 205, 223,
    280, 282, 283, 288–303
  design of 288–303
  rainfall 209
  retention 9, 151, 189, 192, 195
Strength 337
Strip footers 377–380
Structural design of structures 5, 32, 35, 97,
    98, 99, 144, 159, 212, 351–373
  beams 99, 338, 351, 352, 358–359
  columns 99, 212, 217, 223, 229, 360–361
    beam 360
    long/slender 360
    short/nonslender 360
  dead loads 98
  earthquake loads 350–351
  girders 99, 358–359
  lateral loads 338, 368–369
  live loads 9, 98
  load bearing 212
  loads 9, 98, 204, 212, 223
  one way slabs 353–358
  rain loads 340, 341, 346, 347–348
  reinforced concrete 66, 95
  roof loads 346–351
    rain 347–348
    snow 348–350
    wind 347
  slenderness ratio 360
  snow loads 348–350
  steel 5, 66, 95, 99, 126
  two-way slabs 354–358
  wind loads 338, 339, 340, 341, 343–346,
      362, 368

Structural details 371–373
Students 1–14, 17, 18, 21, 22, 23, 24, 26, 27,
    28, 39, 40, 75, 100, 113, 131, 132, 217,
    222, 229
  ABET and 39, 40
Sustainability 4, 8, 10, 24, 143, 144, 154, 157, 190
Sustainable building 154
Sustainable future 2, 24, 149
Sustainable sites 149, 150, 151
SWOT 191
Syntax 75

T

Tables 78, 79–80, 84, 101
Teaming skills 3, 24–28
Technical memoranda 6, 7, 89, 91, 95, 96, 97,
    98–100, 398
Threshold criteria 132, 138, 139, 141
Traffic 9, 21, 187, 189, 197, 199, 205, 288,
    328, 329, 332
Transportation 2, 4, 5, 8, 9, 17, 19, 21, 41, 43,
    49, 57, 95, 96, 99, 102, 116, 144, 151,
    159, 189, 191, 197, 328–333
  design of 328–333
Tributary area 251
Triple bottom line 153–154

U

U.S.C. (Federal Code sections) 157, 159, 160,
    172
U.S. EPA 147–148
U.S. Green Building Council 144, 146, 149–150
USGS 172, 175

V

Value engineering 99, 100, 391
Visual aids 6, 71, 78, 107, 109–112
Voice 21, 22, 25, 26, 85, 107, 113, 127

W

Walls 362
  load bearing 362
  nonload bearing 362
  shear 362–368
Water 6, 8, 9, 19, 20, 32, 49, 64, 65, 95, 96,
    99, 102, 132, 133, 144, 145, 158, 161,
    189, 191, 194, 196, 209, 303–313
  design of 303–313
  distribution 9, 144, 161, 209
  potable 9, 144, 161, 209
  storm 6, 8, 9, 32, 38, 49, 57, 65, 96, 99,
    102, 151, 189, 194, 195, 198, 205, 223
WaterSense® 146, 147, 148, 151
Web 22, 29, 77, 91, 172
Weighting factor 136, 137
Wind loads 338, 343–346, 347, 362, 368
  uplift 343
Work breakdown structure 391
Writing 6, 10, 20, 22, 36, 61, 66, 71–130, 135,
    180
  engineering style 71–72, 86, 88
  persuasive 6, 71, 77–78
  pitfalls 128–129
  references 55, 61, 73, 75–77, 79, 84, 85, 93,
    98, 99, 100, 101, 182
    frame of 134
  technical reports 100–103